METALLOPROTEINS

TOPICS IN MOLECULAR
AND STRUCTURAL BIOLOGY

General Editors:

Watson Fuller
(University of Keele)
and
Stephen Neidle
(University of London
King's College)

METALLOPROTEINS

Part 1: Metal Proteins with Redox Roles

Edited by

PAULINE M. HARRISON
Department of Biochemistry
University of Sheffield

MACMILLAN

First published 1985

Published by
THE MACMILLAN PRESS LTD
Houndmills, Basingstoke, Hampshire RG21 2XS
and London
Companies and representatives
throughout the world

Typeset by Reproduction Drawings Ltd., Sutton, Surrey

British Library Cataloguing in Publication Data
Harrison, Pauline M.
Metalloproteins.–(Topics in molecular
and structural biology, ISSN 0256–4377)
Pt. 1: Metal proteins with redox roles
1. Metalloproteins–Analysis
I. Title   II. Series
574.19'245   QP601

ISBN 978-1-349-06374-1   ISBN 978-1-349-06372-7 (eBook)
DOI 10.1007/978-1-349-06372-7

The Contributors

E. T. Adman,
    Department of Biological Structure,
    University of Washington,
    Seattle, WA98195,
    USA

A. E. G. Cass,
    Centre for Biotechnology,
    Imperial College of Science and Technology,
    London SW7 2AZ,
    UK

P. G. Debrunner,
    Departments of Biochemistry and Physics,
    University of Illinois,
    Urbana, Illinois 61801,
    USA

M. T. Fisher,
    Departments of Biochemistry and Physics,
    University of Illinois,
    Urbana, Illinois 61801,
    USA

C. Greenwood,
    School of Biological Sciences,
    University of East Anglia,
    Norwich NR4 7TJ,
    UK

D. J. Lowe,
    Agricultural Research Council,
    Unit of Nitrogen Fixation,
    University of Sussex,
    Brighton BN1 9RQ,
    UK

R. I. Murray,
    Departments of Biochemistry and Physics,
    University of Illinois,
    Urbana, Illinois 61801,
    USA

S. G. Sligar,
    Departments of Biochemistry and Physics,
    University of Illinois,
    Urbana, Illinois 61801,
    USA

B. E. Smith,
    Agricultural Research Council,
    Unit of Nitrogen Fixation,
    University of Sussex,
    Brighton BN1 9RQ,
    UK

A. J. Thomson,
    School of Chemical Sciences,
    University of East Anglia,
    Norwich NR4 7TJ,
    UK

R. N. F. Thorneley,
    Agricultural Research Council,
    Unit of Nitrogen Fixation,
    University of Sussex,
    Brighton BN1 9RQ,
    UK

Contents

Preface

The last 35 years have seen major developments in structural and molecular biology. With structures of DNA, transfer-RNA and over a hundred proteins known we have the basis for an understanding at the molecular level of some of the biological processes in which these molecules are involved. Aspects of the structural and molecular biology of the nucleic acids form the subject of other volumes in this series and their importance can be taken to be self-evident. It is just 50 years since the first x-ray diffraction photographs of a protein crystal were taken and 26 since the first low resolution structure of a protein was published. This structure was, of course, that of myoglobin, a metalloprotein.

During the last two decades another area of Biochemistry has taken immense strides forward. This is the so-called 'Inorganic' Biochemistry, or the biochemistry of metals and other elements usually considered to be the province of the Inorganic Chemist. It is now clear that metals are involved, in one way or another, in a very large number of biochemical processes, both cellular and extracellular. This volume on Metalloproteins and the one that follows it, illustrate *some* of these processes, namely some of those in which metals are closely associated with proteins.

The structural and molecular biology of metalloproteins is exciting because of its interdisciplinary nature. We need to know about the structure of the protein, the ligands it provides for the metal or metals it binds, the metal site geometry and how this geometry has been selected and evolved. We need to understand the chemistry of the metal and how this chemistry may be modulated by the protein in subtle ways. We need to follow the reactions in which metalloproteins are involved, to derive their mechanisms, to understand how they are regulated in relation to other physiological processes. Hence a wide variety of expertise must be employed to apply a wide variety of techniques and methods including protein chemistry and elemental analysis, x-ray analysis, fast reaction kinetics, and one or more of a whole host of spectroscopies. We need to weld together the approaches and backgrounds of the biologist, the chemist and even the physicist,

ix

to arrive at an understanding of how metalloprotein molecules work, the relationship of their functions to their own structures and to the functions and structures of other biomolecules.

The present volume, which deals with molecules that participate in various intracellular oxidation reduction reactions, is the first of two on metalloproteins. Of necessity the volumes are selective in their approach. Topics have been chosen to illustrate certain aspects of metalloprotein structure and function by concentrating on a few individual molecules or molecular associations and examining them in some depth. The studies encompass both molecules for which three-dimensional structures are already available (in this volume, the blue copper proteins azurin and plastocyanin, superoxide dismutases, several of the ferredoxins, Fe–S cluster model compounds and some other proteins containing iron-sulphur centres and some of the cytochromes c, if not the cytochrome c containing enzymes) and complexes for which structural information is lacking or incomplete. In the case of cytochrome P-450 a three-dimensional electron density map is still under interpretation. The molybdenum and iron containing component of the nitrogenase complex has been crystallised, but its structure determination is only in its early stages. The six chapters of Part I are written by experts or groups of experts in the chosen topics and the approach to the subject matter reflects the special interests of the authors as well as the state of scientific knowledge. Thus analysis of the three-dimensional structures of plastocyanin and two azurins allows speculation in chapter 1 about their possible evolutionary interrelations and about the regions on the azurin surface which react with molecules (cytochrome c_{551} and nitrite reductase) from which it receives or to which it transfers electrons. These proteins are relatively small and each contains a single copper atom in a distorted tetrahedral site which, in its $Cu^{(II)}$ state, confers on the molecule an intense blue colour. Since the cytochromes c have been reviewed extensively elsewhere the author of chapter 2 concentrates on the enzymes cytochrome c peroxidase and nitrite reductase, each of which contains two types of haem group. The haem irons can be distinguished by spectroscopic methods including resonance Raman, electron paramagnetic resonance and variable temperature magnetic circular dichroism. The application of the latter to nitrite reductase provides an elegant example of the power of this technique. The emphasis of chapter 3 is on the iron–sulphur centres and particularly exciting are the discussions of the conflicting structures derived by x-ray crystallography and by Extended X-ray Absorption Fine Structure analysis of the three-iron clusters recently found in aconitase and some ferredoxins, and of the interconversion of the 3Fe and 4Fe clusters. This chapter also includes hydrogenase, the only enzyme in these volumes known to contain nickel. Superoxide dismutases, among the fastest known enzymes, form the subject of chapter 4. These enzymes may contain copper and zinc, manganese or iron. Substitutions of other metals in the cuprozinc enzyme have been used to probe the properties of these sites and kinetic measurements including decay of superoxide produced by pulse radiolysis followed optically, and of water proton relaxation, have provided information relevant to the enzymes's mechanism. The last two chapters on the structure and chemistry of the monoxygenase, cyto-chrome P-450, and of nitrogenase, the enzyme responsible for converting dinitrogen to ammonia, read like detective stories in which evidence of all types is brought to bear on these complex, but fascinating, and potentially economic-ally important enzymes.

The metals appearing in this volume are principally the transition metals copper and iron while molybdenum is associated with iron in one of the proteins of nitrogenase. Changes in the electronic states of the metals can be followed by several methods, Mossbauer spectroscopy, for example, being extensively applied to the iron-proteins. There are no chapters on techniques alone, the chapters being subject-orientated, rather than technique-orientated. However, the reader will be able to build up a picture of the power, usefulness or limitations of the various techniques used in this very active area of research.

I should like to take this opportunity of thanking my contributors for six excellent reviews, which, although written individually, will, I hope, when taken together, provide a coherent picture of some redox roles of metalloproteins.

Sheffield, 1984 P. M. H.

1

Structure and Function of Small Blue Copper Proteins

Elinor T. Adman

I. INTRODUCTION

When one studies the structure of a protein, one is confronted with a plethora of detail: all the reasons for the existence and behaviour of that particular molecule are implicit in its structure. We, as observers, must try and sort out which of those details are relevant to the function, and which to its history or its relatedness to other structures. The structures of the two blue copper proteins, azurin and plastocyanin, provide an opportunity to do just that.

Classifying a group of proteins according to a few distinctive physical properties before a representative structure is known assumes that there will be common structural features responsible for those properties. In fact the three known structures in this class of proteins would appear to have confirmed that assumption. Nevertheless, as we will try to show in what follows, not all is said and done with these few structures: other properties indicate that there are subclasses of blue copper proteins as is the case for the cytochrome family and the ferredoxins.

Blue copper proteins are proteins with intense absorption near 600 nm. They are also paramagnetic in the oxidised form, and have characteristically narrow hyperfine splitting in epr spectra. The proteins which are generally regarded as 'blue copper proteins' and some of their properties are listed in table 1.1a and b. Blue copper *centres* exist in larger proteins such as laccase, ascorbate oxidase and ceruloplasmin. Thus far it appears that the lower molecular weight blue proteins occur only in bacteria and plants.

Most of these proteins were discovered in the process of looking for something else: azurin (Pseudomonas blue protein, later called azurin by Sutherland and Wilkinson (1963)) was discovered by Horio (1958) when he was isolating and purifying cytochromes; stellacyanin (the first blue protein to be discovered)

Table 1.1a Some properties of bacterial cupredoxins

Common Name:	Azurin	Azurin	Azurin	Azurin
Source:	*P. aeruginosa*	*Alcaligines denitrificans*	*Alcaligenes* sp NCIB11015	*Achromobacter cycloclastes*
MW	14 600[1]	14 000[1]	14 000[1]	12 000[12]
His, Cys, Met	4, 3, 6[1]	5, 3, 6[1]	4, 3, 4,[1]	3, 1, 5[12]
Optical Spectra (nm; mM^{-1} cm^{-1})	625, 3.5[3]	–	620, 10.5?[4]	600, 2.0[11]
EPR	axial[3]	–	axial[3]	?
$g_\perp, g_\parallel$	2.052, 2.26	–	2.055, 2.26	?
A (cm^{-1})	0.006	–	0.006	
Redox potential (mV)	330[5]	–	230[4]	245[11]
X-ray	2.7 Å[6]	2.5 Å[7]	xyls[8,9]	–
Composition	seq[1]	seq[1]	seq[1]	seq[12]
	Azurin	Blue protein	Amicyanin	Rusticyanin
Source:	*Paracoccus denitrificans*	*Alcaligines faecalis* S-6	*Pseudomonas* AM1	*Thiobacillus ferroxidans*
MW	13 790[2]	12 000[10]	11 723[13]	16 500[14]
His, Cys, Met	4, 1, 4[2]	3, 1, 5[10]	3, 1, 2[13]	5, 1, 3[14]
Optical Spectra (nm; mM^{-1} cm^{-1})	595, 1.5[2]	593, 2.9[10]	596, 4.5[13]	597, 1.95[14]
EPR	axial[2]	axial[10]	?	rhombic[15]
$g_\perp, g_\parallel$	2.052, 2.29			2.019, 2.064, 2.229
A (cm^{-1})	0.077	0.0055		0.0065, 0.002, 0.0045
Redox potential (mV)	230[2]	?	180?[13]	680[16]
X-ray	no	xyls[10]	no	xyls?[17]
Composition	+[2]		+[13]	seq[18]

1. Ambler (1971).
2. Martinkus *et al.* (1980).
3. Brill *et al.* (1968).
4. Suzuki and Iwasaki (1962).
5. Horio (1958).
6. Adman and Jensen (1981).
7. Norris *et al.* (1983).
8. Strahs (1969).
9. Norris *et al.* (1979).
10. Kakutani *et al.* (1981).
11. Iwasaki (1973).
12. Ambler (1977).
13. Tobari and Harada (1981).
14. Cox and Boxer (1978).
15. Cox *et al.* (1978).
16. Ingledew (1976).
17. Freeman (personal communication).
18. Ambler (personal communication).

was isolated as a blue pigment which copurified (Keilin and Mann, 1940) with the blue oxidase, laccase. Although stellacyanin was originally thought to be free of copper (Omura, 1961), it was later characterised as a blue copper protein at just about the same time that plastocyanin was found (Katoh and Takamiya, 1961). Umecyanin was discovered while isolating peroxidases (Paul and Stigbrand, 1970).

Table 1.1b Some properties of plant cupredoxins

Common name:	Plastocyanin	Umecyanin	Stellacyanin	Plantacyanin (Cusacyanin, cucumber basic protein)	Mavicyanin
Source:	spinach (chloroplast)	Horseradish root	lacquer tree	Cucumber seedlings	Green squash fruit
MW:	10 800[19]	14 600[22]	20 000[29] (107aa)	10 100[27]	18 000[26]
His, Cys, Met	2, 1, 2[20]	3, 3, 4[23]	4, 3, 0[30]	2, 3, 2[28]	5–6, 5, 1–2[26]
Optical Spectra (nm, mM^{-1} cm^{-1})	597, 4.9[19]	610, 3.4[22]	604, 4.0[31]	597, 3.4[28]	600, 5.0[26]
EPR	axial[19]	axial[24]	rhombic[31]	rhombic[28]	rhombic[26]
$g_\perp$, $g_\parallel$	2.053, 2.26	2.05, 2.317	2.03, 2.08, 2.29	2.02, 2.08, 2.207	2.03, 2.08, 2.29
A (cm)	0.005	0.0035	0.0057, 0.0029, 0.0035	0.006, 0.001, 0.0055	0.0057, 0.0029, 0.0035
Redox potential (mV)	340–370[19]	283[25]	184[32]	317[28]	285[26]
X-ray	1.6A[21]	no	no	xyls[27]	no
Composition	seq[20]	+[23]	seq[30]	seq[28]	+[26]

19. Katoh *et al.* (1962).
20. Scawen *et al.* (1975).
21. Freeman and Guss (1981).
22. Paul and Stigbrand (1970).
23. Stigbrand (1970).
24. Stigbrand *et al.* (1971).
25. Stigbrand (1972).

26. Marchesini *et al.* (1979).
27. Colman *et al.* (1977).
28. Murata *et al.* (1982).
29. Omura (1961).
30. Bergman *et al.* (1977).
31. Malmstrom *et al.* (1970).
32. Reinhammar (1972).

New names have abounded as additional blue proteins have been discovered: umecyanin, plantacyanin, cucumber blue protein, rusticyanin, and most recently amicyanin. Perhaps a more descriptive name 'cupredoxin', patterned after the iron–sulphur proteins, the ferredoxins, might instead gain general acceptance. Functions have not been ascribed to all of these proteins as yet, but those that are known involve electron transport systems exclusively, so that the suffix 'redoxin' would be appropriate.

Copper containing proteins have been reviewed previously (Fee, 1975; Holwerda *et al.*, 1976; Boulter *et al.*, 1977) and recently a book reviewing current information on copper proteins has appeared (Spiro, 1981). I hope to complement these by highlighting results reported in the recent literature in the context of the known structures.

II AZURINS

(a) Source

Proteins considered to be azurins have been isolated from several bacterial genera, namely *Pseudomonas, Bordatella, Alcaligines* and *Achromobacter*. They have not been found in *Salmonella, Escherichia* or *Bacillus* (Sutherland and Wilkinson, 1963), but two blue proteins from *Methylomonas J* have been reported (Tobari and Harada, 1981). The sequences of nine azurins from *Pseudomonas, Bordatella* and *Alcaligines*, show these all to be closely related while that from *Achromobacter cycloclastes* differs (Ambler, 1977). Complete sequences are not yet available for the remaining blue proteins in Table 1a.

(b) Structure

The structures of two azurins, one from *Pseudomonas aeruginosa* (Adman *et al.*, 1978; Adman and Jensen, 1981; Adman, unpublished results) and one from *Alcaligines denitrificans* (Norris *et al.*, 1983) have been determined by X-ray diffraction methods.

P. aeruginosa azurin was crystallised from 0.1M acetate buffer, pH 5.5–5.8, 3.2M ammonium sulphate. The structure was determined from X-ray diffraction data to a resolution of 2.7 Å using the multiple isomorphous replacement methods. There are four molecules in the asymmetric unit. The model was derived from an interpretation of an averaged electron density map which resulted from several cycles of density modification. (The latter is a technique which tries to make maximum use of the information implicit in diffraction data without first imposing an interpretation on the density map.) The model has only been partially refined (that is, subjected to the process by which an interpretation of an electron density map is used to calculate the diffraction pattern, and then adjusted by least squares so that the calculated diffraction pattern will agree as closely as possible with the observed pattern). The standard deviations of the coordinates are estimated to be about 0.5 Å: however, since the refinement process frequently includes *reinterpretation* of the electron density maps as phases improve, some residues could eventually be more than 0.5 Å from their present positions.

The model presently consists of coordinates for residues 3-128, and the copper atom. The copper ligands are His46, Cys112, His117 and Met121, with the first three apparently closer (2.0 Å, and nearly coplanar with the copper atom) than the last (about 2.6 Å). The peptide preceding His46 is also close to the copper (about 3.5 Å) if the current interpretation with the carbonyl oxygen towards the copper is correct.

Analysis of the Extended X-ray Absorption Fine Structure spectrum of oxidised *P. aeruginosa* azurin has led to the conclusion that the spectrum could be fitted with either an $N_3 S$ or an $N_2 SS'$ configuration around the copper, but in either case with a short Cu-S bond of 2.10 ± 0.020 Å, short Cu-N bonds, 1.97 Å, and for the latter model, a somewhat longer Cu-S' bond of 2.25 Å (Tullius *et al.*, 1978). A more recent EXAFS investigation of azurin and stellacyanin suggests that the Cu—S bond is more likely to be 2.15-2.20 Å, the earlier value resulting from an artifact of data interpretation (Blumberg and Powers, 1983). Any of the shorter values Cu—N, N, S are consistent with the X-ray structure presently, while a value of 2.7 for the Cu—S' bond is preferred (Adman *et al.*, 1982).

The folding topology of the chain has been described as an eight stranded beta barrel (figure 1.1a) and is like that of superoxide dismutase, whose resemblance to the immunoglobulins has already been noted. This topology is apparently a reflection of a stable folded unit rather than a particular functional unit. It is also present in the structure of some bacterial antitumor proteins (Sieker, 1981). A functional feature common to these structurally similar proteins is that each binds cofactors, such as a metal, or an antigen, or a small organic molecule, but not always in the same relative location.

In azurin, the folding is better described as a beta sandwich consisting of a primarily C-terminal layer and an N terminal layer (see figure 1.1c). The loop between strands G and H of the beta sheet formed primarily from the C-terminal portion of the molecule contains three of the ligands to the copper (Cys112, His117 and Met121). These two strands are flanked on one side by a parallel N terminal strand (B) and on the other side by an antiparallel strand (D) comprised of a few residues to the C terminal side of the fourth ligand, His46. Strand D is then hydrogen bonded in one or two more places to residues 80-84 which forms a fifth strand (E) in the sheet. The beta sheet thus formed has its entire interior surface coated with hydrophobic residues. This sheet is apposed to the other sheet containing two strands (A and C) from the N terminal portion of the molecule, and a third strand (F) from residues just prior to the ligand containing C-terminal strands. The interior side of this layer is also hydrophobic.

The halves of the beta sandwich thus formed are bridged by several morphological features (see figure 1.1e). One of these is a 'dog-leg' in residues 16-20;

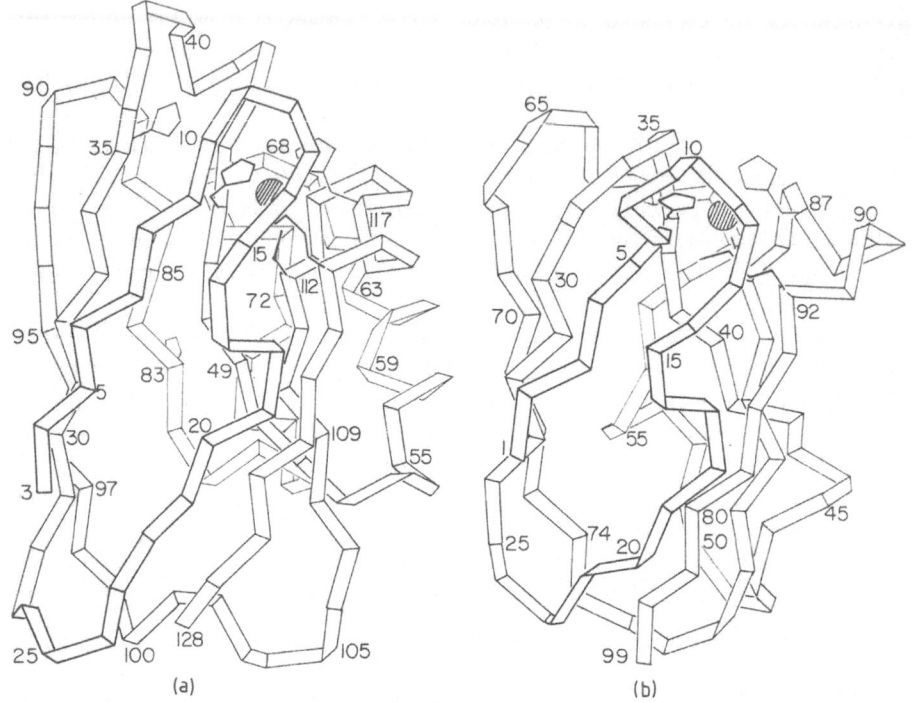

Figure 1.1 (a, b) The polypeptide folding of azurin (a) and plastocyanin (b) with copper atom, copper ligands, and histidine residues indicated. The ribbon drawings for these and subsequent figures were constructed from computer plots of coordinate sets for each of the proteins, rotated to maximise the resemblance between the two (see text). In each of the following parts, azurin is to the left, plastocyanin to the right; each highlights particular features described in the text

another is a closed loop formed from residues 103–108 in which the hydroxyl of Tyr108 is hydrogen bonded to carbonyl O103; the third rather extensive region is the 'flap' which protrudes between two edge strands (D and E) of the C-terminal sheet.

The disulphide, one of the distinguishing differences between azurin and plastocyanin, is quite remote from the copper (>20 Å) and links strand A to the loop (B-C) connecting the two beta sheets. According to our structurally based sequence alignment these sheets are longer in azurin than in plastocyanin (figure 1.1b,c,g and h).

The curvature of the C-terminal sheet is quite extensive: it might be better described as a sheet bent in two along a line passing roughly through the hydrogen bonds from the carbonyl of Tyr 108 to N125 and that from N109 to O51. In contrast, the N-terminal sheet is more extended and regular.

As might be expected nearly all the internal residues are conserved between sequences; however, some surface residues are also conserved. The *hydrophobic*

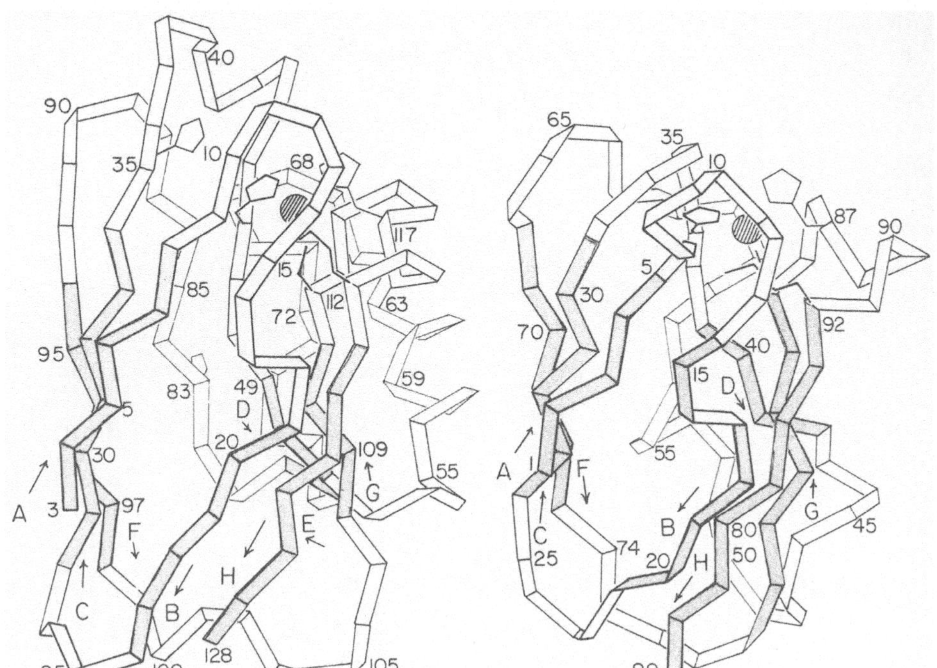

Figure 1.1(c,d)   Beta sheets and labelling of strands

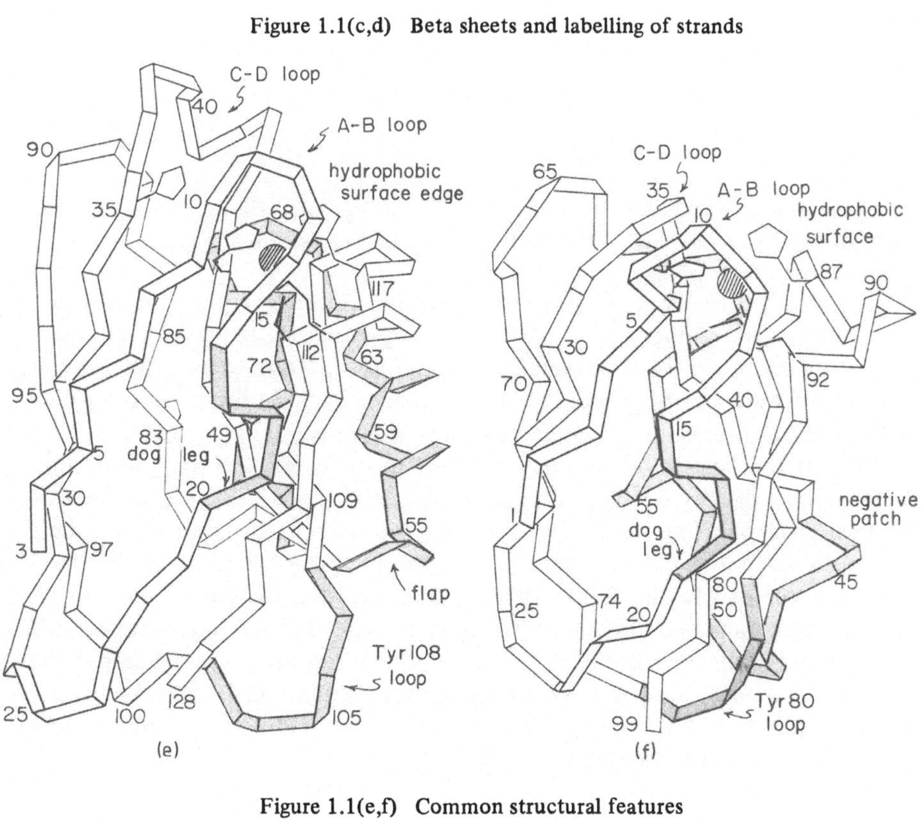

Figure 1.1(e,f)   Common structural features

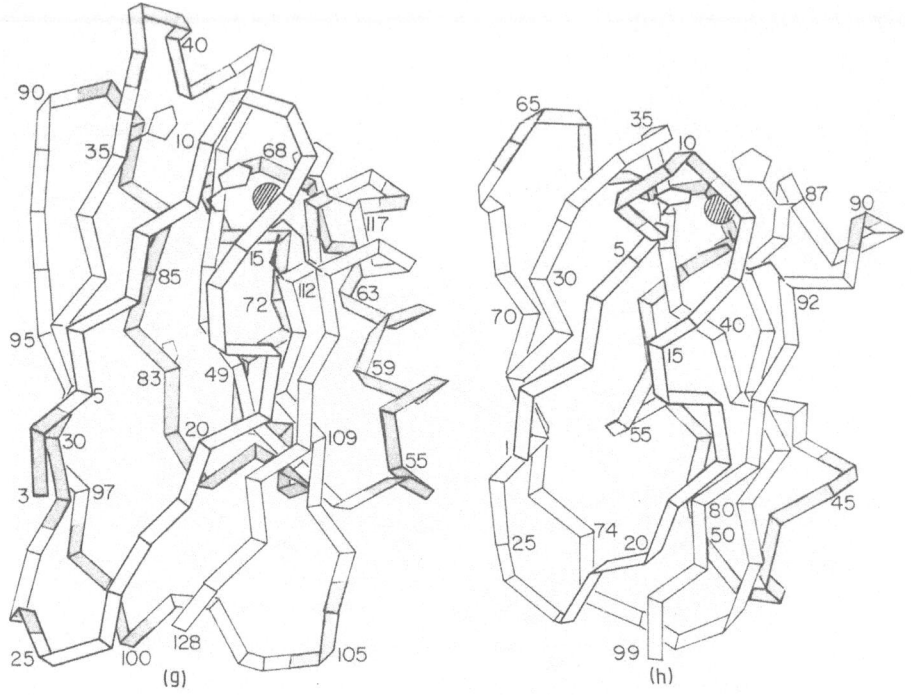

Figure 1.1(g, h)   Structurally inequivalent regions

surface residues which are conserved all lie in a region surrounding the exposed edge of the ligand His117. That surface viewed from the outside in is schematically shown in figure 1.2a.

The closest conserved *charged* residue to the copper is Asp12. The distribution of charged residues on the surface is not particularly remarkable although several are conserved. Most charged residues are in a position to have their charge neutralised by a nearby oppositely charged side chain. At least two conserved acidic residues neutralise Arg79 for example (Adman and Jensen, 1981).

Two of the ligands to the copper appear to be further fixed in position by hydrogen bonds. His46 NE2 hydrogen bonds to O10. Cys112 SG has the peptide nitrogen hydrogen of residue 47 pointing towards it, suggesting a possible hydrogen bond such as the NH. . .S hydrogen bonds observed in the iron-sulphur proteins (Adman *et al.*, 1975). Slightly further away from the cluster, the internal conserved residues Thr113 and Asn47 form a hydrogen bonded pair. Although this pair does not appear to interact directly with ligands of the copper centre, conceivably it serves to provide an additional link between the C-terminal loop (G-H) holding three copper ligands and the remaining strand (D) holding the fourth ligand.

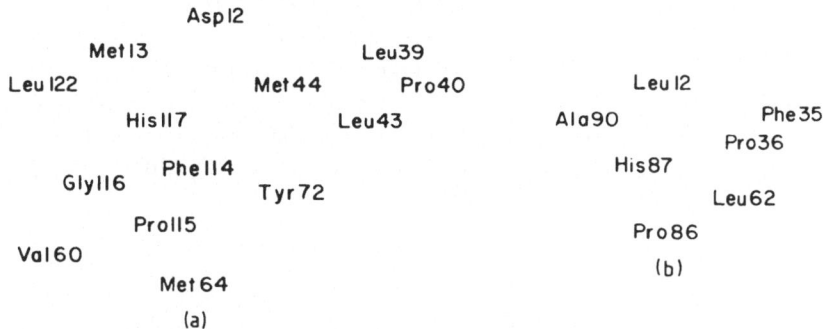

Figure 1.2 Schematic view of the hydrophobic surfaces of (a) azurin and (b) plastocyanin, viewed from the outside of the molecule towards the centre

The Asn47-Thr113 bond lies directly below the Cu centre in the direction of the Cu-S112G bond, His83, which has figured in the interpretation of some kinetic data to be described below, is directly on the other side of this hydrogen bonded pair, adjacent to Asn47, lying on the surface in a shallow cleft formed by the flap and the main part of the molecule.

His35, frequently suggested to be involved in electron transfer activity, lies distal to His46 from the copper. It is conserved, and it is buried. It can be oriented such that depending on its state of protonation it could compete with His46 for O10, or with some rotation from its present position, hydrogen bond to O44. Hydrogen bonds involving main chain atoms of the other ligands are less clear at this time.

The *Alcaligines denitrificans* azurin structure differs little from the *Pseudomonas* one (Norris *et al.*, 1983). It has been determined to 3.0 Å resolution using multiple isomorphous replacement methods, and extended to 2.5 Å resolution by refinement. There are two molecules in the asymmetric unit which have been treated independently. The root mean square difference between the two is 0.8 Å, a value which is expected to decrease as the refinement proceeds. This azurin was crystallised from 0.1M phosphate buffer, pH 6.0, with 60% saturated ammonium sulphate.

Essential features are the same as *Ps. aeruginosa* azurin, including the orientation of the carbonyl oxygen preceding His46. The copper ligands S, N, N are about 2.0-2.2 Å from the copper, and the methionine sulphur still further away, in fact nearly equidistant with the carbonyl oxygen preceding His46 at the present stage of refinement. One possibly important difference detectable at this point appears to be near His35, and involves nearby hydrogen bonds. In the *denitrificans* structure, His35 is hydrogen bonded to O44 and N37. Moreover there is a hydrogen bond between N36 and O9 in this structure (Baker, personal communication) but not in the *Pseudomonas* structure because residue 36 is a proline.

Crystals have been reported for one other azurin, from *Alcaligines* sp. NCIB 11015, formerly *Pseudomonas denitrificans* (Norris *et al.*, 1979) also called Iwasaki blue protein. Tetragonal crystals from unspecified buffer conditions, with cell dimensions of 53 x 53 x 101 Å were originally obtained (Strahs, 1969). Norris *et al.* (1979) reported getting thin plates of the same material from pH 6.0 phosphate buffer, which were not of sufficiently good quality to get diffraction data.

(c) Physical and chemical properties

Studies on blue proteins subsequent to their discovery have centred primarily on characterising the copper centre with various physical techniques (optical, epr and resonance Raman spectroscopy) and characterising the interaction of azurin with small inorganic and with protein redox agents. The spectroscopy has been reviewed previously by Gray and Solomon (1981), and much of the kinetic work by Farver and Pecht (1981c).

1 Spectroscopy

The most salient property of the protein is the blue colour. The absorption at about 600 nm is in fact usually accompanied by two less intense peaks on either side, one at about 440 nm and one at around 780 nm, which in fact are the sum of about eight transitions (Solomon *et al.*, 1980). The current interpretation of the spectra is that the intense blue peak is due to charge transfer between πS and $Cu(d_{x^2-y^2})$ orbitals while the 780 band is due to σ-S and $Cu(d_{x^2-y^2})$ orbitals (Tennent and McMillan, 1979; Gray and Solomon, 1981). The interpretation of the lower wavelength absorption has not been completely agreed upon (Gray and Solomon, 1981). The highly asymmetric environment later confirmed by the structural analysis was hypothesised early in the spectral analysis to account for the relative ordering of energy levels and hence the nature of the observed spectra (Solomon *et al.*, 1976).

Resonance Raman spectra (in which the enhancement of vibrational spectra is due to absorption of energy associated with bonds involving the copper atoms) have been used to confirm the interpretations of electronic spectra (Ferris *et al.*, 1979; Thamann *et al.*, 1982). The vibrational frequencies are unusually high compared to model compounds or non-blue metalloproteins. Woodruff *et al.* (1983) have suggested that the envelope of peaks near 400 cm^{-1} characteristic of the resonance Raman spectra of some blue proteins including azurin, is most likely due to mixing of contributions for both Cu−S and Cu−N stretching frequencies. Thamann *et al.* have further used several specific models for the CU centres including a distorted tetrahedral CuN_2SS' arrangement and a trigonal CuN_2S to predict the force constants and positions of the resonance Raman peaks. They find that a trigonal model fits the observed spectra best, and that the peak at 260 cm^{-1} can be assigned to a hindered torsional rotation along the C−S−Cu−N torsion angle, in contrast to earlier interpretation of this peak as a Cu−Met S stretch (Ferris *et al.*, 1979). Thamann *et al.*, point out that there is also absorption near 260 cm^{-1} in stellacyanin which has no methionine, implying that the azurin peak cannot then be due to the methionine. In view of this and their own unpublished observations that azurin from organisms grown on seleno-methionine show the same resonance Raman spectrum, Thamann *et al.* conclude that the trigonal geometry provides the best description of the copper environment.

The characteristically small epr hyperfine splitting constants are also a consequence of the asymmetry of the copper environment (Gray and Solomon, 1981). Analysis of the azurin epr parameters with a view towards determining the vibronic contributions to activation parameters for electron transfer showed that the spectrum could be best explained by a summation over a distribution of states around a distorted tetrahedral one (Brill, 1978). The distribution was used to obtain an effective force constant for that distortion which corresponded to a small spread in the states, equivalent to about 0.05 Å movement of the ligand positions (Brill, 1978). Normal thermal motion of proteins more than covers displacements of that magnitude.

NMR spectra reveal features of the entire molecule but as a result are quite complex. Their usefulness depends on the ability to assign specific resonances to specific residues, a difficult task. The work until 1978 largely focussed on identification of the near neighbours of the Cu (those residues whose resonances are so broadened by the paramagnetic copper in the oxidised form as to be invisible), and of residues which were affected by changing pH or temperature (Ulrich and Markley, 1978). Now that the structure is known, much of this work takes on new meaning, for the behaviour of various resonances as a function of environment can be interpreted in light of the structure.

Three features of the NMR spectra are most interesting in view of other studies. One is the presence in ^{13}C-NMR spectra of a peak denoted 'x' in the reduced structure, and which has an unusually large chemical shift (Ugurbil *et al.*, 1977). This was originally ascribed to a carbonyl group ligand to the copper, but then qualified in a review of NMR work (Ulrich and Markley, 1978) because the newly published structure of plastocyanin indicated the presence of only four ligands, exclusive of carbonyls. It now appears that a suitable structural feature for this resonance is the carbonyl preceding the ligand His46 which in the oxidised structure may be as close as 3.5 Å, albeit still too far to be a ligand. (*Alcaligenes* results may reopen this question.)

A second feature is the pH behaviour of resonances assigned to histidines. In the aromatic region of the ^{1}H-NMR spectrum two pairs of resonances change from pH 4.9 to 9.0. For one pair, the first peak decreased at the same position, while the second increased. This pair was assigned to a CD proton of a histidine with slow exchange behaviour, having a pK^* = 6.9–7.3 (Hill and Smith, 1979). The second pair gradually shifted towards a smaller chemical shift as the pH increased, and was assigned to the CD2 and CE1 protons of a histidine with normal exchange behaviour (Hill *et al.*, 1976; Ugurbil and Bersohn, 1977). Since the resonances associated with the first histidine were absent in the spectrum of the oxidised form of the molecule it is close to the copper, but because it was titratable it is not a ligand. (Resonances could be identified as due to histidines which did not titrate and which are the ligands.) Resonances for the second histidine could be seen in both oxidised and reduced spectra with different pKs: pK = 7.57 in reduced, 7.35 in oxidised. The ^{13}C-NMR spectrum interpretation is consistent with these values, moreover it suggests that the first histidine deprotonates at the ND1 position while the second histidine deprotonates at NE2 (Ugurbil *et al.*, 1977). The fact that its resonances are present in both oxidised and reduced azurin clearly suggests that the second histidine corresponds to His83, whose centre is some 14.7 Å from the copper and thus the first histidine is His35, whose centre is about 8 Å from the copper.

The pH behaviour of the NMR resonances is understandable in light of the location of the histidines in the structure. His83 is on the surface, and could move to become hydrogen bonded to Asn47 via ND1, which is entirely consistent with the unusual deprotonation at NE2 mentioned above. His35 is internal as described above, and in its protonated form (which is probably the form occurring in the structures done at pH less than 6) *could* be hydrogen bonded to O44 and O9 (Adman, unpublished observations). In its unprotonated form the ND1 hydrogen, the most accessible to the surface, could have been lost, as suggested by ^{13}C NMR data. Then perhaps ND1 could compete with O10 as an acceptor for N46 NE2 hydrogen, so that His35 can remain hydrogen bonded in both forms, explaining its slow relaxation time. *Alcaligenes denitrificans* azurin which was crystallised at pH 6.0, has His35 hydrogen bonded to O44 and N37, implying that ND1 is *not* protonated at that pH (Norris *et al.*, 1983).

Subsequent analysis of ^{1}H-NMR of reduced *P. aeruginosa* azurin as a function of pH and temperature identified a peak corresponding to the methionine ligand methyl group which was shifted far upfield relative to most methionines (Adman *et al.*, 1982). The upfield shift may be the result of that methionine methyl group being affected by the ring current of Phe15 against which it is packed. Upon increasing the pH, or raising the temperature, that resonance decreases in intensity suggesting increased motion under these conditions.

The change of hydrogen bonding of His35 described above has been suggested (Adman *et al.*, 1982) as a possible explanation for the NMR behaviour of that methionine resonance: if the hydrogen bond of His46 to O10 were relaxed (i.e. ND1 of His35 competed successfully as an acceptor) the loop of residues containing Phe15 would be free to move more, relaxing the packing constraint on the methyl group of Met121. One might then expect other resonances to show changes too, such as those associated with Phe15. A complete assignment of ^{1}H-NMR resonances is in progress, and such details will soon be available (Hill, personal communication). The precise details of the environment of His35 must await higher resolution X-ray data.

A recent paper (Blaszak *et al.* (1982) on high resolution NMR (360 and 470 MHz) of Ni(II) azurin raises some interesting points. The aliphatic resonances for Ni(II) azurin resemble those of Cu(I) azurin, hence one may conclude that the metal has not affected these resonances. In contrast to the paramagnetic broadening observed with Cu(II) azurin, the paramagnetic Ni(II) gives narrow contact shifted resonances far from the normal aromatic and aliphatic regions. One of seven such resonances was assigned to Met-121 inasmuch as the peak integrated to three protons. When these resonances were examined as a function of pH (pH = 5.3 to 6.8 at 25°) the methionine resonance shifted downfield with decreasing pH. Out of the seven resonances, 6 were affected by the pH change. The pK of these changes is associated with a pK (= 6.0) which also is that of two resonances associated with His35. Thus there appears to be somewhat greater fluxionality at the metal site with Ni bound, which tends to amplify the effects of protonation-deprotonation of His35. Two questions come to mind: is this evidence for the equivalent of a His35 conformation change equivalent to oxidised azurin or reduced azurin? Is the lower pK a result of a looser metal binding site and hence easier protonatability?

2 Redox chemistry

Studies on the chemical behaviour of azurin include measurement of its kinetic behaviour with small inorganic oxidants and reductants and a novel exploration of the electron transfer pathway with redox inert Cr(III), as well as a kinetic analysis of the reactions of azurin and plastocyanin with various other protein redox partners. Much of this work has been well reviewed by Farver and Pecht (1981c). The various sites of chemical interaction to be described are schematically summarised in figure 1.3a for those sites for which it has been possible to identify a specific locus.

The relative 'kinetic accessibility' of plastocyanin, stellacyanin, high potential iron protein, cytochrome c and azurin, with respect to certain oxidants has been determined by using a Marcus theory formalism for interpreting the kinetic rate constants of the reactions of these proteins with $Ru(NH_3)_5py^{3+}$ (Cummins and Gray, 1977). The basic assumption of this formalism is that kinetic factors are separable from inherent driving forces of reactions, so that rates of reactions can be compared for pairs of reactants that have different redox potentials (Holwerda *et al.*, 1976). If the derived 'self exchange' rates are truly constant, one may conclude that a single reaction pathway is used; if they vary from one system to another then one may infer different reaction mechanisms. Cummins and Gray observed that the order of reactivity against the ruthenium compound was stellacyanin > plastocyanin > cytochrome c > Hipip > azurin, and that the self exchange rate constants determined for each of these versus the oxidants $Fe(EDTA)^{2-}$, $Co(phen)_3^{3+}$, $Ru(NH_3)_5py^{3+}$ and $Ru(NH_3)_6^{3+}$ varied. The authors suggested that the spread of self exchange constants represented the number of different mechanisms of electron transfer employed by a protein and could be considered a measure of the 'accessibility' of the electron transfer centre. Hence the large spread for azurin reflected minimal accessibility. A simple view would be one where all reagents react at a single surface of the molecule, and accessibility reflects the 'penetrating power' or ability of the oxidant to get close to the

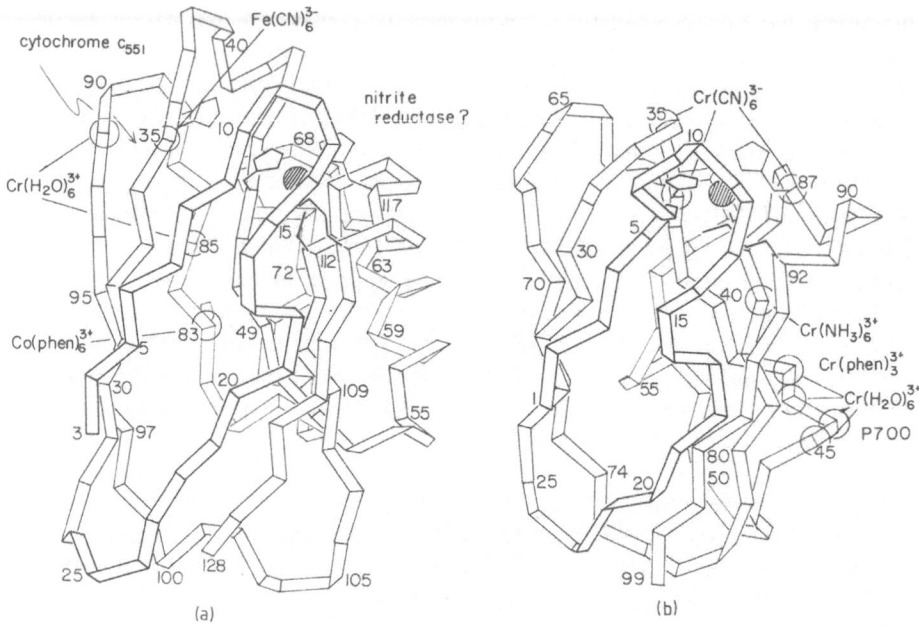

Figure 1.3 Identifiable sites of interaction of redox agents with (a) azurin and (b) plasto-cyanin

copper centre. However it is not likely to be that simple, inasmuch as other evidence suggests multiple sites of interaction at least for other oxidants.

Marcus theory has also been used to interpret the rates of interaction of azurin (and plastocyanin) with various cytochromes (Wherland and Pecht, 1978). Two important conclusions came from this work: (1) the inherent self exchange rates were among the highest ever measured and (2) the values were nearly constant for each protein implying that each protein always used the same pathway with respect to the particular partners used in the study. It would be interesting to know if this sort of analysis would be useful in comparing inherent reactivities of azurin with cytochrome oxidase as well as the small cytochromes, inasmuch as other evidence suggests one site of interaction for cytochromes and another for cytochrome oxidase (Farver *et al.*, 1982a).

Sykes and coworkers have compared rates of electron transfer of several inorganic oxidants of azurin at a variety of pHs and in the presence of other reactants (Lappin *et al.*, 1979a). The pH dependence of the second order rate constant for the oxidation of reduced azurin by $Co(phen)_3^{3+}$ and for $Fe(CN)_6^{3-}$ showed that the rate increases with increasing pH for $Co(phen)_3^{3+}$ and decreases for $Fe(CN)_6^{3-}$ with pKs of 7.6 and 7.1 respectively. Because the pKs resemble those of the two histidines described above, Lappin *et al.* suggest that $Co(phen)_3^{3+}$ and $Fe(CN)_6^{3-}$ interact via His83 and His35 respectively. There is some discrepancy with the ^{1}H-NMR results: the pH$_{mid}$ of the slowly titrating

His35 is somewhat lower than the pK determined from the rate constants; however, qualitatively the assignments agree. (These differences may be due to the different experimental conditions required.) There is yet another site of interaction for the third oxidant used (Co(4,7-DPSphen)$_3^{3-}$). Its reaction exhibits little pH dependence, and does not affect the reaction with Fe(CN)$_6^{3-}$. Since the DPSphen has more hydrophobic character it may seek out a more hydrophobic binding surface, such as the face of the molecule. Its low effectiveness in transferring an electron may then be the result of its bulk, or non-optimal orientation with respect to the copper.

Lappin *et al.*, caution that self exchange 'constants' may be suspect since they rely on a measured redox potential. If a particular redox couple interacts at a pH dependent site, the apparent redox potential will vary with pH, and hence the self exchange constants may vary with pH. Lappin *et al.* point out that the redox potential of azurin varies from 360 mV at pH 5.0 to 300 mV at pH 9.1, when measured with the Fe(CN)$_6^{3-}$ couple, which binds to azurin. They suggest that the redox potential might be found to be constant with pH if Co(4,7-DPSphen)$_3^{3-}$ were used.

The pathway involving His35 has also been suggested by the use of substitutionally inert Cr(III) complexes. This approach has been used by Farver and Pecht on a number of blue copper proteins and had been used a number of years ago on cytochrome c (Grimes *et al.*, 1974). A basic premise is that the chromium remains at the site of electron transfer. In point of fact, although determination crystallographically of the location of the chromium on cytochrome c has never been reported (possibly because phosphate buffer is used in the crystallisation of cytochrome c), the chemical analysis showed it to be at the site which later was shown crystallographically (Takano and Dickerson, 1981) to exhibit the largest differences between oxidised and reduced forms. So there *is* evidence that the chromium in fact can label the active site.

In azurin the Cr(III) is found to label a tryptic peptide containing residues 80-92, and a chymotryptic peptide 84-95 (Farver and Pecht, 1981a,b,c). The possible chromium binding side chains common to both of these are Lys85 and Glu91, both of which are conserved in all except *A. faecalis* azurin. According to Pecht (personal communication) *A. faecalis* azurin is not labelled with Cr(III). Farver and Pecht suggested that the transfer of an electron from chromium to the copper was mediated by water (which would get in to His35), and by nearly parallel orientation of the plane of His35 (and hence its π orbitals), with His46. In spite of the preliminary state of the coordinates on which this hypothesis is based, the hypothesis is useful. Two further observations were extremely interesting. The Cr(III) is lost if the protein is reoxidised with Fe(CN)$_6^{3-}$ suggesting that Fe(CN)$_6^{3-}$ acts at the same site (which is not inconsistent with the results of Lappin *et al.* described above), but binds more strongly, or that a product of CrFe(CN)$_6$ (Farver, personal communication) forms and then the protein is oxidised with excess Fe(CN)$_6^{3-}$. The reaction with cytochrome c$_{551}$ is slower if Cr(III) is already bound, suggesting that this is also the site used by cytochrome

c_{551}. The reaction with cytochrome oxidase is unaffected, suggesting that that reaction occurs elsewhere.

Silvestrini *et al.* (1981) have studied the pH dependence of the interaction of azurin and cytochrome c_{551} using temperature jump methods and observed two relaxation processes, one slow and one fast previously described by Pecht and Rosen (1973), Wilson *et al.* (1975) and Rosen and Pecht (1976). The fast process is attributed in the most recent paper to the transfer of electrons from reduced azurin to oxidised cytochrome c_{551}, the slow one to a previously hypothesised conversion between an active and inactive form of reduced azurin. A maximum in the amplitude of the slow relaxation event was found near pH 7, and protons are released at higher temperatures. Thus, since they previously observed that higher temperatures favoured the *inactive* form, Silvestrini *et al.* concluded that the active form is protonated, which is consistent with the observations of Lappin *et al.* on the pH behaviour of $Fe(CN)_6^{3-}$ interaction with His35. Since the time constant for the slow step is approximately the same as that observed by NMR to be associated with a slowly titrating histidine, Silvestrini *et al.* proposed that the switch from inactive to active form was correlated with the slowly titrating histidine 35 and is due to two environments for His35 in the protonated and unprotonated forms, both being hydrogen bonded.

Farver *et al.* (1982) subsequently showed from the kinetics of oxidation of the chromium labelled azurin with cytochrome c_{551} that labelled azurin exhibited the same two relaxation processes, one fast and one slow. The slow step was again attributed to two different conformers associated with His35. Interestingly they found the rate of conversion between conformers to be sensitive to both buffer type and concentration (slower in the HEPES buffer than in phosphate). At comparable buffer concentrations relaxation amplitudes for both the slow and fast steps differed considerably, between native and Cr-labelled azurin while relaxation times did not. In contrast to Silvestrini *et al.* these workers find a negative heat of reaction for the active to inactive conversion (the slow relaxation) for both the Cu-labelled azurin and the unlabelled while observing that hydrogen ions were released as the temperature was increased and hence were led to the conclusion that the active form is a *deprotonated* form.

Azurin from *A. faecalis* exhibits only a single fast relaxation time in its reaction with cytochrome c_{551} (Rosen *et al.*, 1981; Wherland and Pecht, 1978). Recently Mitra and Bersohn (1982) looked for NMR resonances associated with His35 in *A. faecalis* azurin in both oxidised and reduced forms, and in Co-substituted and Hg-substituted azurin. Consistent with the temperature jump studies (Pecht) no titratable resonances associated with His35 could be found, although the titratable ones assigned to His83 behaved just as they did with *P. aeruginosa* azurin. Mitra and Bersohn concluded that His35 could not be protonated. They suggested that since His35 lies in a cleft, it is perhaps less accessible in the *Alcaligenes faecalis* azurin possibly because residue 34 is a lysine for both *Alcaligenes faecalis* and *Alcaligenes spp.* azurins which show similar behaviour.

Figure 1.4 schematically illustrates the region around His35 viewed from the surface of the molecule which I believe is responsible for this effect. The cleft described by Mitra and Bersohn probably corresponds to the region between strands A and C. The sequence changes illustrated in the figure suggest a rationale for this. Residue 34 is lysine, but the neighbouring surface residue (residue 8) is glutamic acid for *A. faecalis* azurin and could be salt linked. Ser89, whose side group is hydrogen bonded to the carbonyl of Gly37 in *P. aeruginosa*, is Gly in *A. faecalis*. The combined effect of the loss of the hydrogen bond between strands C and F and gain of the salt link between C and A could be to close off access to His35 by permitting chain C to move to the right. Since the pH effect is also absent in *Alcaligenes spp.* azurin (Wherland and Pecht, 1978), the OG89-O37 bond must be the more important interaction since although residue 34 is a Glu, residue 8 is a Ala, and there is no possible salt link.

This hypothesis is testable: *A. denitrificans* should show the same non-titratable histidine since it has the same amino acid changes as *A. faecalis* azurin. The structure of *A. denitrificans* azurin shows His35 to be hydrogen bonded to N37 and O44. This implies that His35 is unprotonated at pH 6 and hence lends support to the observations of Mitra and Bersohn that His35 is not titratable in *A. faecalis* azurin, with a similar environment. Both changes observed in *Alcaligenes faecalis*, namely Gly replacing Ser89 and Lys and Glu replacing Ser34 and Gln8 respectively, are present in *Alcaligenes denitrificans*. Possibly as a result of this, N36 (strand C) is then close enough to hydrogen bond to O9 (strand A). In point of fact, N36 is hydrogen bonded to O9, effectively both closing off the cleft, and preventing His35 ND1 from hydrogen bonding to O9 as postulated

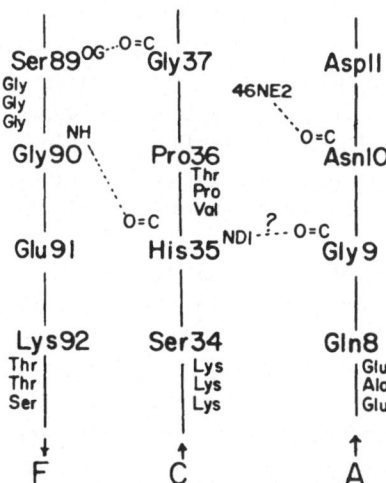

Figure 1.4 Comparison of sequences of residues near His35, and which may be involved in protonatability of His35. The first is from *P. aeruginosa*, the second, *Alcaligenes faecalis*, the third, *Alcaligenes spp.* and the fourth *Alcaligenes denitrificans.* Sequences are from Ambler (1971)

from the *P. aeruginosa* X-ray coordinates. In *P. aeruginosa* azurin residue 36 is Pro rendering this hydrogen bond impossible and making possible the His35 O9 bond. The combined effects of the hydrogen bond between 89OG and O37, and the Pro in position 36 in *P. aeruginosa* also prevent N37 from being available to hydrogen bond to His35; hence making His35 ND1 available for protonation in the *aeruginosa* structure, but not the others. The kinetic experiments need to be done on *Alcaligenes denitrificans* azurin.

Two forms of reduced azurin have also been postulated to be necessary to account for the kinetics of the interaction of unlabelled azurin with the oxidase (Brunori *et al.*, 1975; Parr *et al.*, 1977). Nevertheless the rate of reduction of cytochrome oxidase does not appear to be affected by Cr(III) (Farver *et al.*, 1982). If these two forms are the same as those postulated for the interaction with cytochrome c_{551} (which has been assumed because the time constants are similar (Brunori *et al.*, 1975)), then cytochrome oxidase does not interact with azurin at the same site as c_{551}, for if it did, the bound Cr should affect its reaction in the same way.

Cytochrome peroxidase will also react with reduced azurin (Ronnberg *et al.*, 1981); two interconvertible forms of the *peroxidase* are required to explain its kinetics. However, the two pseudo first order rate constants are affected by azurin concentration. The authors dismissed the possibility that the azurin inter-conversion affected the peroxidase interconversion because the latter was also seen (quantitatively?) for reduction with $Fe(CN)_6^{3-}$. Moreover, the reaction was carried out at pH 6, where the azurin would be expected to be more in the active form. pH studies carried out on the peroxidase reaction would be helpful.

The pH profile of the rate of reconstitution of apoazurin suggests that a residue with a pK of 6.75 is involved in the reconstitution (which is faster at higher pH). The similarity of far uv ORD and Trp fluorescence emission spectra for apo- and holo-azurin suggest that these structures differ little (Marks and Miller, 1979). Visual inspection of the structure shows His46 to be buried to about the same extent as His35. It seems likely that His46 is the residue detected in the reconstitution, and that the slight conformational change is the replace-ment of a hydrogen bond from His46 to His117 by the copper as it becomes ligated.

In summary, azurin must have at least two surfaces through which biological molecules may interact: one which cytochrome c_{551} uses, which is also used by Cr, and probably $Fe(CN)_6^{3-}$, and one used by nitrite reductase (cytochrome oxidase). $Co(phen)_3^{3+}$ probably interacts via His83, and $Co(4,7\text{-DPSphen})_3^{3-}$ through yet another, pH independent, pathway. It is not clear where the redox agents used by Gray and coworkers interact. The invariant residue His35 is involved in one internal path to the copper centre inasmuch as it has been identified by several different methods to be involved in the kinetics of azurin reactions.

Recently, a chemically crosslinked complex between *Pseudomonas aeruginosa* azurin and *Ps. aeruginosa* cytochrome c_{551} has been reported (Marks, 1982) which when fully characterised will surely clarify the nature of the interaction between the two.

The higher reactivity at low pH vs high pH of cytochrome c with its lysine charges chemically modified (Augustin *et al.*, 1983) parallels the pH behaviour of $Fe(CN)_6^{3-}$ with azurin, and indirectly suggests that cytochrome c also acts through the $Fe(CN)_6^{3-}$ site. The enhanced reactivity at lower pH, with a pK of 6.9 would seem to be more consistent with the postulation of Silvestrini *et al.* that the *protonated* His35 form of azurin is more reactive, and inconsistent with the conclusion of Pecht *et al.* that the unprotonated form is more reactive.

III PLASTOCYANIN

There is an excellent review of the origins and properties of plastocyanins (Boulter *et al.*, 1977) which should be consulted for complete references to the literature. Although copper was recognised as an essential element of plants in 1937, plastocyanin was not characterised until 1961, by Katoh and Takamiya. Plastocyanins are found in nearly all chloroplast containing organisms, including some blue-green bacteria. There are some cases reported however in which a cytochrome c_{553} can be induced instead of plastocyanin. Other small blue copper proteins in plants include stellacyanin, umecyanin, cucumber basic blue protein (plantacyanin) and mavicyanin. These are not all chloroplast associated. Some of their properties are listed in Table 1b. There are also larger plant proteins laccase and ascorbate oxidase with blue centres not covered in this review.

(a) General properties

Plastocyanins are generally smaller than bacterial cupredoxins, and their characteristic blue absorption maximum is at a slightly shorter wavelength, 597 nm instead of 625 nm. Their epr spectrum is like azurin, usually axial, with small hyperfine splitting constants. Purification is sometimes complicated by the presence of bound ferricyanide (Boulter *et al.*, 1977), which is used to keep the protein oxidised. There may be carbohydrate associated with the protein—this feature is still problematical. Plastocyanin generally lacks arginine and tryptophan and has only one cysteine. The algal plastocyanins are generally longer than the plant ones and may be less acidic. The complete sequences of 12 plastocyanins are given in Boulter *et al.*, 1977; Freeman (1981) refers to 67 known sequences.

Both Boulter *et al.* (1977) and Ryden and Lundgren (1976, 1979), used the homologies among the sequences of plastocyanins and azurins to predict the probable ligands to the coppers, the latter group correctly predicting the two histidines, a cysteine, and a methionine. Boulter *et al.* include a sequence for 'dock' (*Rumex obtusifolius*) plastocyanin which at the time had a leucine at the methionine position. There is extensive homology among all the remaining residues. However, that residue is in fact a methionine according to the more recent complete report of the sequence (Haslett *et al.*, 1978). Apparently the peptide containing that methionine could never be isolated in sufficient quantity to sequence it, so that, as pointed out by the authors, the evidence for its being a methionine rests on amino acid composition and cleavage specificity.

(b) Structure

The X-ray structure of plastocyanin from poplar leaves (the first that was crystallisable after many attempts with other plastocyanins) has been reported at 2.7 Å resolution (Colman *et al.*, 1978), and some details of the elegant work at 1.6 Å have appeared (Freeman, 1981; Guss and Freeman, 1983). It was crystallised at pH 6.0 from phosphate buffer and ammonium sulphate. The structure was solved using the multiple isomorphous replacement method, one of the derivatives was created by replacing the copper with Hg(II). The metal ligand distances determined from the refined oxidised structure at 1.6 Å are Cu–N (His37) 2.04 Å, Cu–S (Cys83) 2.13 Å, Cu–N (His87) 2.10 Å, and Cu–S (Met92) 2.90 Å (Freeman, 1981). The folding topology is summarised in figure 1.1b which has been oriented to maximise the resemblance between azurin and plastocyanin. Figure 1.1d illustrates the two beta sheets. Figure 1.1f illustrates other features similar to those of azurin including a 'dog leg', the 'tyr loop', and highlights the difference in the strand which corresponds to the 'flap' in azurin.

Comparison of the two structures, by visual identification of spatially equivalent residues, leads to identification of the insertions/deletions illustrated in figure 1.1g and 1.1h. Figure 1.5 gives the alignment of composite sequences (after Ryden and Lundgren, 1976) purely from this qualitative structural viewpoint.

According to this alignment there are nine regions which have additions and deletions. There are deletions in three adjacent strands at the bottom of the azurin molecule leading to the rather compact plastocyanin structure. There is also an *insertion* in the A-B loop as well as deletions in the C-D loop and G-H loop, with the remaining deletions in strand E and the E-F loop (see figure 1.1g and 1.1h).

Chothia and Lesk (1982) have also compared the three dimensional structures of plastocyanin and azurin to understand their evolutionary relationship. We concur with their alignment of the C-terminal sheet, but believe that the N-terminal sheet is out of register by two residues. Their observation remains true that the interior surfaces of the beta sheets contain different residues and therefore shift relative to each other because of differences in the bulk of the residues which have to pack together, but the sense of the shift differs in our comparison.

It is useful to consider the residues which have been equivalenced by our comparison in seven classes summarised in table 1.2 and figure 1.6a and b: the ligands (I), the hydrophobic residues on the interior of the beta sheet which pack together (II), those which pack next to ligand residues (III, IV, V), a pair

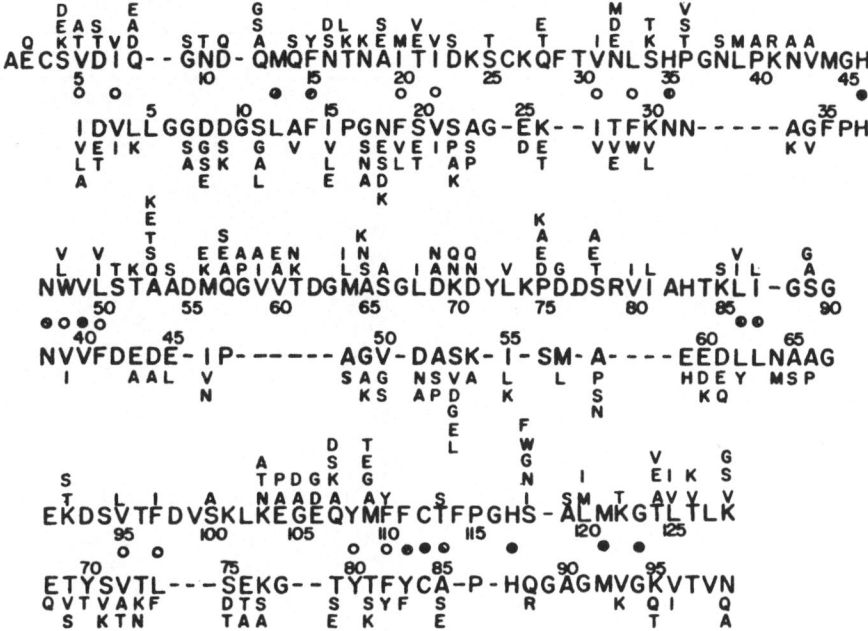

Figure 1.5 Sequence alignment of azurin and plastocyanin, using sequences of Ryden and Lundgren (1976), but aligned according to structure

of residues external to the barrel (VI) and two invariant glycines (VII). The ligands are of course the two histidines, the cysteine and methionine. Group II residues have been described before by Adman and Jensen (1981) and Chothia and Lesk (1982).

Group III, IV and V residues are perhaps the most interesting to consider, for they effectively enlarge what we consider to be the electron transfer centre area. Within group (III) are residues Met13 and Phe15 in azurin, Met12 and Phe14 in plastocyanin, which close pack on either side of the methionine ligand. Morgan *et al.* (1978) have observed a clustering of sulphur containing and aromatic side chains in several globular proteins, which others argue represent an energetically favourable situation. Morgan (1982, personal communication) pointed out to us that these (group III) residues form part of such a clustering in azurin, which he feels may illustrate a good opportunity for extensive delocalisation of electrons.

Asn47 or 38 (group IV) is invariant in both structures and is part of a rather intricate hydrogen bonded network which helps maintain the Cys environment. Thr113 (Ser85) and main chain atoms of residue 61 in plastocyanin are also involved. These hydrogen bonds combine with hydrogen bonds from O46 (or O37) to N87 (or N63) and result in NH47 (or NH38) being directed towards the cysteine sulphur, probably forming an NH . . . S hydrogen bond.

Table 1.2 Environmentally equivalent regions in azurin and plastocyanin

Group	Function	Residues	
		Azurin	Plastocyanin
I	ligands	46, 112, 117, 121	37, 84, 87, 92
II	internal	see figure 1.6a and b	
III	methionine neighbours	Met 13 Phe 15	Leu 12 Phe 14
IV	cysteine neighbours	Asn 47 Thr 113 O46...N87 N47...S112	Asn 38 Ser 85 O37...N63 N38...S84
V	histidine neighbours	His 35 Ile 87 Ile 86 O10...N46E2	Asn 31 Ile 63 Ile 62 O33...N37E2
VI	???	Phe 111 Val 49 Tyr 108	Tyr 83 Val 40 Tyr 80
VII	glycines	Gly 123 Gly 90	Gly 94 Gly 67

The environment of His46 is affected by group V residues: His35 and Asn31. Although to my knowledge Asn31 has not been implicated in any kinetic behaviour of plastocyanin, because it is invariant, and does correspond in relative position to the aforementioned His35, it would not be surprising to find it to be kinetically important. The environment of His46 may be further affected by the beta sheet holding the cluster, which is extended by the hydrogen bonds from main chain atoms of residue 87(63) to those of 46. The consequent relative orientation of Ile87 (Leu63) to His35 (Asn31) are the same, as are those of Az-Ile86 and Pc-Ile62 to Az-Asn47 and Pc-Asn38 respectively.

Group VI residues Az-Phe111 (Pc-Tyr83) and Az-Va149 (Pc-Va140) which project towards the outer molecular surface of the C-terminal beta sheet, may be part of an extended environment not affecting the ligands directly but never-theless required for the function of the molecule. As we shall describe below, Tyr83 of plastocyanin turns out to be quite important for its function. Finally, Group VII: the conserved glycine 123 is necessary for the twist in the beta sheet at that point, while that at 90 and 67 is necessary for the 3_{10} turn at that point.

Tyr108 and Tyr80 are both internal and hydrogen bonded to a carbonyl oxygen six and four residues preceding them, respectively. The function may be purely architectural, a way of keeping the crossover connection between the

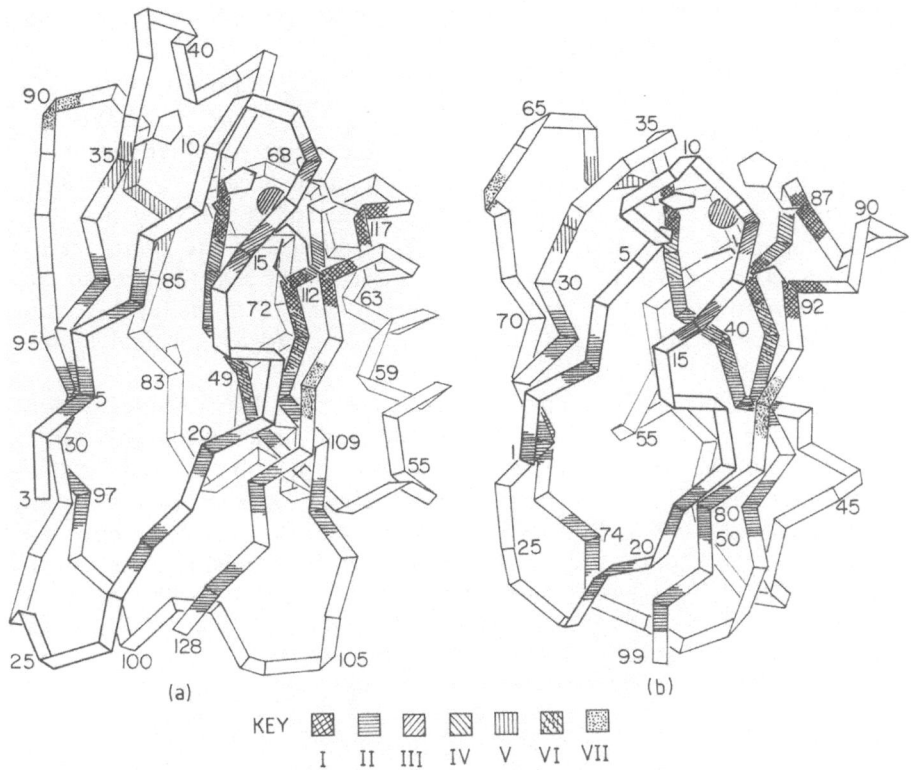

Figure 1.6(a,b) Structurally equivalent residues of azurin and plastocyanin shaded according to grouping in table 1.3

two sheets relatively compact. The dog leg illustrated in figure 1.1c and d is also a crossover connection between sheets, but despite the similar path it is comprised of entirely different residues for each protein.

The differences in detail, where similarities are expected are also noteworthy. One feature which aids in maintaining the relative orientation of the ligands to the copper is the hydrogen bond from His46 (His37) to main chain carbonyl oxygens. In azurin the acceptor is O10, a carbonyl oxygen in the A-B loop, the loop which is shorter in azurin. In plastocyanin, the acceptor is O33, which is in the C-D loop, which is a shorter loop in plastocyanin.

The 'hydrophobic patch' has been suggested as a possible site of electron transfer interaction for both proteins (Colman *et al.*, 1978; Adman *et al.*, 1978). The 'hydrophobic patch' is less extensive in plastocyanin (figure 1.2a and b): although there is a rather extensive surface from which charged residues appear to be excluded. The copper in plastocyanin is much more accessible to the surface than it is in azurin, primarily due to the absence of Phe114. A very important feature of plastocyanin which is absent in azurin is the negative patch consisting of residues 42–45, a patch which can be extended to define a ring including conserved residues 59, 60 and 51 (figure 1.7).

The X-ray structure of the reduced form of poplar plastocyanin has been determined at several pHs. The interesting result is that at low pH the His87 to Cu distance increases much more than would be expected. Freeman *et al.* propose that in fact the histidine becomes protonated since a very nice correlation between Cu-N distance and pH can be made, and in the X-ray structure one is looking at an average of a population of protonated and unprotonated molecules. At pH 6.0 there is very little difference in the structure around the copper between the oxidised and reduced forms, whereas in the low pH form His87 becomes protonated and the copper becomes trigonal, bound to the methionine, histidine and cysteine.

Other crystalline plastocyanins have been reported; that from pea (*Pisum sativum*) and corn (*Zea mays*) give excellent diffraction photographs (Chirgadze *et al.*, 1977). Perhaps a more interesting one with respect to establishing evolutionary relationships is that from a green algae, *Enteromorpha prolifera* (Yoshizaki *et al.*, 1981). The algal sequences, especially that from *anabaena* differ quite a bit from the plant ones, although all the internal residues remain much the same. The remarkable negative patch consisting of residues 42–45 is hydrophobic in the *anabaena* sequence. The sequence of the *Enteromorpha* plastocyanin has not been reported, nevertheless should a structure be made available, functional comparisons of the two would be quite informative.

Recently, preliminary crystallographic data for a cucumber plastocyanin and possibly a second crystal form of *Enteromorpha prolifera* algal plastocyanin have been obtained (Freeman *et al.*, 1983), and the sequence of cucumber Pc reported (Ramshaw, 1982).

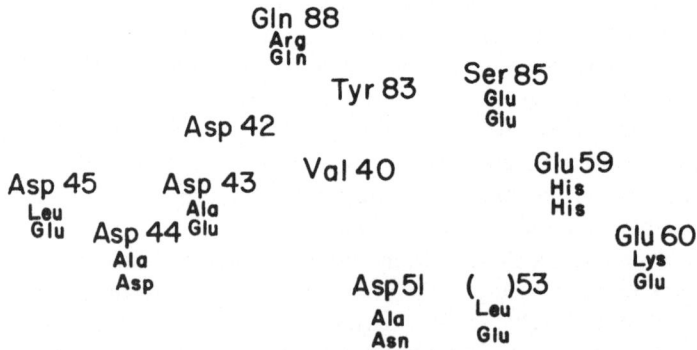

Figure 1.7 Surface residues of plastocyanin in the vicinity of the negative patch, the proposed site of cation effect on the interaction with P700 (see text). The first sequence is poplar, the second and third from *Anabaena* and *Chlorella* respectively (Boulter *et al.*, 1977)

(c) Physical and chemical properties

1 Spectroscopy
The polarisation ratios of optical absorption observed in the crystal can be correlated with the four positions of the copper centre found in the crystal and thus clearly confirm the assignment of the major 597 band to charge transfer from the cysteine sulphur to the Cu (Penfield *et al.*, 1981). Moreover, the orientation of the $d_{x^2-y^2}$ Cu orbitals containing the reducing electron is such that the normal to the plane containing those orbitals is about 5° off from the Cu-S(Met) bond. Figure 1 in Penfield *et al.* (1981) suggests that two adjacent lobes contain each of the histidine nitrogens while the other two lobes then would lie between the Cys and each His, an observation which may prove important in understanding how reducing electrons enter the protein. The reasons for the axial nature of the epr spectra were also explored with single crystal epr spectra, and it was concluded that ligand field strength of ligands is as important as describing the site geometry as 'C_{3v} with significant rhombic distortion'. Unfortunately in all of the analyses of spectra of blue copper proteins (Gray and Solomon, 1981) it is not clear to me what structural feature is likely to correlate with the characteristically shorter absorption wavelength for plastocyanin over azurin.

The length of the Cu—MetS bond of 2.9 Å has led workers to question whether or not it is in fact a bond, and as discussed earlier for azurin, a Cu—MetS bond was not important in assigning resonance Raman spectra. In a further EXAFS study intended to explore the reasons for the weak or nearly absent contribution of this bond to the spectrum, Scott *et al.* (1982) examined single crystal EXAFS spectra in orientations such that the Cu—S' bond should contribute maximally. The contributions of the remaining ligands were predictably orientation dependent, while the Cu—S' bond persisted in lacking a contribution. A possible explanation offered was that since the bond was weak, there was little *correlated* motion between the Cu and the Met S; hence the back scattering on the average was small. Clearly one must conclude that the absence of a feature in the EXAFS spectrum does not necessarily mean that the neighbour is absent in the environment. As pointed out by the authors, the complementarity of the two techniques needs to be emphasised.

NMR studies have been carried out on French bean, spinach, *Anabaena* and cucumber plastocyanin. The earlier studies summarised in Ulrich and Markley (1978) focussed on assignment of residues in the neighbourhood of the copper, primarily by identifying those resonances which are most sensitive to oxidation of the reduced form. Histidine proton resonances could be identified in the proton spectra; in ^{13}C-NMR fifteen resonances were affected by the form of the copper, and were tentatively identified according to distance from the copper. In addition five resonances present in the oxidised form but not reduced are indicative of small localised conformational differences between the two structures since they must be shifted in the reduced form rather than absent.

2 Redox chemistry

Some of the initial clues to numbers of reactive sites were provided by studies of the rates of reactions of plastocyanin with small inorganic oxidants. Three sites of interaction depending on the charge of the reactants and the nature of the ligands were inferred from the kinetic experiments. $Co(4,7\text{-DPSphen})_3^{3-}$ oxidises reduced plastocyanin independently of pH from 5.2 to 7.5, and binds strongly (Lappin *et al.*, 1979b). When it is bound it does not affect the rate of electron transfer from $Fe(CN)_6^{3-}$ (it oxidises plastocyanin(I) about 500 times more slowly than $Fe(CN)_6^{3-}$) indicating at least two sites for electron transfer. $Co(phen)_3^{3+}$ oxidises plastocyanin and has a strong pH dependence (increasing with increasing pH); its rate of reaction *is* affected by the presence of redox inert $Cr(phen)_3^{3+}$ whose binding in turn is affected by ionic strength. The pH dependence of this reaction suggests a third site, the pK of which is about 6.2

According to another NMR study chemically inert $Cr(CN)_6^{3-}$ binds near the copper and affects resonances of His87, His37 and Phe35 (Cookson *et al.*, 1980). These effects can be eliminated by raising the pH to 7.0 from 6.3. Positively charged oxidants $Cr(NH_3)_6^{3+}$ affect the Tyr83 resonance and some resonances in the aliphatic region.

The $Cr(phen)_3^{3+}$ site probably corresponds to the $Cr(NH_3)_6^{3+}$ site determined by NMR to affect the region containing the invariant acid residues 42–45 (see figure 1.7. (The $Fe(CN)_6^{3-}$ site, according to NMR, is the ligand His87. The $Co(4,7\text{-DPSphen})_3^{3-}$ site is unknown.) The Cr site has also been indicated by reacting Pc with Cr(II) and isolating the peptide containing the substitution inert bound Cr(III) (as was done earlier with azurin). The peptide was found to contain residues 42–45, which are all acidic (Farver and Pecht, 1981a). The presence of the chromium label was found to affect tyrosine fluorescence emission, a further indication that it is near Tyr83. It was observed that the presence of Cr(III) did not influence the rate of photoreduction with chloroplasts, but did slow the photo-oxidation of Pc-Cr(III) by isolated photosystem I reaction centres. The reduced system could be reoxidised with $IrCl_6^{3-}$ with no loss of Cr(III), indicating that $IrCl_6^{3-}$ acts on a different pathway (Farver and Pecht, 1981a,b,c).

The kinetics of the reactions of Cr(III) labelled plastocyanin with chloroplasts and pure Photosystem I particles have been described for both unlabelled and Cr(III) labelled plastocyanin(II) in terms of rates of electron transfer from chloroplasts directly to plastocyanin and between a 'dead-end' complex of plastocyanin(II)-cytochrome f and plastocyanin(II) (Farver *et al.*, 1982). These rates are unaffected by chromium labelling. A similar analysis of electron transfer to photosystem I particles yielded two rate constants for the oxidation of plastocyanin(I) by P700 and again a slow rate from an abortive complex. What is important is that the rate for oxidation of reduced plastocyanin was considerably slowed by the presence of chromium whereas the rate for reduction of plastocyanin(II) by chloroplasts was unaffected. This then suggests that since the Cr(III) label goes on the 'negative patch' at residues 42–45 this is the pathway from which an electron passes to another molecule, and further that this is not the site used by cytochrome f.

The rate of reduction of spinach plastocyanin by ascorbate is also pH dependent in the range from 5 to 10, going up by orders of magnitude after pH 7.0. This however is presumed to be not a protein effect, but because divalent ascorbate is the species involved. The rate of electron transfer from ascorbate to plastocyanin is so poor that it is felt to not be a physiologically important reaction (Takabe *et al.*, 1980a).

Plastocyanin undergoes autoreduction at higher pH (Takabe *et al.*, 1980b). The observation that the rate of reduction by $Fe(CN)_6^{4-}$ is unchanged at higher pH *again* suggests multiple sites for electron transfer. Restoration of blue colour with $Fe(CN)_6^{3-}$ indicates that the decolorisation is truly reduction and not loss of copper. The epr spectrum indicated that the environment of the remaining oxidised plastocyanin was different at higher pH. The fact that the protein remained intact is indicated by the unchanged fluorescence spectra, and the reversibility of the reaction. A possible explanation (Takabe *et al.*, 1980b) of the autoreduction at higher pH was that the redox potential was sufficiently increased *for that pathway* that the equilibrium shifts to reduced protein. This means that a deprotonated species of plastocyanin would prefer to be in a reduced form, and a population which is not deprotonated (reflected by the pK of whatever is protonated) still reacts with $Fe(CN)_6^{3-}$. One possible explanation could be deprotonation of His37 (NE2) so that there would be no longer a hydrogen bond to the protein backbone at O33. Azurin also undergoes autoreduction at higher pH (Brill *et al.*, 1968). It is highly speculative, but conceivable that the loss of His46 hydrogen bond to O10 could do the same thing.

This is not the same as the existence of the protonated form of plastocyanin at low pH which Katoh *et al.* (1962) proposed to explain an increase in redox potential below pH 5.5. In that case, His87 is protonated forcing Cu to a three coordinate environment which is redox inactive as described above.

Another study on the interaction of plastocyanin with P700 in spinach photosystem I particles (Davis *et al.*, 1980) also implicates this pathway. In this study the effect of cations on the rates of reactions of an acidic (plant) plastocyanin and basic (algal) plastocyanin and several cytochromes which are known to replace plastocyanin, on P700 was examined. Cations enhance the rate of reaction of the acidic plastocyanins and depress the basic ones. Davis *et al.* point out that a major difference between algal and plant plastocyanin is the replacement of the ring of negative charges which includes the Cr binding site, by uncharged alanine in 43–44, Leu in 45; and a His-Lys for Gln-Glu 59 and 60. We should also point out that Gln88 is replaced by Arg, and Asp51 by Ala, which is also likely to affect this region. However, a neighbouring residue 85 becomes Glu in *Anabaena*, keeping at least one negative charge near 59–60 (see figure 1.7). Davis *et al.* suggest that the apparent evolution from an efficient to a less efficient system (one which does not require cations to one which does) means that in the eukaryotic system there is yet another factor between plastocyanin and P700. The extent of the surface changes implicates a much larger interaction surface than has been suggested previously by the studies with inorganic redox agents, and suggests some reason why 59 and 51 are also conserved, inasmuch as they may be involved in specific orientation effects.

Tyr83 is adjacent to the ring of negative charges, and as described above, is the site associated with perturbation of NMR signal when $Cr(NH)_3^{3+}$ is bound. Nevertheless, when Tyr83 (probably) is nitrated (Davis and San Pietro, 1979), it does not affect the electron transfer between plastocyanin and P700. (It is not likely that the other tyrosine (80) is nitrated because it is hydrogen bonded internally and thus far less accessible to outside reagents.) This would suggest that nitration does not affect the properties of Tyr83 that are relevant to electron transfer even though Tyr83 is right in the middle of the surface presumably affected by the cations. Nevertheless the nitrated Tyr has not been identified.

Finally, four carboxyl groups can be chemically modified with a water soluble carbodiimide which effectively neutralises the negative charge and hence abolishes the cation effect on the reaction of plastocyanin with P700 (Burkey and Gross, 1981): the rate of reaction is greatly enhanced without added cations. The redox potential (measured with $Fe(CN)_6^{3-}$) increased to 423 from 393 mV as a result of the modification. The effect on the redox potential is not due to changing binding properties of $Fe(CN)_6^{3-}$, for there is no concentration effect. Therefore it must reflect an increase in stabilisation of Cu(I)— analogous to the effect of increasing pH. Some of this effect could be abolished by an increase in ionic strength.

These results are reminiscent of similar modification studies carried out on cytochrome c (Aviram *et al.*, 1981). Increasing numbers of lysines in this molecule were maleated, and the changes of the redox potentials and rates of reaction with ascorbate measured as a function of ionic strength for each set of derivatives. In this case, the redox potentials decreased with an increase in the number of negatively charged groups (from 280 mV to 80 mV, when extrapolated to zero ionic strength for zero to thirteen modified lysines) in a manner expected from modifying purely electrostatic interactions.

Apparent electrostatic effects could be looked upon as acting in two possible ways: modifying the intrinsic redox potential associated with a redox centre by actually modifying the structure, or acting in accordance with Debye–Huckel theory in which differences in energy are associated merely with the fact that there are differences in charges. If the differences in redox potentials could be adequately described by the latter, there would be no need to speculate on the former. Nevertheless, in the case of blue copper proteins possibly both effects are important, particularly because the network of next nearest neighbour interactions around the copper centre is so complex and hence susceptible to change.

The question is, do the carboxyl modifications affect the structure or are only charge differences important? In the plastocyanin study (Burkey and Gross, 1981) it is unfortunate that the specific residues which were chemically modified were only partially identified. At most only two of five carboxyls were modified in residues 31–54, and one to two in the peptide containing residues 55–71. All of the carboxyl groups project from the surface of the molecules, so that modification would not be likely to affect internal structures and hence the copper environment and hence the redox potential. There are possibly two exceptions, one rather remote from the negative patch that is involved in the cation effect. In the poplar structure residue 68 is hydrogen bonded to the E-F loop, and is adjacent to Asn31. If ethylenediamine changed residue 68 to a positively charged one, that hydrogen bond would disappear perhaps then enabling residue 68 to interact with Asn31 which in turn affects its interaction with His37. The second modification possibly affecting the redox potential would be Asp61, which could perhaps change the network of hydrogen bonds around Asn38 which it is near and hence affect the complex. It seems more likely however that the effect is not on structure but purely electrostatic.

(d) Biological activity

Plastocyanin has rather clearly been established as the electron acceptor from cytochrome f, and an electron donor to the P700 reaction centres in chloroplasts (Wood, 1978). In some algae (*Scenedesmus*) the amount of plastocyanin present is dependent on the copper available during growth (Bohner and Boger, 1978; Bohner *et al.*, 1980a). In this organism, plastocyanin can be completely replaced by a cytochrome c_{553}. Another algae, *Bumilleriopsis* contains no plastocyanin, while a third cannot synthesise the cytochrome c_{553} (Bohner *et al.*, 1980b). Interestingly, in the absence of copper, *Scenedesmus* can be stimulated by silver ions to produce immunodetectable plastocyanin. This Ag-plastocyanin production decreased the production of the inducible cytochrome c_{553}. Silver will replace the Cu in plastocyanin *in vitro*, even with Cu in three to four fold excess, but in the cell excess copper could decrease incorporation of silver suggesting a specific carrier protein supplying the metal to apoplastocyanin (Bohner *et al.*, 1981).

There is a cytochrome c_{552} inducible in both in *Chlamydomonas reinhardtii*, and a mutant (*C. reinhardtii*-208), and which is present in another species, *C. mondati* which lacks plastocyanin completely. This cytochrome will replace plastocyanin in the plastoquinone–cytochrome f-plastocyanin-P700 chain (Wood, 1978). As noted by Wood, this situation is analogous to ferredoxin and flavodoxin being able to replace one another (Mayhew and Ludwig, 1975). Wood also suggests that since cytochrome c_{552} and plastocyanin both interact with cytochrome f, surface features on plastocyanin and c_{553} should be similar, a directly testable hypothesis, when and if the structures of both components from the same organism become available.

Although cytochrome f has been identified as the specific electron donor to plastocyanin, it is part of a larger complex which has been isolated from spinach chloroplasts. The complex contains five polypeptides, including a Rieske iron protein, two cytochrome b_6 and cytochrome f molecules, and catalyses electron transfer to spinach plastocyanin and to Scenedesmus cytochrome c_{552} (Hurt and Hauska, 1981). Notably, it is twice as reactive with its own plastocyanin as with cytochrome c_{552} from algae.

From studying the reduction of P700 while still in chloroplasts with flash photometry, Haehnel *et al.* (1980) conclude that plastocyanin transfers electrons to P700 best as a complexed form with a transfer half time of $20\,\mu s$ although a slower transfer (about $200\ \mu s$) also is observed when followed by changes in absorbance of the P700. The amplitude of the fast phase is increased by $MgCl_2$ or sorbitol and decreased by the plastocyanin inhibitors KCN and $HgCl_2$, while the half time is not affected. The authors interpret these observations to mean that there is a pool of plastocyanin which is complexed to P700 with a dissociation rate slow compared to the electron transfer reaction time. When complexed to P700, plastocyanin is not accessible to inhibitors, hence leaving the half time the same. Formation of the complex is enhanced by salt or low effective volumes available to plastocyanin within the thylakoid membranes. The authors further suggest that the binding constant must not be very high and that surface charge compensation is important in the binding. The latter view is consistent with the cation effect as described by Burkey and Gross (1981).

In light of the known structure, and the evidence pointing to the surface near the ring of negative charges being the region interacting with P700, the lack of an inhibitory effect by the plastocyanin inhibitor remains puzzling. If one envisages a complex involving the negative patch in plastocyanin it would seem that the inhibitory agent might still have access to the Cu centre. Clearly both the nature of the inhibitor and the nature of the complex have to be explored further. Since the results suggest that the binding site involves the negative patch, further studies with algal chloroplasts might be illuminating.

Lockau (1979) reported inhibition of photosynthesis in isolated chloroplast particles by low ionic strength and/or low osmotic strength. The site of inhibition was ascertained to be within Photosystem I since PS-2 was deliberately blocked. Because impermeable salts or osmotic agents had little effect on the total amount of light utilisation and because previous experiments with plastocyanin antibodies had localised plastocyanin within the thylakoid membrane, Lockau suggested that plastocyanin was the site of inhibition. It is not clear if plastocyanin is the only member of Photosystem I which is inside the membrane and which could behave in this manner, nevertheless the results are consistent with those of Haehnel described above.

Very little has appeared in the literature regarding the biosynthesis of these molecules. A very interesting study of the question of what proteins are transported into the chloroplast organelle demonstrated that plastocyanin can be synthesised *in vitro* as a precursor which is then transported into the chloroplast (Grossman *et al.*, 1982). The plastocyanin precursor is 15 000 Da larger than plastocyanin itself.

In short, one may conclude that there are at least two biologically important surfaces of interaction for plastocyanin and at least three for inorganic oxidants. The first involves the negative patch, and it is utilised by P700, Cr and may involve Tyr83. The second is used by cytochrome f, probably near His87 (Farver *et al.*, 1982b). $Fe(CN)_6^{3-}$ interacts near His87, and $Co(4,7\text{-DPSphen})_3^{3-}$ acts at a pH independent site. $Cr(phen)_3^{3+}$ probably uses the negative patch also.

IV OTHER BLUE PROTEINS

(a) Bacterial blue proteins

The bacterial blue copper proteins listed in table 1.1a as a group have lower redox potentials than that of *P. aeruginosa* azurin (except for rusticyanin) and smaller extinction coefficients as well as absorption maxima shifted more towards that of plastocyanin. Amicyanin and *A. faecalis* blue protein have lower molecular weights. The amino acid compositions of *P. aeruginosa* azurin, and the last five members of table 1.1a are given in table 1.3a.

The blue protein from *Achromobacter cycloclastes* (Iwasaki and Matsubara, 1973) is of interest because it turns out that its sequence, which has been

Table 1.3a   Bacterial cupredoxin amino acid compositions

	Az	Achr. Az	Par. Az	BP	Amic.	Rc
Lys	13	13	9	10	10	14
His	4	3	3	3	3	5
Arg	1	1	3	1	2	0
Asx	16	10–18	10	10	7	13
Glx	9	7–11	14	11	15	8
Thr	9	7	6	4	8	19
Ser	12	2	3	4	6	6
Pro	2	8	7	4	8	14
Gly	13	14	9	9	10	19
Ala	9	16	12	14	11	14
Cys	3	1	1	1	1	1
Met	4	5	4	5	2	3
Val	12	13	10	10	13	15
Ile	7	4	5	9	3	6
Leu	5	6	6	5	3	5
Tyr	2	3	2	4	2	6
Phe	6	5	5	3	4	9
Trp	1	0	0	0	1	2
	128	124	109	107	109	159

1. *Pseudomonas aeruginosa* azurin (Ambler (1971)).
2. *Achromobacter cycloclastes* azurin (Ambler (1982)).
3. *Paracoccus denitrificans* azurin (Martinkus *et al.* (1980)).
4. *Alcaligenes faecalis* S-6 Blue Protein (Kakutani *et al.* (1981)).
5. Amicyanin (Tobari and Harada (1981)).
6. Rusticyanin (Cox and Boxer (1978)).

Table 1.3b   Plant cupredoxin amino acid compositions

	Pc	Uc	Sc	CBP	Mc	Pc(algal)
Lys	6	10	10	6	14	4
His	2	3	4	2	5	2
Arg	0	3	4	3	4	1
Asx	13	17	18	10	19	15
Glx	10	8	4	5	13	6
Thr	2	11	9	5	10	6
Ser	11	5	5	6	13	4
Pro	5	7	3	5	10	5
Gly	10	13	8	13	19	13
Ala	7	8	3	6	10	12
Cys	1	3	3	3	5	1
Met	2	4	0	2	1	2
Val	9	10	13	9	14	11
Ile	6	6	5	4	7	7
Leu	6	4	3	4	11	4
Tyr	2	4	7	4	5	3
Phe	7	5	5	6	10	4
Trp	0	4	3	2	4	1
	99	125	107	96	174	101
CHO		3	40%	?	7%	

1. Plastocyanin (poplar) (Freeman (1981)).
2. Umecyanin (Stigbrand (1970)).
3. Stellacyanin (Bergman *et al.* (1977)).
4. Cucumber Blue Protein (Murata *et al.* (1982)).
5. Mavicyanin (Marchesini *et al.* (1979)).
6. Algal Plastocyanin (*Enteromorpha prolifera*, Yoshizaki *et al.* (1981)).

partially completed, in some respects resembles plastocyanin's more than azurin (Ambler, 1977) (figure 1.8), as does its spectrum. It has a -GHNV- stretch corresponding to the first His ligand and a -CTPH- stretch which includes two potential ligands. It lacks the characteristic two aromatic residues preceding the cysteine for both plastocyanin and azurin although there are enough residues preceding the cysteine to allow residue 74 to be matched with a F and 80 to be matched with a Y. In order to maximise the alignment of ligands one must align the sequences so that the region containing Az-His35/Pc-Asn31 is missing, as well as the neighbouring E-F loop containing the invariant gly90/67. The 'flap' containing strand E is greatly shortened over plastocyanin; a region corresponding to strand F can be aligned with the plastocyanin sequence and beyond the second His ligand through a stretch corresponding to strand H, with 30 additional residues beyond that. The fragments of matching residues and their relative locations are extremely intriguing– but really will be understood only when three dimensional structures are available and when more is known about the genetic relationships.

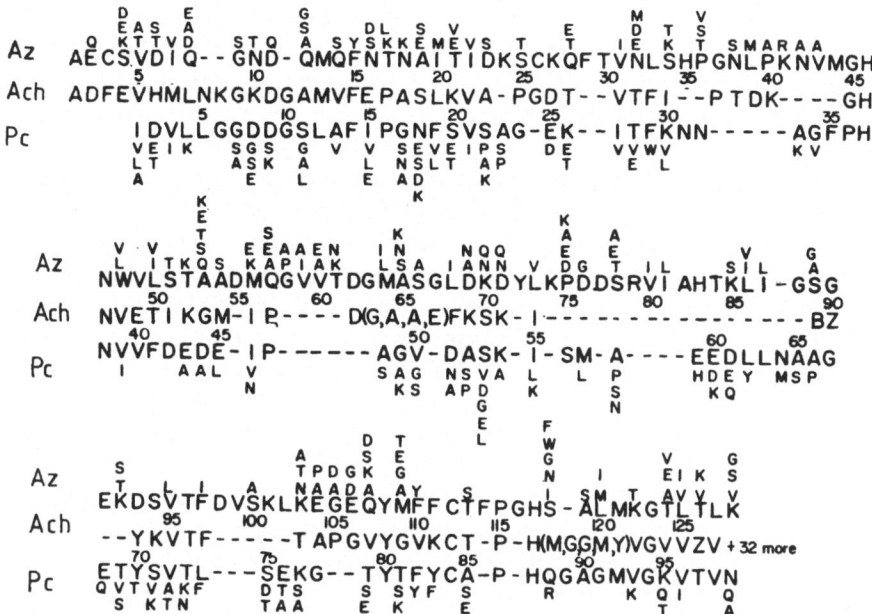

Figure 1.8   Possible sequence alignment of *Achromobacter cycloclastes* azurin with sequence of azurin and plastocyanin in figure 1.5. Sequence is from Ambler, 1977

Paracoccus denitrificans azurin has been characterised by Martinkus *et al.* (1980), and has properties differing slightly from that of *Pseudomonas* azurin, tabulated in table 1.1a and 1.2a. Its role as an electron donor to a cytochrome cd-nitrite reductase has been questioned because it is isolated in much lower yields than normal azurins, and it exhibits much greater reactivity with *Paracoccus* membrane fragments from cells grown aerobically and therefore presumably lacking nitrite reductase than it does with purified nitrite reductase.

Amicyanin is notable for several reasons. It appears to lack a disulphide, and has a molecular weight closer to plastocyanin. It was isolated as an electron acceptor from methylamine dehydrogenase from methylamine grown *Pseudomonas* AM1, and it donates electrons to cytochrome c from the same organism (Tobari and Harada, 1981). Methanol grown *Pseudomonas* AM1 and *Methylamonas* J both contain a blue protein which cannot be reduced by methylamine dehydrogenase and can be reduced by reduced cytochrome c_{551} as normal azurin is. *Thus these organisms switch between at least two different blue proteins depending on growth conditions* (Tobari and Harada, 1981).

The blue protein from *A. faecalis* S-6, whose structure is under investigation in our laboratory, has an amino acid composition similar to amicyanin, also lacking a disulphide. It is reported to specifically inhibit a copper containing nitrite reductase by being *required* to reduce that molecule in the presence of oxygen and hence to repress nitrite respiration (Kakutani *et al.*, 1981). Its normal function is probably to reduce the nitrite reductase in the normal nitrite respiration pathway. Apparently no blue protein is detected in aerobically grown S-6 cells, although other strains of *A. faecalis* do produce a normal azurin.

Rusticyanin is unique in its function. It is a primary electron acceptor from Fe2+ in *Thiobacillus ferroxidans* (Ingledew *et al.*, 1977) and is unusual because it is stable at pH 2.0. Moreover it is reported to have an extremely high redox potential (Ingledew, 1976), nearly as high as the blue copper in laccase. Although larger than other bacterial azurins, it lacks a disulphide. Its epr spectrum is somewhat different (and more like stellacyanin) in that it is rhombic, but still has small hyperfine splitting constants (Cox *et al.*, 1978). Partial sequence information is available (Ambler, personal communication): of the five histidines none has neighbouring residues like -FPHNV- of plastocyanin or -MGHNW- of azurin, nor does the single Cys have flanking residues like those of azurin and plastocyanin. In particular FFC is not present, although there is a -YY- fragment. Thus it appears that its structure will be quite different. The surroundings of one histidine, -EVHDK-, resemble VVHEK of bovine superoxide dismutase, but no other sequence stretches do. Small crystals of rusticyanin have been grown by Freeman's group in Sydney, but none large enough for a structure determination yet (Freeman, personal communication).

(b) Plant blue proteins

Algal plastocyanins have been described above. Several other proteins, one also associated with chloroplasts, the others not, have been isolated and characterised to varying extents (see Table 1.1b). The mung bean and rice bran blue protein apparently have not been worked on since 1975 (Fee, 1975). The function of these other proteins is not known, although a possible function for stellacyanin will be considered below. Available amino acid compositions are summarised in Table 1.3b.

The physical properties of umecyanin have been studied, but little is known about its functions. It has been isolated from horseradish root (*Armoracia lapatifolia*, Gilib) (Paul and Stigbrand, 1970). Its molecular weight is very similar to azurin; it apparently has three to four moles of carbohydrate associated with it. It may dimerise at higher concentrations.

Its CD spectrum suggests more beta structure than alpha helix which would be consistent with an azurin type folding. At high pH its 610 nm absorption band is shifted to 580 nm, as described for mavicyanin; however its epr spectrum is axial, not rhombic. It transfers electrons to laccase at about one-fourth the rate that azurin does. Its optical spectrum differs somewhat in that the 800 nm peak is not really a peak, but a gradually increasing absorption from about 750 nm up to 1000 nm.

Stellacyanin comes from the latex of lacquer tree, also the source for one of the polyphenol oxidases, laccase. It has also been isolated from cucumber peelings, along with three other blue proteins (Aikazyan and Nalbandyan, 1979).

The sequence of stellacyanin confirms the fact that stellacyanin lacks methionine (Bergman *et al.*, 1977). The absence of Met indicated some time ago that there must be more than one way to form a blue copper site.

The sequences of stellacyanin and cucumber basic protein (Murata *et al.*, 1982) resemble each other to a certain extent, more each other than either azurin or plastocyanin. There is an -HN- fragment, all that remains of the much longer conserved region around that ligand in plastocyanin and azurin. The similarity of the behaviour of NMR to pH titration to that of plastocyanin and azurin suggests that there are two histidine ligands as well (Hill and Lee, 1979). The intense blue is suggestive of the Cys ligand. In stellacyanin there is a disulphide between the two C-terminal cysteines, leaving the other Cys as the 'blue' copper ligand although Hill and Lee dispute that because of lack of homology with azurin and plastocyanin. The identity of the fourth ligand is in question (Ferris *et al.*, 1978). McMillan and Morris (1981) have suggested that the disulphide provides the fourth ligand, and argue that absorption by a $Cu-(S-S)$ bond might be expected to occur in the same place as the transition assigned to $Cu-Met$.

Knaff *et al.* (1981) have attempted to probe the environment of the stellacyanin copper with chloromercurinitrophenol and with a dansyl group. Both of these rendered the protein no longer reconstitutable with Cu, implying that each occupies the Cu site and is bound to the cysteinyl sulphhydryl. The lack of enhancement of the fluorescence of the dansyl reagent and the lack of a change in the pK of the chloromercury derivative both suggested that their environments were *not* hydrophobic, and hence that of the copper in stellacyanin was not either, possibly because it is close to the molecular surface. Experiments in which azurin, plastocyanin and stellacyanin were compared with the same oxidants (Cummins and Gray, 1977) did indicate that the Cu centre in stellacyanin was relatively more accessible. The plastocyanin centre seems so accessible, it is hard to imagine how stellacyanin could be more so. Perhaps this merely confirms what all the earlier experiments suggest: kinetic access is not directly via the most exposed portion of the centre.

Holwerda *et al.* (1981) have proposed that stellacyanin may function as a quinol reductase in a regulatory capacity with laccase which produces quinols. However, the kinetic study indicated that stellacyanin reacts rapidly with such molecules, but shows no particular specificity for them, or enhancement of reactivity over what would be expected. Therefore they feel that possible quinol reductase activity is still moot.

Stellacyanin can be reduced by Cr(II) with a single Cr(III) remaining bound, as has been done with azurin and plastocyanin (Morpurgo and Pecht, 1982). The residues affected by this have not yet been identified, but it is known that Cr-labelled stellacyanin can be reduced again by another Cr.

Plantacyanin, cusacyanin and cucumber basic blue protein all appear to be the same protein, although rather different molecular weights and amino acid compositions have been reported by two groups (Markossian *et al.*, 1974;

Colman *et al.*, 1977). Plantacyanin has been isolated from cucumber peelings, spinach leaves, beet leaves and asparagus stems (Aikazyan and Nalbandyan, 1981) and although associated with chloroplasts it will not replace plasto-cyanin in its reactions and hence must have a different function. The cucumber basic blue protein was originally isolated by Vickery (1971) as a by-product in the preparation of indole 3-ethanol oxidase from light grown cucumber seedlings. The epr spectra of both show rhombic environments, small hyperfine splitting constants, and an extinction coefficient at 597 of $3.4 \, mM^{-1} \, cm^{-1}$. The X-ray structure determination is underway in Freeman's laboratory. Since the sequence (figure 1.9) shows both a disulphide and a Met available to bind the Cu, it will be very interesting to see what the ligands are.

```
          5      10      15      20      25      30
Sc   T V Y T V G D S A G W K V P F F G D V D Y D W K W A S N K T F H I G D
CBP  A V Y V V G G S G G W T - - - F - N T - - - E S W P K G K R F R A G D
          5      10              15        20      25


          40      45      50      55      60      65
Sc   V L V F K Y D R R F H N V D K V T Q K N Y Q S - C N D T T P I A S - Y
CBP  I L L F N Y N P X M H N V V V V N Q - G G F S T C N - T P A G A K V Y
        30      35      40      45      50      55      60


      70      75      80      85      90      95      100     105
Sc   N T - G N N R - - I N L K T V G Q K Y Y I C G V P K H C D L G Q K V H I N V T V R S
CBP  - T S G - - R D Q I K L - P K G Q S Y F I C N F P G H C Q S G M K - - I A V N A L
          65      70      75      80      85      90      95
```

Figure 1.9 Sequence of stellacyanin and cucumber blue protein aligned as in Murata *et al.*, 1983

Mavicyanin is another blue protein isolated from the fruit of green squash, which has a rhombic epr spectrum and molecular weight of 18 000 (Marchesini *et al.*, 1979). Inasmuch as the larger blue protein, ascorbate oxidase, is also isolated from the same fruit, the possibility that this was a fragment or subunit of that protein was tested, and found not to be the case. Interestingly, on increasing pH from 7 to 11 or 12, the epr spectrum of mavicyanin looks very much like stellacyanin in that it becomes a square planar type of spectrum, and the optical absorption shifts to 570 nm. The midpoint potential is 285 at pH 7.0. Indeed the epr spectrum can be fitted with stellacyanin parameters. The reasons this is *not* considered a stellacyanin are that the carbohydrate content is 7%, compared with 20% for stellacyanin, and that it has a much lower redox potential. Nevertheless, the fact that Aikazyan *et al.* report isolation of a stella-cyanin from cucumber peelings which they call a stellacyanin because of its characteristic epr, makes one suspect that mavicyanin too is a stellacyanin.

V CONCLUDING REMARKS

On the basis of sequence comparisons, Ryden and Lundgren (1979) proposed that plastocyanin evolved from an azurin and that stellacyanin and the blue copper oxidase ceruloplasmin evolved from the plastocyanin branch. The partial *Achromobacter cycloclastes* sequence suggests that there may have been an earlier precursor than the proposed azurin-plastocyanin split; the fact that a strain of *Alcaligines faecalis* and *Methylomonas* can produce an azurin type and a plastocyanin type suggests that possibly genes for both have been carried along in some species.

Again on the basis of sequence homologies it has been suggested that sub-unit II of bovine cytochrome c oxidase is also a blue copper protein (Steffens and Buse, 1979), and that Cu_{a_3} is the Type I copper of the two coppers found in the oxidase. Brudvig and Chan (1979) tried to test this hypothesis, reasoning that a sulphhydryl reagent, Hg or Ag should bleach the absorption near 600 nm (as these reagents will do with plastocyanin) if there is indeed a blue copper centre. They found little effect of the reagents. On the other hand, Darley-Usmar *et al.* (1981) used iodoacetamide to label free sulphhydryls and found that the two cysteines in sub-unit II were not labelled, presumably *because* they are bound to the copper and therefore inaccessible to the reagent. In addition, modification of sub-unit II with $HgCl_2$ by these same workers resulted in some loss of redox activity, further implicating these same cysteines. Powers *et al.* (1981) have proposed a model for the Cu_{a_3} site of cytochrome oxidase, based on EXAFS studies of carefully defined redox intermediates. The model includes two nitrogen ligands, a sulphur ligand which is shared by heme a3, and a more remote (2.8 Å) S, N, or O ligand. The model resembles one they have proposed for stellacyanin.

In spite of Buse and Steffens' sequence comparisons, there appears to be insufficient homology to actually enable one to suggest that the cytochrome oxidase sub-unit folds like azurin or plastocyanin. It is possible, with molecular models, to arrange the potential ligands around a copper with reasonable tetrahedral geometry. As in the case of the *Achromobacter* sequence alignment, a large deletion must occur between the first His ligand and the first Cys ligand which, in effect, eliminates the 'flap' and the E and F strands of the beta barrel structure. It then becomes highly speculative to contemplate how the remaining sequence folds.

Are blue proteins really a class of proteins? If so, there are also at least four subclasses as shown in table 1.4. The three known structures (two in class I, one in class III) show more structural homology than one might have been led to expect considering all the possible differences in properties. Clearly classes II and IV will be further subdivided as more information is made available about the members of each. but meanwhile this may serve as a useful way of thinking about these very interesting molecules.

Table 1.4 Suggested classification of cupredoxins

Properties	I	II	III	IV
Absorption	>600 nm	<600 nm	<600 nm	600 + 4
Extinction	>3.5	<3.5	>4.0	3–5
Epr	axial	axial	axial	rhombic
Hyperfine	small	small	small	small
MW	14 000	12–14 K	10 K	10–20 K
AA comp	3 Cys	1 Cys	1 Cys	>1 Cys
Structure	beta sand	?	beta sand	?
Cu ligands	His2, Cys1, Met1	?	His2, Cys1, Met1?	?
Function	interact with c551 and cyto cd-nitrite red	interact with e.t. agents other than c551	photosystem I e.t. from cyto f to P700	unknown
Members	azurins from P. aeruginosa, A. denitrific., A. faecalis umecyanin?	amicyanin, Az. from Achromobacter cycloclastes, Paracoccus denitrificans Blue protein A. faecalis S-6	plastocyanin from plants, blue-green algae	stellacyanin, mavicyanin, cucumber blue protein, rusticyanin?

ACKNOWLEDGMENTS

I am grateful for many fruitful discussions with my colleagues, particularly Dr Larry Sieker who started me in blue copper proteins; to those many colleagues who shared preliminary publications, particularly Dr Ted Baker, Dr Israel Pecht, Dr Allen Hill, and Dr Richard Ambler; to Dr Israel Pecht and Dr Ole Farver for careful critical reading of this manuscript; to Dr Lyle Jensen who has supplied much encouragement and support over the years, and to NIH for support on grant GM13366.

REFERENCES

Adman, E. T., Canters, G. W., Hill, H. A. O. and Kitchen, N. A. (1982) *FEBS Lett.*, **143**, 287

Adman, E. T. and Jensen, L. H. (1981) *Israel J. Chem.*, **21**, 8

Adman, E. T., Stenkamp, R. E., Sieker, L. C. and Jensen, L. H. (1978) *J. mol. Biol.*, **123**, 35

Adman, E. T., Watenpaugh, K. D. and Jensen, L. H. (1975) *Proc. Nat. Acad Sci.*, **72**, 4854

Aikazyan, V. Ts. and Nalbandyan, R. M. (1979) *FEBS Lett.*, **104**, 127

Aikazyan, V. Ts. and Nalbandyan, R. M. (1981) *Biochim. Biophys. Acta*, **667**, 421

Ambler, R. P. (1971) In: *Recent Developments in the Chemical Studies of Protein Structures* (ed. A. Previero, J.-F. Pechere and C. Previero), Inserm, Paris, p. 289

Ambler, R. P. (1977) In: *The Evolution of Metalloenzymes, Metalloproteins and Related Materials*, Proceedings of a Symposium of the Inorganic Biochemistry Discussion Group of the Chemical Society, University of Sussex (ed. G. S. Leigh), Symposium Press, London, p. 100–118

Ambler, R. P. (1982) Personal communication

Augustin, M., Chapman, S. K., Davies, D. M. Sykes, G. A., Speck, S. H. and Margoliash, E. (1983) *J. Biol. Chem.*, **258**, 6405

Aviram, I., Myer, Y. P. and Schejter, A. (1981) *J. biol. Chem.*, **256**, 5540

Bergman, C., Ganvik, E-K., Nyman, P. O. and Strid, L. (1977) *Biochem. Biophys. Res. Comm.*, **77**, 1052

Blaszak, J. A., Ulrich, E. L., Markley, J. L. and McMillin, D. R. (1982) *Biochemistry*, **21**, 6253

Blumberg, W. E. and Powers, L. (1983) *Fed. Proc.*, **41**, 862

Bohner, H. and Boger, P. (1978) *FEBS Lett.*, **85**, 337

Bohner, H., Bohme, H. and Boger, P. (1980a) *Biochim. Biophys. Acta*, **592**, 103

Bohner, H., Merkle, H., Kroneck, P. and Boger, P. (1980b) *Eur. J. Biochem.*, **105**, 603

Bohner, H., Sandmann, G. and Boger, P. (1981) *Biochim. Biophys. Acta*, **636**, 65

Boulter, D., Haslett, B. G., Peacock, D., Ramshaw, J. A. M. and Scawen, M. D. (1977) *Plant Biochemistry II, International Rev. Biochem.*, **13**, 1

Brill, A. S. (1978) *Biophys. J.*, **22**, 139

Brill, A. S., Bryce, G. F. and Maria, H. J. (1968) *Biochim. Biophys. Acta*, **154**, 342

Brudvig, G. W. and Chan, S. I. (1979) *FEBS Lett.*, **106**, 139

Brunori, M., Parr, S. R., Greenwood, C. and Wilson, M. T. (1975) *Biochem. J.*, **151**, 185

Burkey, K. O. and Gross, E. L. (1981) *Biochemistry*, **20**, 5495

Chirgadze, Y. N., Garber, M. B. and Nikonov, S. V. (1977) *J. mol. Biol.*, **113**, 443

Chothia, C. and Lesk, A. M. (1982) *J. mol. Biol.*, **160**, 309

Colman, P. M., Freeman, H. C., Guss, J. M., Murata, M., Norris, V. A., Ramshaw, J. A. M. and Venkatappa, M. P. (1978) *Nature, Lond.*, **272**, 319

Colman, P. M., Freeman, H. C., Guss, J. M., Murata, M., Norris, V. A., Ramshaw, J. A. M., Venkatappa, M. P. and Vickery, L. E. (1977) *J. mol. Biol.*, **112**, 649

Cookson, D. J., Hayes, M. T. and Wright, P. E. (1980) *Biochim. Biophys. Acta*, **591**, 162

Cox, J. C., Aasa, R. and Malmstrom, B. G. (1978) *FEBS Lett.*, **93**, 157

Cox, J. C. and Boxer, D. H. (1978) *Biochem. J.*, **174**, 497

Cummins, D. and Gray, H. B. (1977) *J. Am. chem. Soc.*, 99, 5158

Darley-Usmar, V. M., Capaldi, R. A. and Wilson, M. T. (1981) *Biochem. Biophys. Res. Comm.*, 103, 1223

Davis, D. J., Krogmann, D. W. and San Pietro, A. (1980) *Plant Physiology*, 65, 697

Davis, D. J. and San Pietro, A. (1979) *Anal. Biochem.*, 95, 254

Farver, O., Blatt, Y. and Pecht, I. (1982) *Biochemistry*, 21, 3556

Farver, O. and Pecht, I. (1981a) *Proc. nat. Acad. Sci.*, 78, 4190

Farver, O. and Pecht, I. (1981b) *Israel J. Chem.*, 21, 13

Farver, O. and Pecht, I. (1981c) In: *Copper Proteins* (ed. T. Spiro), John Wiley, New York, p. 151

Farver, O., Shahak, Y. and Pecht, I. (1982b) *Biochemistry*, 21, 1885

Fee, J. A. (1975) *Structure and Bonding*, 23, 1

Ferris, N. S., Woodruff, W. H., Rorabacher, D. B., Jones, T. E. and Ochrymowycz, L. A. (1978) *J. Am. chem. Soc.*, 100, 5939

Ferris, N. S., Woodruff, W. H., Tennent, D. L. and McMillin, D. R. (1979) *Biochem. Biophys. Res. Comm.*, 88, 288

Freeman, H. C. (1981) In: *Coordination Chem. Rev.* (ed. F. P. Laurent) Pergamon Press, Oxford, 21, p. 29

Freeman, H. C., Garret, T. P. J., Guss, J. M., Murata, M., Yoshizaki, F., Sugimara, Y. and Shimakoriyama, M. (1983) *J. mol. Biol.*, 164, 351

Freeman, H. C. and Guss, M. (1981) Unpublished work referred to in Freeman (1981)

Goldberg, M. and Pecht, I. (1976) *Biochemistry*, 15, 4197

Gray, H. B. and Solomon, E. I. (1981) In: *Copper Proteins* (ed. T. Spiro), John Wiley, New York, p. 1

Grimes, C. J., Piskiewicz, D. and Fleisher, E. B. (1974) *Proc. nat. Acad. Sci.*, 71, 1408–1412

Grossman, A. R., Bartlett, S. G., Schmidt, G. W., Mullet, J. E. and Chua, N-H. (1982) *J. biol. Chem.*, 257, 1558

Guss, J. M. and Freeman, H. C. (1983) *J. mol. Biol.*, 169, 521

Haehnel, W., Propper, A. and Krause, H. (1980) *Biochim. Biophys. Acta*, 593, 384

Haslett, B. G., Bailey, C. J., Ramshaw, J. A. M., Scawen, M. D. and Boulter, D. (1978) *Phytochemistry*, 17, 615

Hill, H. A. O. and Lee, W-K. (1979) *J. inorg. Biochem.*, 11, 101

Hill, H. A. O., Leer, J. C., Smith, B. E., Storm, C. B. and Ambler, R. P. (1976) *Biochem. Biophys. Res. Comm.*, 70, 331

Hill, H. A. O. and Smith, B. E. (1979) *J. inorg. Biochem.*, 11, 79

Holwerda, R. A., Knaff, D. B., Gray, G. O. and Harsh, C. E. (1981) *Biochemistry*, 20, 4336

Holwerda, R. A., Wherland, S. and Gray, H. B. (1976) *Ann. Rev. Biophys. and Bioeng.*, 5, 363

Horio, T. (1958) *J. Biochem. (Tokyo)*, 45, 195

Hurt, E. and Hauska, G. (1981) *Eur. J. Biochem.*, 117, 591

Ingledew, W. J. (1976) *Proc. Soc. Ital. Bioch.*, Oct. 1976, 84

Ingledew, W. J., Cox, J. C. and Halling, P. J. (1977) *FEMS Microbiol. Lett.*, 2, 193

Iwasaki, H. and Matsubara, T. (1973) *J. Biochem.*, 73, 659

Kakutani, T., Watanabe, H., Arima, K. and Beppu, T. (1981) *J. Biochem.*, 89, 463

Katoh, S., Shiratori, I. and Takamiya, S. (1962) *J. Biochem. (Tokyo)*, 51, 32

Katoh, S. and Takamiya, A. (1961) *Nature, Lond.*, 189, 665

Keilin, D. and Mann, T. (1940) *Nature, Lond.*, 145, 304

Knaff, D. B., Harsh, C. E. and Holwerda, R. A. (1981) *Biochemistry*, 20, 4333

Lappin, A. G., Segal, M. G., Weatherburn, D. C., Henderson, R. A. and Sykes, A. G. (1979a) *J. Am. chem. Soc.*, 101, 2302

Lappin, A. G., Segal, M. G., Weatherburn, D. C. and Sykes, A. G. (1979b) *J. Am. chem. Soc.*, 101, 2297

Lockau, W. (1979) *Eur. J. Biochem.*, 94, 365

Malmstrom, B. O., Reinhammar, B. and Vanngard, T. (1970) *Biochim. Biophys. Acta*, 205, 48

Marchesini, A., Minelli, M., Merkle, H. and Kroneck, P. M. H. (1979) *Eur. J. Biochem.*, 101, 77

Markossian, K. A., Aikazyan, V. Ts. and Nalbandyan, R. M. (1974) *Biochim. Biophys. Acta*, 395, 47

Marks, R. H. L. (1982) *Fed. Proc.*, **41**, 890
Marks, R. H. L. and Miller, R. D. (1979) *Arch. Biochem. Biophys.*, **195**, 103
Martinkus, K., Kennelly, P. J., Rea, T. and Timkovich, R. (1980) *Arch. Biochem. Biophys.*, **199**, 465
Mayhew, S. G. and Ludwig, M. L. (1975) In: *The Enzymes* (ed. P. D. Boyer), Vol. 12, Academic Press, New York, p. 57
McMillin, D. R. and Morris, M. C. (1981) *Proc. nat. Acad. Sci.*, **78**, 6567
Mitra, S. and Bersohn, R. (1982) *Proc. nat. Acad. Sci.*, **79**, 6807
Morgan, R. S., Tatsch, C. E., Gushard, R. H., McAdon, J. M. and Warme, P. K. (1978) *Int. J. Prot. Pept. Res.*, **11**, 209
Morpurgo, G. and Pecht, I. (1982) *Biochem. Biophys. Res. Comm.*, **104**, 1592
Murata, M., Begg, G. S., Lambrou, F., Leslie, B., Simpson, R. J., Freeman, H. C. and Morgan, F. J. (1982) *Proc. nat. Acad. Sci.*, **79**, 6434
Norris, G. E., Anderson, B. F. and Baker, E. N. (1983) *J. mol. Biol.*, **165**, 501
Norris, G. E., Anderson, B. F., Baker, E. N. and Rumball, S. V. (1979) *J. mol. Biol.*, **135**, 309
Omura, T. (1961) *J. Biochem. (Tokyo)*, **50**, 394
Parr, S. R., Barber, D., Greenwood, C. and Brunori, M. (1977) *Biochem. J.*, **167**, 447
Paul, K. G. and Stigbrand, T. (1970) *Biochem. Biophys. Acta*, **221**, 255
Pecht, I. and Rosen, P. (1973) *Biochem. Biophys. Res. Comm.*, **50**, 853
Penfield, K. W., Gay, R. R., Himmelwright, R. S., Eickman, N. C., Norris, V. A., Freeman, H. C. and Solomon, E. I. (1981) *J. Am. chem. Soc.*, **103**, 4382
Powers, L., Chance, B., Ching, Y. and Angiolillo, P. (1981) *Proc. nat. Acad. Sci.*, **77**, 928
Ramshaw, J. A. M. (1982) *Phytochemistry*, **21**, 1317
Reinhammar, B. R. M. (1972) *Biochim. Biophys. Acta*, **275**, 245
Ronnberg, M., Araiso, T., Ellfolk, N. and Dunford, B. (1981) *J. Biol. Chem.*, **256**, 2471
Rosen, P. and Pecht, I. (1976) *Biochemistry*, **15**, 775
Rosen, P., Segal, M. and Pecht, I. (1981) *Eur. J. Biochem.*, **120**, 339
Ryden, L. and Lundgren, J-O. (1976) *Nature, Lond.*, **261**, 344
Ryden, L. and Lundgren, J-O. (1979) *Biochemie*, **61**, 781
Scawen, M. D., Ramshaw, J. A. M. and Boulter, D. (1975) *Biochem. J.*, **147**, 343
Scott, R. A., Hahn, J. E., Doniach, S., Freeman, H. C. and Hodgson, K. O. (1982) *J. Am. chem. Soc.*, **104**, 5364
Sieker, L. C. (1981) Ph.D. thesis, University of Washington, Seattle
Silvestrini, M. C., Brunori, M., Wilson, M. T. and Darley-Usmar, V. M. (1981) *J. inorg. Biochem.*, **14**, 327
Solomon, E. I., Hare, J. W., Dooley, D. M., Dawson, J. H., Stephens, P. J. and Gray, H. B. (1980) *J. Am. chem. Soc.*, **97**, 5260
Solomon, E. I., Hare, J. W. and Gray, H. B. (1976) *Proc. nat. Acad. Sci.*, **73**, 1389
Spiro, T. (Ed.) (1981) *Copper Proteins*, John Wiley, New York
Steffens, G. J. and Buse, G. (1979) *Hoppe Seyler's Z. Physiol. Chem.*, **360**, 613
Stigbrand, T. (1970) *Biochim. Biophys. Acta*, **236**, 246
Stigbrand, T. (1972) *FEBS Lett.*, **23**, 41
Stigbrand, T., Malmstrom, B. G. and Vanngard, T. (1971) *FEBS Lett.*, **12**, 260
Strigbrand, T. and Sjoholm, I. (1972) *Biochim. Biophys. Acta*, **263**, 244
Strahs, G. (1969) *Science, N.Y.*, **165**, 60
Sutherland, I. W. and Wilkinson, J. F. (1963) *J. gen. Microbiol.*, **30**, 105
Suzuki, H. and Iwasaki, H. (1962) *J. Biochem. (Tokyo)*, **52**, 193
Takabe, T., Niwa, S., Ishikawa, H. and Miyakawa, M. (1980a) *J. Biochem.*, **87**, 111
Takabe, T., Niwa, S. and Ishikawa, H. (1980b) *J. Biochem.*, **87**, 1335
Takano, T. and Dickerson, R. E. (1981) *J. mol. Biol.*, **153**, 95
Tennent, D. L. and McMillin, D. R. (1979) *J. Am. chem. Soc.*, **101**, 2307
Thamann, T. J., Frank, P., Willis, L. J. and Loehr, T. M. (1982) *Proc. nat. Acad. Sci.*, **79**, 6396
Tullius, T. D., Frank, P. and Hodgson, K. (1978) *Proc. nat. Acad. Sci.*, **75**, 4069
Tobari, J. and Harada, Y. (1981) *Biochem. Biophys. Res. Comm.*, **101**, 502
Ugurbil, K. and Bersohn, R. (1977) *Biochemistry*, **16**, 3016
Ugurbil, K., Norton, R. S., Allerhand, A. and Bersohn, R. (1977) *Biochemistry*, **16**, 886
Ulrich, E. L. and Markley, J. C. (1978) *Coordination Chem. Rev.*, **27**, 109

Vickery, L. E. (1971) Ph.D. thesis, University of California, Santa Barbara

Wherland, S. and Pecht, I. (1978) *Biochemistry*, 17, 2585

Wilson, M. T., Greenwood, C., Brunori, M. and Antonini, E. (1975) *Biochem. J.*, 145, 449

Wood, P. M. (1978) *Eur. J. Biochem.*, 87, 9

Woodruff, W. H., Norton, K. A., Swanson, B. I. and Fry, H. A. (1983) *J. Am. chem. Soc.*, 105, 657

Yoshizaki, F., Sugimara, Y. and Shimokoriyama, M. (1981) *J. Biochem.*, 89, 1533

2
Cytochromes *c* and Cytochrome *c* Containing Enzymes

C. Greenwood

INTRODUCTION

Porphyrin-containing compounds fulfil many different roles in biological systems. Broadly speaking, they fall into two main categories. There are the carrier molecules in which the substance carried is either oxygen, as in the case of the haemoglobins and myoglobins, or electrons, as in the cytochromes. Iron porphyrin containing enzymes also carry out oxidation reduction reactions on bound substances with the prosthetic group acting to bind the substrate via the central iron and at the same time, either alone or in concert with non-binding iron porphyrin groups or other redox active centres, act as a source of electrons for the coupled redox events.

Thinking about cytochromes *c* has been very largely dominated by the most familiar member of the class, cytochrome *c* from the mitochondrial respiratory chain which contains a single haem group per molecule of 103–113 amino acids and a redox potential of + 260 mV. However, other members of the class have multiple haems, and redox potentials covering the range −290 to +400 mV and can act as passive redox carriers or enzymes. The term cytochrome *c* is a structural and spectral classification concerned mainly with the haem and its specific attachment to the polypeptide backbone rather than to the protein which surrounds it. The structure of haem *c* is given in figure 2.1 with the conventional numbering of the pyrrole rings and shows the feature, unique to cytochromes *c*, of the thio-ether bridge to two (or in some exceptional cases, one) cysteine residues in the protein. The iron atom in cytochrome *c* alternates between the oxidised state in which the iron is ferric (Fe III) and has a single, unpaired electron, and the reduced or ferrous state where the iron is ferrous (Fe II) and there are no unpaired electrons. Associated with these stable redox states are characteristic absorption spectra and these are shown in figure 2.2, the typical

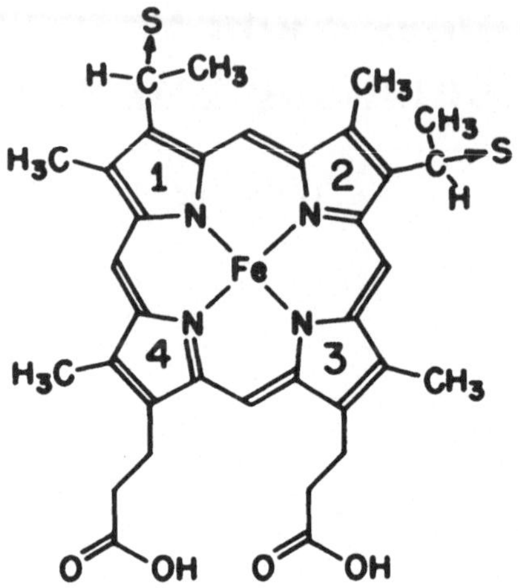

Figure 2.1 Haem c, with the conventional numbering of pyrrole rings. The sulphur atoms are from cysteine side chains on the protein

three banded (in the visible region) of reduced cytochrome c having an α band at 550 nm, a β band at 520 nm and a Soret or γ band at 415 nm. Some variation in the form of the spectra and position of the reduced maxima occurs throughout the group with the range being, typically, α band 550–558 nm, β band 520–528 nm and Soret band 415–423 nm; the higher values being found in cytochromes c with only a single thio-ether bridge (Euglena and Crithida). The principal classes of cytochromes c are summarised in table 2.1, taken from Dickerson and

Table 2.1 Principal classes of cytochrome c

Type and examples	$E_0{'}$	Haems	MW/chain
I. High potential: eukaryotic and prokaryotic photosynthesis and respiration $- c, c_1, c_2, c_4, c_5, c_6$ (=f), c_{551}, etc.	+200 to +390	1	9000–81 000
II. High-spin iron: c' from some pseudomonads, *Alcaligenes* and several purple photosynthetic bacteria	0 to +130	1	13 000–16 000
III. Flavocytochromes			
c_{553} from *Chlorobium*	+98	1	50 000
c_{552} from *Chromatium*	+10	2	72 000
IV. Low potential			
c_{553} from *Desulphovibrio*	−100 to 0	1	9 000
c_3 from *Desulphovibrio*, c_7	−200	3 or 4	12 000
c_{551} from purple non-sulphurs, others	−150 to −250	1 or 2	16 000–21 000

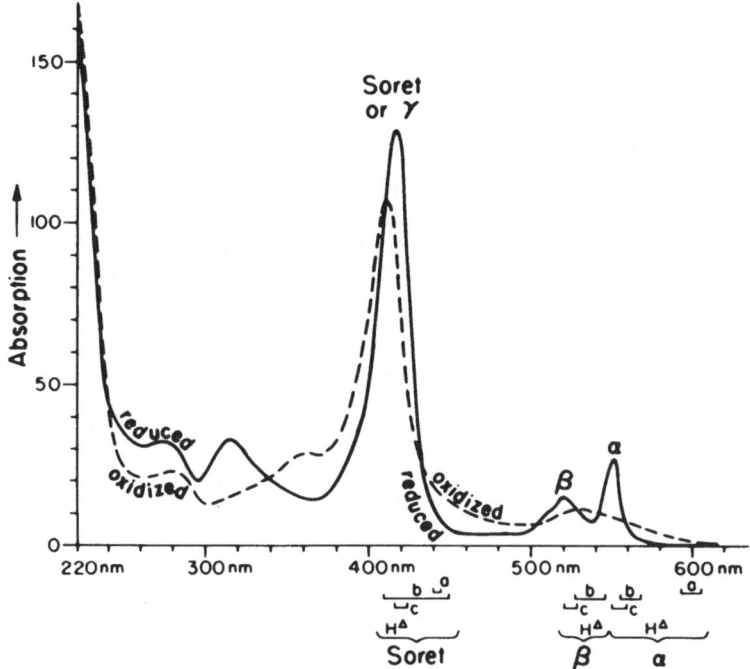

Figure 2.2 Absorption spectra of oxidised and reduced horse heart cytochrome c. Wavelength ranges for the α, β, and Soret bands in cytochrome a, b, and c, and haemoglobin are indicated at the bottom

Timkovich (1975), in addition to which a large number of excellent reviews of the structure, function and history of cytochromes c can be found in Margoliash and Schejter (1966), Keilin (1966), Lemberg and Barrett (1973), Salemme (1977), Timkovich (1979) and Ferguson Miller *et al.* (1979).

CYTOCHROME *C* CONTAINING ENZYMES

The main role for cytochromes of the c group thus appears to be one of transferring electrons and since the electron is not usually considered to be a substrate these proteins are not enzymes in the strictest sense. Furthermore, with the exception of cytochromes of the c^1 type, they appear to be specifically designed with a buried active site (the haem c group) which is chemically constrained to prevent its reaction with potential substrates/ligands. A number of proteins containing haem c covalently bound to their polypeptide chain, which are arguably enzymes capable of reacting with and converting substrate molecules to products, do, however, exist amongst certain groups of bacteria. These enzymes, which are associated with anaerobic respiratory activity in the organism concerned, are often repressed by the presence of oxygen and, as in the case of the nitrite reductases, are not expressed unless nitrate is present in the growth medium.

CYTOCHROME c PEROXIDASE

In the early 1950s Lenhoff and Kaplan (1953, 1956) demonstrated the presence of cytochrome c peroxidase (cytochrome c : H_2O_2 oxidoreductase EC 1.11.1.5) in *Pseudomonas fluorescens* and described a method for the partial purification of the enzyme. Since then a number of laboratories have published details of purification procedures which have resulted in the isolation of electrophoretic-ally pure protein from a number of bacterial sources (Ellfolk and Soininen, 1970; Coulson and Oliver, 1979; Foote *et al.*, 1982a). Most work has been done using the enzyme from *Pseudomonas aeruginosa* (Ellfolk and Soininen, 1970; Foote *et al.*, 1982a) which, although apparently water soluble after isolation, is normally associated with the membrane fraction as prepared by lysozyme EDTA treatment of whole cells (Soininen *et al.*, 1970).

Spectra

The enzyme shows a typical c type spectrum (figure 2.3) with a ratio A_{407}/A_{280} of 4.2, a ratio which provides a ready test of purity in the late states of purifi-cation of the enzyme. The spectra of *Pseudomonas* cytochrome c peroxidase (Ps. CCP) differ considerably from those of yeast cytochrome c peroxidase (Y. CCP) and from those of peroxidases like horse radish peroxidase (HRP). Similar peroxidases with α and β bands have been reported and classified by Morita and Kameda (1957) as b group or low spin (Yamazaka *et al.*, 1968) (ferrihaemochrome type). However, PsCCP does not form the alkaline pyridine ferrohaemochrome with an α band at 557 nm as does protohaemin but one with an α band at 550 nm. An identical band was recorded after reduction of a denatured preparation of PsCCP indicating that the haem in PsCCP is present as a haemochrome in the native enzyme (Ellfolk and Soininen, 1971). Iron and haem analysis indicated that the enzyme contains two haem c components in a single polypeptide chain, the total molecular weight being approximately 44 kDa (Soininen and Ellfolk, 1973a). In spite of the low spin character of the absorp-tion spectrum, only one of the haems is low spin whilst the other is high spin (Ronnberg *et al.*, 1979) suggesting that the enzyme might be classified as a cytochrome cc^1 type.

The spin state assignment and assumed haem inequivalence emerged initially from circular dichroism studies carried out by Ronnberg *et al.* (1979) on the oxidised, reduced and carbonmonoxy derivative (see figure 2.4). The CD spectra in the Soret region indicate that, in the oxidised enzyme, the two haems are degenerate, whereas they are perturbed differently in the reduced state and one haem appears to be non-degenerate. Furthermore, the formation of a carbon-monoxy derivative of the reduced enzyme causes changes in the optical activity

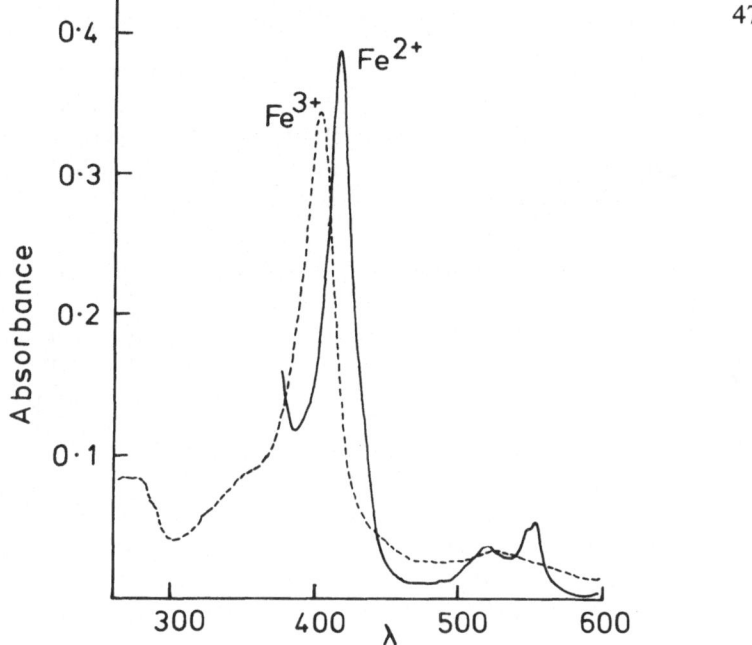

Figure 2.3 Absorption spectra of oxidised and reduced cytochrome *c* peroxidase from *Ps. aeruginosa*

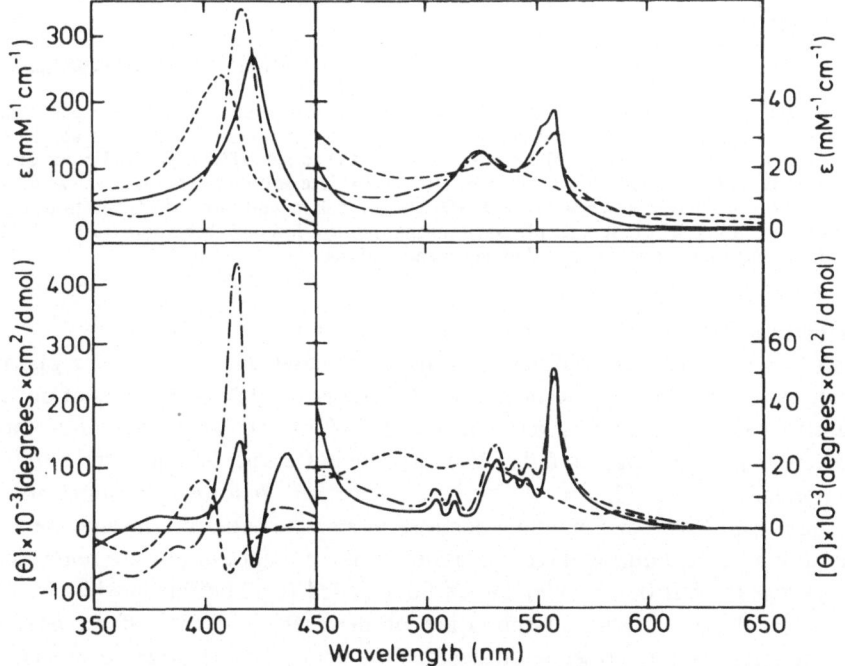

Figure 2.4 Soret and visible absorption (upper) and CD (lower) spectra of ferrous (——), ferric (-----) and ferrous-carbonyl (-.-.-.-) forms of *Pseudomonas* cytochrome *c* peroxidase in sodium phosphate buffer (pH 6.0, *I* = 0.1). (With permission of *Biochim. Biophys. Acta* and author)

indicative of a spin state conversion and consistent with the idea that one haem
is high and the other low spin. Experiments in which the reduced enzyme was
titrated with carbon monoxide show clearly that the enzyme binds CO in a 1 : 1
molar ratio (see figure 2.5) and this must mean that only half the haem present
in the enzyme is responsive to this strong field ligand.

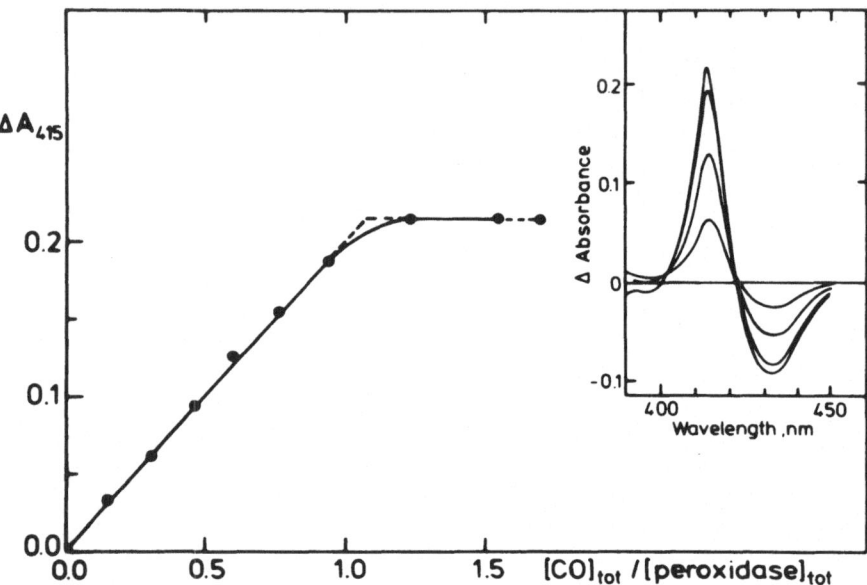

Figure 2.5 Carbon monoxide titration of ferrous *Pseudomonas* cytochrome *c* peroxidase
(1.8 μM) in sodium phosphate buffer (pH 6.0, I = 0.1). Absorbance increments at 415 nm
are plotted against the ratio of the concentration of CO to peroxidase. The end point of the
titration is that measured from a CO-saturated peroxidase solution. The inset shows difference spectra observed between 390 and 450 nm on making additions of CO solution to the
sample cell. Some of the spectra obtained during the titration have been omitted for clarity.
(With permission of *Biochim. Biophys. Acta* and author)

Stability of PsCCP

PsCCP is unstable during degassing procedures designed to replace the gaseous
environment of the enzyme in solution (Foote *et al.*, 1982). The loss of optical
density has been found to be as much as 70% of the Soret peak after three
successive gas exchanges in 0.025 M phosphate buffer, pH 6.0, in which buffer it
is normally stable. The instability is more marked at lower pH values and is
accompanied by a corresponding decrease in enzyme activity. The enzyme may
be fully protected during these operations by the presence of low concentrations
of detergent, thus 0.01% (v/v) Tween 80 is added to all buffers in which PsCCP
is to be degassed. This concentration of detergent does not affect enzyme
activity nor does its presence throughout the course of the isolation procedure
increase the enzyme yield.

Resonance Raman spectra

The resonance Raman spectrum of the ferric form of PsCCP taken from Ronnberg *et al.* (1980) can be seen in figure 2.6. Both polarisations, $I_\perp$ and $I_\parallel$, of the scattered intensity are shown over the wavelength range 1300–1700 cm^{-1} using an exciting wavelength of 514.5 nm. The two sets of Raman lines indicate that one of the haems is high spin and one is in the low spin form, and the anomalously polarised band at 1313–1315 cm^{-1} indicates that the two haems are mesohaems (Adar, 1975). The band at approximately 1500 cm^{-1} is known to be a spin state marker and also indicates the position of the iron relative to the haem plane (Verma and Bernstein, 1974). Thus the 1485 cm^{-1} band seen in the oxidised enzyme correlates closely with that of myoglobin fluoride at 1484 (Verma and Bernstein, 1974), suggesting a high spin haem with an out-of-plane iron. The bands at 1496, 1580 and 1640 cm^{-1} confirm the presence of a low spin haem.

The resonance Raman spectra for the reduced and reduced carbonmonoxy complex of the enzyme, again over the range 1300–1700 cm^{-1}, are presented in

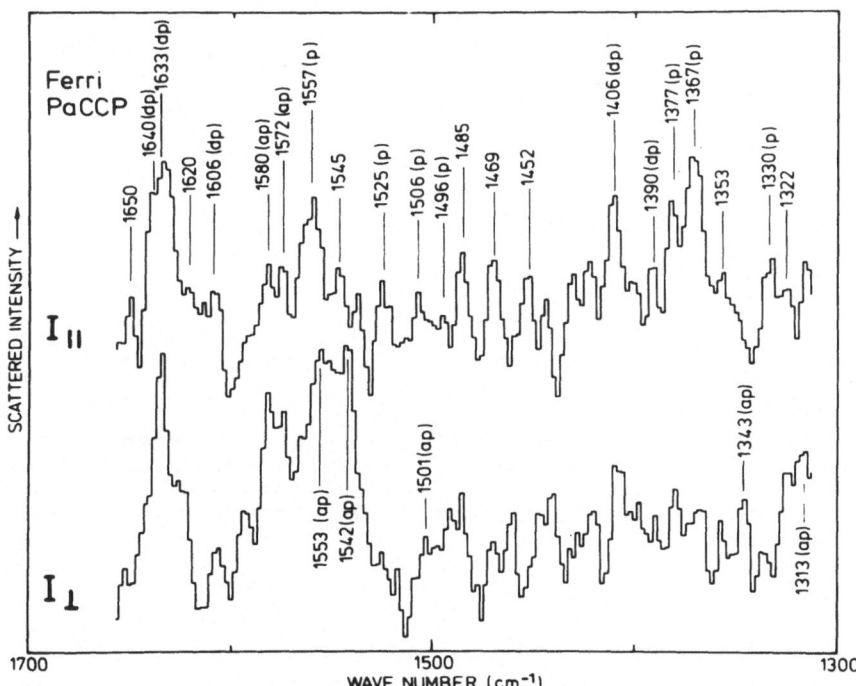

Figure 2.6   The polarised resonance Raman spectra of ferric *Pseudomonas* cytochrome *c* peroxidase (Ferri PaCCP). 130 μM enzyme in 50 mN phosphate buffer, pH 6.0, was measured with Ar$^+$ laser excitation at 514.5 nm (incident power 160 mW, slit width 250 μm, scan rate 20 cm^{-1} min^{-1}, integrating time 3 s). (With permission of *Biochim. Biophys. Acta* and author)

figure 2.7. Bands at 1475, 1555 and 1602 cm^{-1} seen in the spectrum correspond closely to bands at 1473, 1553 and 1607 cm^{-1} which have been assigned
to a high spin, five co-ordinate ferrous iron complex with the iron displaced
from the porphyrin plane for deoxymyoglobin and deoxyhaemoglobin. Thus, it
is concluded that the high spin haem in reduced PsCCP also has a five co-ordinate
out-of-plane structure. The reduced enzyme shows an additional set of bands
at 1363, 1490 and 1618 cm^{-1} characteristic of the low spin haem component
and closely similar to the respective bands in the spectrum of cytochrome c
which are situated at 1363, 1493 and 1620 cm^{-1}.

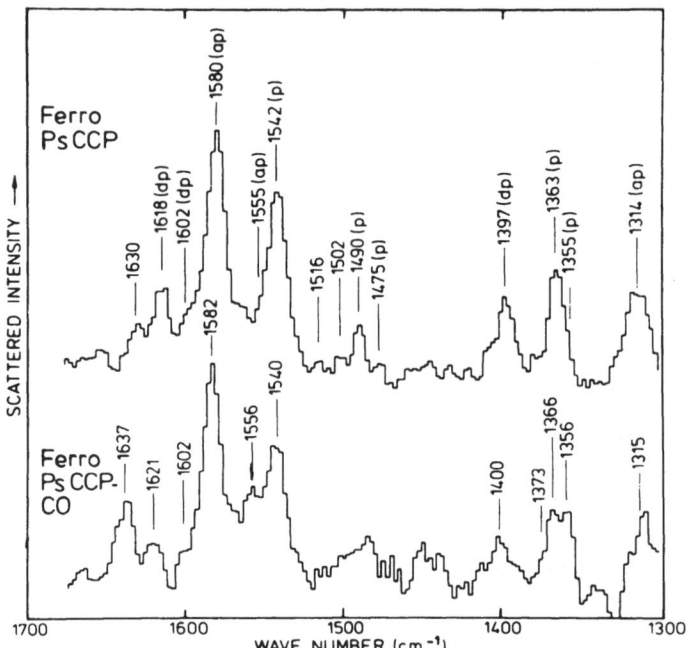

Figure 2.7 The unpolarised resonance Raman spectra of ferrous *Pseudomonas* cytochrome
c peroxidase (Ferro PsCCP) and its carbonyl compound (Ferro PsCCP-CO). Spectra were
obtained with an incident power of 90 mW and a scan rate of 50 cm^{-1} min^{-1}. Other
experimental details as in the legend to figure 2.6. (With permission of *Biochim. Biophys.
Acta* and author)

Electron paramagnetic resonance

The epr spectra of high and low spin haem compounds are quite distinct. High
spin compounds have $g_\parallel = 2$, $g_\perp = 6$ and low spin complexes have three g values,
$g_x \approx 1.7$, $g_y \approx 2.2$ and $g_z \approx 2.8$. Furthermore, in high spin spectra the signals
may show evidence of asymmetry or splitting whilst those from low spin ferric
haemoproteins are often broad and weak even at extremely low temperatures
(liquid helium). Mixed spin situations are detectable by epr but because the

position of the spin equilibrium is often temperature dependent information obtained at liquid helium temperatures may not be relevant to the situation at room temperature. With these reservations, epr is clearly a very useful technique to apply in the case of haemoproteins and especially in the case of PsCCP.

The epr spectrum of oxidised PsCCP recorded at 15 K is presented in figure 2.8, and an assignment to two low spin species consistent with the sum of the squares of all three g values is indicated as stick spectra. All the lines with g values less than 4, except for the small structures around $g = 2$, can be interpreted as arising from two low spin ferric species. Integration of both spectra indicates that both low spin species correspond to 0.8–1.0 haem per molecule. The high spin peak between g 5–6 is estimated to correspond to an Fe^{3+} concentration of 5% of that of the enzyme. Partial reduction of the enzyme, in which one haem is converted to the ferrous state, apparently induces changes in the epr signal of the remaining ferric haem to g_z 2.83, g_y 2.35 and g_x 1.54 consistent with an interaction between the two haems resulting in the deprotonation of a histidine from that haem which is axially coordinated to two histidines (Aasa *et al.*, 1981).

It is evident that the change to low spin character must occur on cooling below 77 K, since the room temperature absorption spectra and resonance Raman clearly show one haem to be high spin. A similar dependence of spin state on temperature has been reported for chloroperoxidase (Hollenberg *et al.*, 1980) and yeast cytochrome *c* peroxidase (Yonetani *et al.*, 1966). Recent X-ray crystallographic studies on the latter protein by Paulos and Kraut (1980) suggest that the sixth ligand position of the haem iron is occupied by water but with a histidine residue close by. It seems possible that the spin state change observed

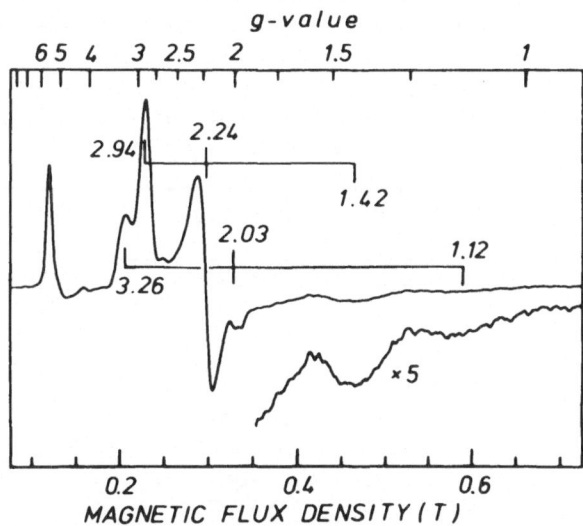

Figure 2.8 EPR spectrum of oxidised *Pseudomonas* cytochrome *c* peroxidase. Enzyme concentration, 410 μM, in 10 mM sodium phosphate buffer, pH 6.8. Temperature 15 K, microwave frequency 9245 MHz and power 2 mW. (With permission of *Biochim. Biophys. Acta* and author)

in this case on cooling may be due to this histidine binding to the distal coordination position of the haem iron and a similar coordination by a histidine may conceivably occur in PsCCP.

Endogenous haem ligands

Although a definitive answer to the problem of what protein residues provide the ligands to the haem iron of PsCCP must await a crystallographic analysis, it is nevertheless possible to infer what these might be from a variety of spectroscopic markers and by analogy with other cytochromes of the c group whose structures are known. The occurrence of the positive CD band at 263 nm in all derivatives of the enzyme implicates a histidine residue as the fifth ligand of the haem iron, as proposed for cytochrome c by Flatmark and Robinson (1968). The band at 640 nm, present in the oxidised enzyme and titrating with a pK of 9.4, is similar to that encountered in myoglobin. For native myoglobin the pK is 8.9 whilst for mesomyoglobin with a prosthetic group similar to PsCCP it is 9.3. It seems reasonable to conclude that, as in met-myoglobin, the sixth ligand of the high spin ferric haem of PsCCP at room temperature is water. On the basis of the 705 nm absorption and by analogy with the 695 nm band seen in oxidised cytochrome c, it appears likely that the sixth ligand of the low spin haem is a methionyl sulphur. That this ligand should be replaced by a nitrogenous group, either histidine or lysine, on reduction of the low spin iron seems rather unlikely but is apparently indicated from Raman studies (Ronnberg $et\ al.$, 1980) where a line at 1542 cm^{-1} appears which is identical to that of reduced cytochrome c_3 and reduced alkylated cytochrome c.

The g_z values of 3.26 and 2.94 found in epr experiments indicate that, at low temperatures, both haems are low spin with one haem, presumably c, having imidazole N and methionyl S coordination whilst the other, c^1, has a bis imidazole N structure. Partial reduction of the enzyme at the level of the c haem results in a deprotonation of one of the imidazole nitrogens of the c^1 haem.

A scheme summarising the iron ligation patterns in PsCCP is given below:

Redox State of Iron	Room Temperature		<77K	
	Low spin heam	High spin haem	All low spin	
	c	c^1	c	c^1
Oxidised (Fe^{3+})	$^{His}N-Fe^{3+}-SMet$	$^{His}N-Fe^{3+}$ with H_2O, $O-C-$, O	$^{His}N-Fe^3-SMet$	$^{His}N-Fe^{3+}-N_{His}$
Reduced (Fe^{2+})	$^{His}N-Fe^{2+}-N^{His}$ or ^{Lys}	$^{His}N-Fe^{2+}$		
Partially reduced ($c\ Fe^{2+}\ c^1 Fe^{3+}$)			cFe^{2+}	$^{His}N-Fe^{3+}-N^{His^-}$

Ligand binding

Ferri PsCCP

The oxidised enzyme reacts slowly with high concentrations of cyanide and azide, the difference spectrum between oxidised and the ligand-bound forms showing characteristics of the spectrum of reduced PsCCP with well formed α and β bands and a red shifted Soret maximum (Soininen and Ellfolk, 1973). There was no apparent reaction between ferric PsCCP and fluoride. The extremely slow binding of cyanide and azide suggests that access to the high spin haem is hindered, perhaps through the presence of an adjacent anionic group like carboxylate and/or the need to displace water. It is not ruled out that cyanide and azide might also react with the low spin haem, although this requires the replacement of an endogenous ligand (presumably methionyl sulphur). Both ligands inhibit the peroxidative oxidation reaction catalysed by the enzyme. Cyanide inhibition was of a mixed type, intermediate between competitive and noncompetitive with respect to hydrogen peroxide, suggesting that several phenomena are involved in the interaction of enzyme and inhibitor. Azide inhibits the peroxidative reaction non-competitively, showing that it reacts as a different site from that reacting with peroxide, although the possibility of irreversible inhibition giving a non-competitive plot cannot be ruled out on the basis of the studies to date.

Ferro PsCCP

The affinity of the reduced enzyme for cyanide is very low as judged from the fact that only 30% of the spectral change seen in the presence of 0.46 M cyanide was elicited by 4.6 mM cyanide whilst in 0.1 mM cyanide no change was observed.

Carbon monoxide reacts rapidly with ferro PsCCP to give a light sensitive carbon monoxide derivative with a characteristic low spin spectrum. The rate of combination measured by stopped flow and flash photolysis showed a linear variation over a wide range of CO concentration with a second-order rate constant of $1.5 \times 10^6 \ M^{-1} . \ s^{-1}$ at pH 6.0 (Foote *et al.*, 1983). However, the reaction measured by flow was not simple and a slower process showing first-order kinetics was observed. It seems likely that two forms of the unbound reduced enzyme are in equilibrium, only one of which can combine with CO directly and the slow, strictly first-order process, is a measure of the interconversion as represented below:

$$
\begin{array}{ccc}
A & + \ CO & \xrightleftharpoons[k_{-2}]{k_2} \quad A-CO \\
k_1 \Big\updownarrow k_{-1} & & \\
A^* & &
\end{array}
\qquad \ldots (1)
$$

The fact that the fast process represents a substantial portion of the observed reaction amplitude implies that the equilibrium between A and A* lies towards A and hence k_{-1} is less than k_1. Both k_1 and k_{-1} are much slower than k_2 with the result that following photolytic removal of CO there is negligible equilibration of A with A* and hence no slow process is observed in photolysis experiments. No direct estimate of k_{-2}, the CO off reaction is presently available but the form of the CO titration curve (figure 2.5) suggests that PsCCP has a very high affinity for this ligand.

Carbon monoxide does not inhibit the peroxidative activity of the enzyme, behaviour characteristic of other peroxidases and which is taken to suggest that the haem iron at the active site does not become reduced to the ferrous state during turnover, but instead cycles between ferric and ferryl forms (see later).

Redox reactions

PsCCP as isolated is in the oxidised state, both haem irons being ferric. Addition of sodium dithionite results in the formation of the fully reduced form which has peaks at 420 nm, 524 nm and at 551 nm, with a shoulder at 557 nm. The enzyme will accept electrons from a number of reductants, and whilst azurin and ferricytochrome c_{551} can be considered natural partners, ferrocyanide, ascorbate, N N N' N' tetramethyl phenylenediamine (TMPD) and NADH will also act as reductants.

Static measurements

The titration of oxidised PsCCP with reduced azurin (Cu^+) revealed a progressive reduction of the enzyme with successive additions of reduced azurin. The results of such a titration can be seen in figure 2.9, which is taken from Ronnberg et al. (1981b). The isosbestic point at 413 nm is good evidence of the presence of only two spectrally distinct forms and the absorbance differences at 407 nm when plotted against the concentration of azurin indicates a 1 : 1 molar ratio for the reaction between PsCCP and azurin. Since the enzyme is known to contain two haems c, this suggests that only one of these is readily reduced by azurin. Indeed, Foote et al. (1982) have found a similar state of affairs to be the case with other reductants like ferrocyanide and ascorbate.

Kinetic studies

Anaerobic stopped-flow experiments in which oxidised PsCCP was mixed with reduced azurin have been performed in a number of labs. Ronnberg et al. (1981) report that, at low azurin concentrations under pseudo first-order conditions, a biphasic reduction of the enzyme occurs, the slower phase becoming dominant at low (5 μM) azurin. Using double reciprocal plots of their data over a limited range of azurin, they obtain good linear plots which, extrapolated to infinite azurin, suggest a value for the maximal rate constant of 0.60 s^{-1} for the fast phase and 0.12 s^{-1} for the slower phase with dissociation constants of 6.6 μM and 7.0 μM respectively. Unfortunately, these conclusions are not borne out in practice since at high azurin concentration the reaction was observed to be

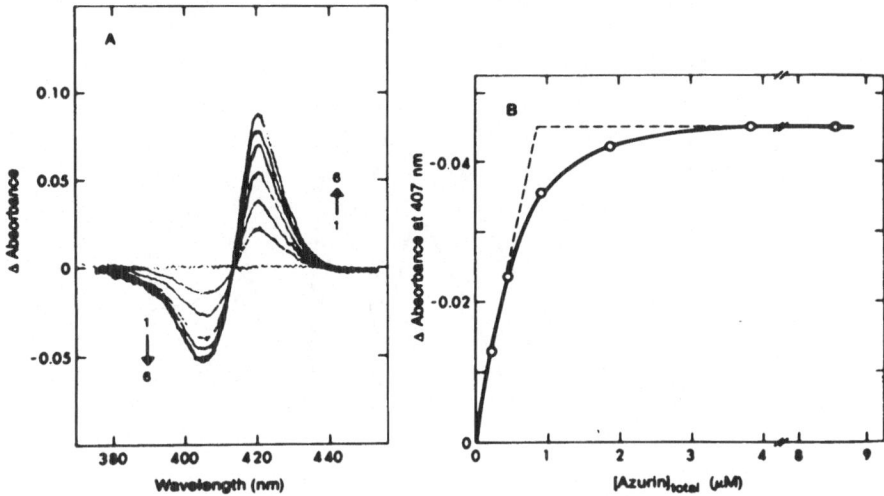

Figure 2.9   Reduction of *Pseudomonas* cytochrome *c* peroxidase by reduced azurin. (A) Difference spectra showing progressive reduction of the enzyme (0.9 μM) upon each addition of azurin. Concentration of azurin was increased from 0.2 to 8.5 μM (measurements 1 to 6). The spectra were measured using the 0 to 0.5 slide wire. (B) Absorbance differences at 408 nm are plotted against the concentration of total azurin after each addition. The static titration indicates a 1:1 molar ratio for the reaction between the peroxidase and azurin. (With permission of *J. biol. Chem.* and author)

essentially a single exponential with a pseudo first-order rate constant at infinite azurin of $5 \, s^{-1}$ (Foote *et al.*, 1982). A simple explanation for the biphasic nature of the reaction of PsCCP with azurin would be that two, inequivalent haem *c* moities are being reduced. This model, however, is not consistent with the static titration data which indicate only one haem *c* to be involved. Neither can the biphasic character be explained in terms of two forms of reduced azurin, as has been suggested in the case of the azurin-Ps. nitrite reductase (Brunori *et al.*, 1975), since similar observations have been made with ferrocyanide reduction whilst at high concentrations of azurin the progress curves are simple exponentials (Foote *et al.*, 1982). A more likely explanation is that two conformational states of the oxidised enzyme are in equilibrium and that the position of this equilibrium is influenced by a change in azurin concentration. Ronnberg *et al.* (1981) claim that the influence of reduced azurin on the enzyme is demonstrated by CD studies on the half-reduced enzyme in the Soret region. The positive ellipticity of ferric haem *c* increases markedly in the half-reduced enzyme (figure 2.10) due to a reduction in the negative component in the Soret transition seen in the fully oxidised enzyme. They suggest that azurin specifically induces the formation of a molecular species with a new geometric position of the ferric haem *c* in the protein pocket. This argument would be more convincing if CD studies on the half-reduced form of the protein in the absence of azurin had been made and had been found to be different.

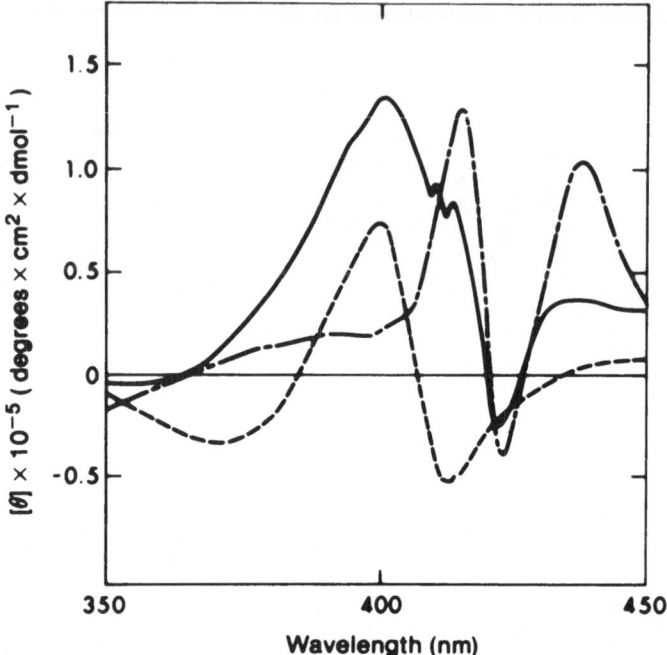

Figure 2.10　Soret region CD spectra of *Pseudomonas* cytochrome *c* peroxidase in 0.05 M sodium phosphate buffer, pH 6.0. ----, totally ferric enzyme; .–.–. totally ferrous enzyme reduced with dithionite; ——, enzyme partially reduced with reduced azurin. Concentrations used: enzyme, 41.0 μM; and azurin, 206.5 μM in sodium phosphate buffer, pH 6.0. (With permission of *Biochim. Biophys. Acta* and author)

It is quite clear that the reduction of the enzyme by azurin is complex and more data will be required before the process can be fully understood. On the basis of the present results, the minimum mechanism suggested by Ronnberg *et al.* (1981) is as follows:

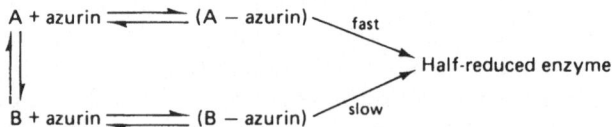

A and B represent different conformational forms of the enzyme and the rate of attainment of the equilibrium between these forms must be slower than the rate of complex formation but faster than the electron exchange process.

Anaerobic stopped-flow experiments using ferrocyanide as reductant have revealed a similarly complex picture, the reduction process occurring in two distinct phases, neither of which showed a clear linear dependence upon ferrocyanide concentration. Estimates of the maximal rate constant for each phase yielded values of 4.3 s^{-1} for the fast and 0.2 s^{-1} for the slow phase (Foote *et al.*, 1982).

Mechanism

The catalytic mechanism has been studied by Ronnberg *et al.* (1981) using a variety of rapid spectrophotometric techniques coupled to more conventional steady-state kinetics. The high spin haem *c* appears to be accessible to hydrogen peroxide although the reaction between the ferric form of the enzyme and peroxide was slow, $k = 7 \text{ s}^{-1}$ and independent of peroxide concentration. Following the formation of the initial complex (A) further very slow changes lead to a secondary complex (B). Neither of these complexes is capable of peroxidative oxidation of added ferrocytochrome c_{551} at any significant rate and therefore the oxidised complexes are considered 'abortive' although there is some evidence from steady-state kinetic measurements that they may become reactivated.

In sharp contrast to the fully oxidised enzyme, partially reduced peroxidase reacts rapidly with hydrogen peroxide to form an intermediate which appears to be active in the peroxidative cycle. The formation of the intermediate (compound I) was maximal when the ratio of reduced azurin to enzyme was 1 : 1, higher levels of reduction resulting in a decrease in the extent of compound I formation. The rate of compound I formation is $1.2 \times 10^8 \text{ M}^{-1} \text{ s}^{-1}$, a value approaching that of diffusion control whilst its decay in the presence of reduced azurin was found to be linearly dependent on the reductant concentration with a second-order rate of $2 \times 10^7 \text{ M}^{-1} \text{ s}^{-1}$. The observed pseudo first-order rates in excess of 10^3 s^{-1} are a remarkable testimony to the skill of the authors of this work. Taken together the results seem to show that the mechanism can be characterised as a modified ping-pong form similar to that accepted for other peroxidases. The following scheme taken from Ronnberg *et al.* (1981) summarises the observations:

'Abortive' complexes

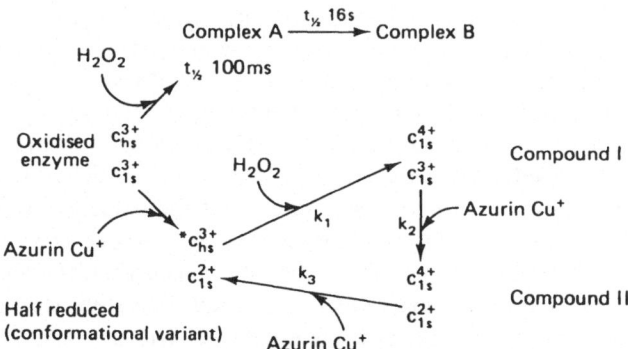

The cycle between half-reduced compound I and compound II is not unlike that proposed for horseradish peroxidase (HRP) where the two equivalent oxidation of the native enzyme leads to the formation of compound I in which the iron is ferryl Fe^{4+} and the extra electron is provided by the porphyrin or a protein group, e.g. tryptophan, resulting in the formation of a radical cation $(R\cdot)$. In PsCCP it is proposed that the extra electron is supplied by the reduced low spin haem c. By analogy with other peroxidases compound II is suggested to contain ferryl iron, although it is not excluded that it is $c_{hs}^{3+}c_{ls}^{3+}$. The formal valence and spin state of intermediates in the mechanistic cycle of the enzyme must await further clarification.

NITRITE REDUCTASE

The ability to reduce nitrate to nitrite is widespread in the bacterial world and two types of nitrate reductase have been distinguished, A and B (Pichinoty, 1964). Nitrate reductase B is found in the soluble fraction of cell extracts and functions in the reductive assimilation of nitrate nitrogen through nitrite and hydroxylamine to ammonia, thus providing a source of nitrogen for biosynthetic processes. In contrast, nitrate reductase A performs a respiratory function, initiating a series of reactions which successively use all the oxidation states of nitrogen as terminal electron acceptors in a form of nitrogen respiration or dissimilatory process. The dissimilatory nitrate reducing bacteria are almost exclusively aerobic bacteria and constitutive nitrate respiration has not yet been observed in wild-type organisms. The synthesis of the nitrate respiratory system occurs only in cells grown anaerobically or at low oxygen tension and is maintained only whilst these conditions persist. Aeration results in the rapid destruction of nitrate reductase and cells rapidly develop their capacity to respire aerobically (Payne, 1973).

The nitrite reductase isolated from *Pseudomonas aeruginosa* grown anaerobically in the presence of nitrate is undoubtedly the most studied and well characterised of the enzymes involved in the dissimilatory process. Horio (1958a) purified four soluble redox proteins from this organism: a blue copper-containing protein, later called azurin, cytochrome c_{551} and c_{554} and a further cytochrome which, because of its greenish-brown colour, he termed cytochrome$_{GB}$. Later work by Horio (1958a) and Horio *et al.* (1958) established that cytochrome$_{GB}$ was autoxidisable and could catalyse the oxidation of reduced azurin and cytochrome c_{551}; hence the enzyme was named *Pseudomonas* cytochrome oxidase. Yamanaka *et al.* (1960) discovered the nitrite reductase activity of the enzyme, thus providing a rationale for the observation that the oxidase was produced in large amounts by cells growing anaerobically in the presence of nitrate. Unfortunately, at the time of writing, the enzyme still remains classified as ferrocytochrome c_{551}: O_2 oxido reductase, EC 1.9.3.2 and is also on occasion referred to as Pseudomonas cytochrome cd_1 for reasons which will later become apparent.

The initial purification procedure devised by Horio *et al.* (1961a) and slightly modified by Yamanaka *et al.* (1963) yielded highly pure crystalline material but proved difficult to reproduce in other hands. As a result a number of other laboratories (Kuronen and Ellfolk, 1972; Gudat *et al.*, 1973; Parr *et al.*, 1976) have all independently devised purification methods which produce good yields of pure nitrite reductase. Molecular weight determinations give values of 119 kD (Kuronen and Ellfolk, 1972) and 121 kD (Gudat *et al.*, 1973) whilst SDS/ polyacrylamide gel electrophoresis has shown the molecule to be dimeric, apparently composed of two identical sub-units. The dimer is stable over the pH range 4–10 and in 3 M NaCl and 6 M urea, suggesting that ionic bonds have no crucial role to play in stabilising the dimeric structure (Kuronen *et al.*, 1975). A covalent dimer produced by chemically cross-linking the sub-units with dimethyl suberimidate was found to have a molecular weight of 130 kD as determined by SDS polyacrylamide gel electrophoresis (Silvestrini *et al.*, 1979). Iron analysis found a minimum molecular weight of 33.5 kD per iron (Kuronen and Ellfolk, 1982), i.e. four iron atoms per dimer and this has been confirmed in later studies (Silvestrini *et al.*, 1979). Electron micrographs using negative staining techniques with uranyl oxalate (pH 6.0) stain confirm the idea of the dimeric structure of the enzyme and a rough model of the enzyme, in which the sub-units are identical and sterically equivalent, was constructed based on different projections of the molecule seen in the electron micrographs (Saraste *et al.*, 1977).

Spectral properties

The spectra of purified *P.* nitrite reductase in its oxidised and ascorbate reduced forms is shown in figure 2.11. The first comprehensive description of the spectral properties was given by Horio *et al.* (1961), who reported that the observed spectrum resulted from the presence of two distinct chromophores in the molecule, one an unusual type of haem *c*, which exhibited a split alpha band in the reduced form of the enzyme, and a second chromophore, which was classified as haem a_2. The haem *c* was found to be covalently bound to the protein but the haem a_2 could be easily removed by treatment with acid acetone or simply by the addition of HCl to give a pH of 0.7–1.0, a process which, performed in the presence of ~0.8 M imidazole, yields the imidazole ferrihaem complex on adjustment to alkaline pH (Walsh *et al.*, 1980). Haem a_2 was reclassified as haem *d* (Enzyme Nomenclature Recommendations, 1965) although the spectral differences between the acid-haemin spectrum of haem *d* and the haem from *P.* nitrite reductase were noted by Lemberg and Barrett (1973) along with the fact that haem *d* is lipophilic, whereas the haem extracted from *P.* nitrite reductase is hydrophilic. It is now generally accepted that the acid-extractable haem of the enzyme is a chlorin type referred to as d_1 and although its structure has yet to be established unequivocally it has been suggested by Lemberg and Barrett (1973) that the conjugation of the tetrapyrrole ring is modified by the introduction of hydroxyl groups. Successful reconstitution of the apo-protein with haem d_1 has been achieved in which 90–100% recovery of enzyme activity was observed. However, only 5% recovery was found when haem *a* was substituted and proto, meso, deutero and haematohaem were all ineffective in yielding a product with enzyme activity (Hill and Wharton, 1978). The apoprotein preci-

C. Greenwood

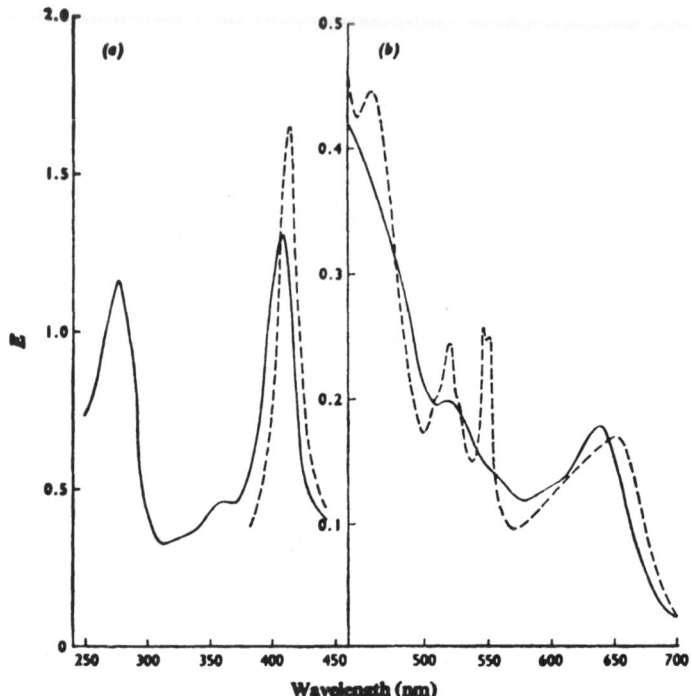

Figure 2.11 Absorption spectra of oxidised and ascorbate-reduced *Pseudomonas* cyto-chrome oxidase (nitrite reductase). —, oxidised; -----ascorbate-reduced *Pseudomonas* cytochrome oxidase in 0.04 M-potassium phosphate buffer, pH 6.9. (With permission of *Biochem. J.* and author.) The enzyme concentration was 8.9 μM and the spectra were recorded at room temperature (20°C), under N_2, in a Thunberg cuvette with a light-path of 1 cm

pitated out in the initial stages of haem d_1 removal can be re-dissolved in aqueous solution at pH values above 10.0. The spectrum of the apoprotein has the form typical of a low spin cytochrome c although this and the epr spectrum are some-what variable – presumably because of partial denaturation caused by the extremes of pH used. Although the apoprotein precipitates at pH values below 10.0, a large proportion of the precipitate can be dissolved in the narrow pH range 1-1.5. The spectrum of this material with Soret maximum of 395 nm and alpha peak at 500 nm corresponds to that of a high spin, five coordinate haem c.

The native enzyme crystallises with great facility although the thinness of the crystalline plates means that they are not ideal for X-ray diffraction studies. However, preliminary studies using this technique (Takano *et al.*, 1979) have shown that the molecule possesses a two-fold axis of symmetry. This infor-mation combined with the electron microscopy studies referred to earlier (Saraste *et al.*, 1977) has recently been used to interpret the results of emission spectroscopy involving fluorescence lifetime measurements of singlet tryptophan, triplet tryptophan and dansyl labelled enzyme in terms of a model of the enzyme in which the haems are all at one end of the molecule (Mitra and Bersohn, 1980) (see figure 2.12).

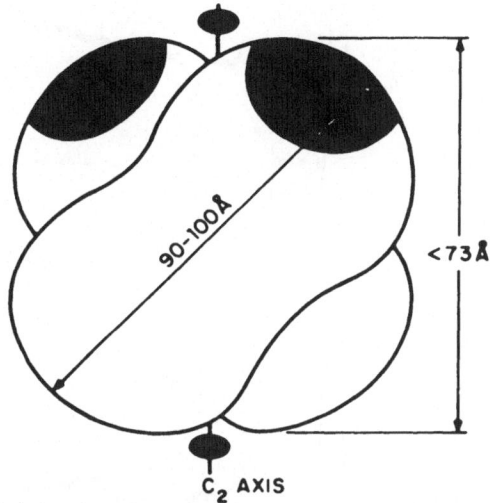

Figure 2.12   The shape, dimension and symmetry of *P. aeruginosa* cytochrome cd_1 oxidase as obtained from electron microscopy and preliminary X-ray investigation (see text). The shaded portion at the end of the molecule shows roughly the location of haems as inferred from emission studies. (With permission of *Biochemistry* and author)

Magnetic properties

Epr

The low temperature epr studies of Gudat *et al.* (1973) indicated that both haems c and d_1 are in the low spin state in the oxidised enzyme and, somewhat surprisingly, the addition of CN^- was said to abolish the signals assigned to haem d_1. In their epr spectrum Gudat *et al.* assigned the peaks at $g = 2.93, 2.31$ and 1.4 to the haem c component and the features at 2.46 and 1.71 to the haem d_1 on the basis that removal of the haem d_1 component of the enzyme resulted in the loss of the latter two resonances. Confirmation of these arrangements has come from the work of Walsh *et al.* (1979) who abolished the $g = 2.53$ and the $g = 1.73$ by complexing the haem d_1 with NO, thus obviating the need of drastic structural modification of the protein. The low temperature epr spectra of oxidised *Ps.* nitrite reductase and its cyanide complex taken from Walsh *et al.* (1979) are presented in figure 2.13. The spectrum in figure 2.13a agrees well with that reported by Gudat *et al.* (1973); the features at $g = 4.2$ and 2.08 are most probably due to adventitious iron and copper respectively. The absence of signals in the $g \sim 6$ region of the epr spectrum of the enzyme suggests that no

C. Greenwood

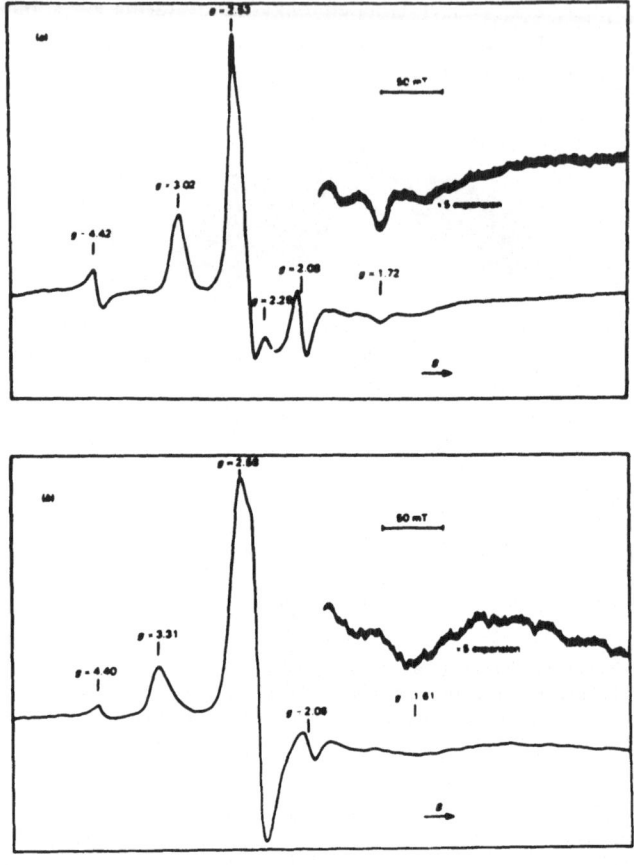

Figure 2.13 Low-temperature epr spectra of oxidised *Pseudomonas* cytochrome oxidase and its cyanide complex. (a) The epr spectrum of oxidised *Pseudomonas* cytochrome oxidase. The spectrum was run at 10 K on a 63 μM-protein sample in 0.04 M-potassium phosphate buffer at pH 7.0. The microwave power was 5.0 mW, field modulation amplitude 0.63 mT, microwave frequency 9.155 GHz and receiver gain 2.5×10^5. B is the magnetic field strength (expressed in T). (b) The epr spectrum of the cyanide complex of oxidised *Pseudomonas* cytochrome oxidase. The spectrum was run at 10 K on a 70 μM-protein sample in 0.04 M-potassium phosphate buffer, pH 7.0, and in the presence of 100 mM-KCN. The microwave power was 20 mW, field modulation amplitude 0.63 mT, microwave frequency 9.150 GHz and receiver gain 2.0×10^5. (With permission of *Biochem. J.* and author)

high spin components are present. The epr spectrum in figure 2.13b is totally at variance with the earlier report (Gudat *et al.*, 1973) on the effect of cyanide and confirms that both haem c and d_1 bind this ligand and remain low spin ferric.

While epr studies are useful in investigations of haem electronic states, they are generally restricted to ferric haem complexes. Variable temperature magnetic circular dichroism (mcd) spectroscopy can provide an optical probe into the magnetic properties of both ferric and ferrous states of haemoproteins (Springall *et al.*, 1976). Previous workers (Orii *et al.*, 1977; Vickery *et al.*, 1978) have published the room temperature mcd spectra of *Ps*. nitrite reductase but reached no conclusions about the magnetic states of haem d_1. The power and facility of variable temperature mcd is nicely illustrated in figure 2.14, which shows the room temperature absorption spectrum and low temperature mcd spectra of ascorbate reduced nitrite reductase. The spectral region between 500 and 550 nm is dominated by the sharp α and β bands of the haem c moiety and the temperature independence of the mcd throughout this range confirms the diamagnetic character of the low spin ferrous haem c. By contrast, the mcd between 550 nm and 700 nm are strongly temperature dependent and since only haem d_1 contributes in the region it is concluded that haem d_1 is paramagnetic in the ascorbate reduced protein. The Soret region confirms these conclusions but is more complex, since bands from the haems c and d_1 overlap; the arrows at 413 and 424 nm in figure 2.14b indicate the position of the positive and

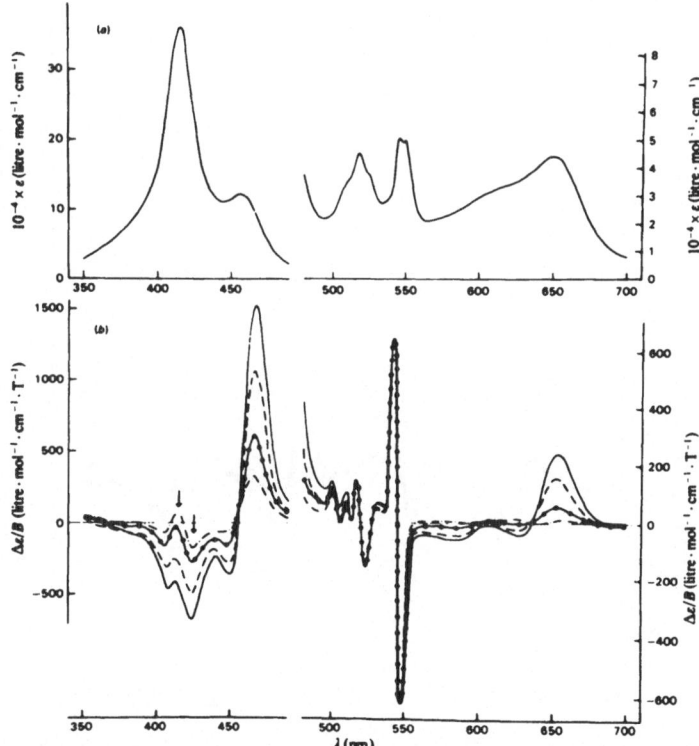

Figure 2.14   Room-temperature absorption spectrum (a) and low-temperature mcd spectra (b) of ascorbate-reduced *Pseudomonas* cytochrome oxidase. The mcd spectra were run at temperatures of 16 K (—), 27 K (----), 53 K (•-•) and 95 K (−.−.) on a sample of enzyme (in 0.04 M-potassium phosphate buffer, pH 7.0) that had been saturated with sucrose to give a final protein concentration of 29 μM. (With permission of *Biochem. J.* and author)

negative peaks of the temperature independent mcd spectrum of haem c which
are however underlain by a broad temperature dependent negative peak belong-
ing to haem d_1. The form of the mcd spectrum of reduced haem d_1 in the Soret
region is similar to that of deoxymyoglobin (Springall *et al.*, 1976), the a_3
component of reduced mammalian cytochrome oxidase (Thomson *et al.*, 1977),
suggesting that haem d_1 has the high spin (S = 2) ferrous form in the reduced
enzyme. Addition of CN^-, which is known to bind only to haem d_1 in the
reduced enzyme (Barber *et al.*, 1978a), results in a completely temperature
independent mcd spectrum showing that haem d_1 has switched to the low spin
ferrous form on binding cyanide. The variable temperature mcd spectra of
oxidised *Ps.* nitrite reductase are rich in detail and all the peaks show strongly
temperature dependent intensities (see figure 2.15). The mcd of the Soret region
between 350 and 450 nm which can be assigned to haem c is very similar to the

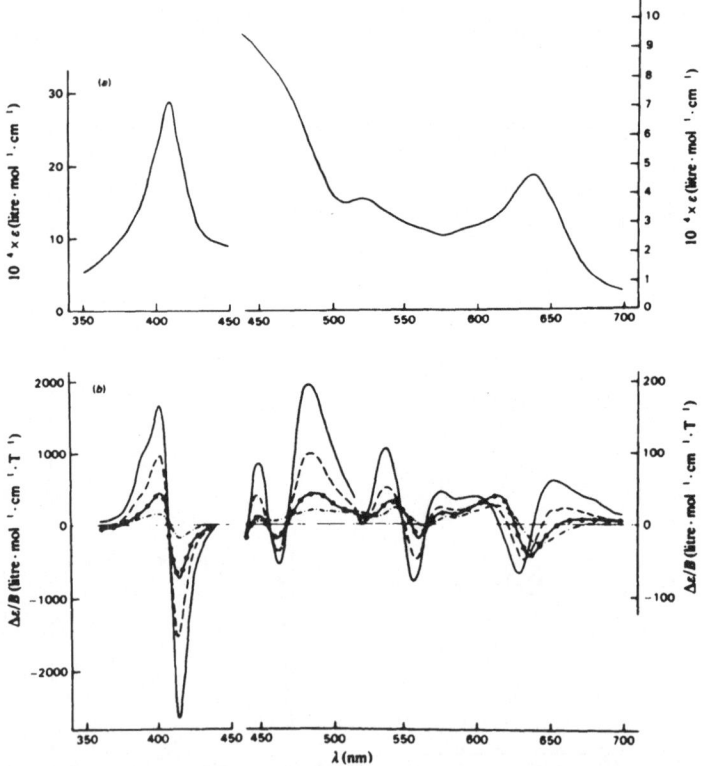

Figure 2.15   Room-temperature absorption spectrum (a) and low-temperature mcd spectra
(b) of oxidised *Pseudomonas* cytochrome oxidase. The mcd spectra were run at temperatures of 16 K (—), 34 K (-----), 85 K (●-●) and 293 K (−.−.−) on a sample of enzyme
(in 0.04 M-potassium phosphate buffer, pH 7.0) that had been saturated with sucrose to give
a final protein concentration of 29 μM. (With permission of *Biochem. J.* and author)

mcd spectra of low spin ferric haems such as metmyoglobin cyanide (Springall *et al.*, 1976) and haem *a* in mammalian cytochrome oxidase (Thomson *et al.*, 1977) confirming its assignment as a low spin moiety. The wavelength region above 600 nm which can be confidently assigned to haem d_1 also shows mcd peaks which are strongly temperature dependent indicating that haem d_1 is paramagnetic. Since high spin ferric haems have only weakly temperature dependent spectra, the inescapable conclusion is that haem d_1 is low spin. This is a rather surprising finding since the d_1 haem of the reduced protein is high spin and very probably five coordinate. If such a geometry were to be maintained in the oxidised state, it seems likely that a high spin or mixed spin form of haem d_1 would result. Possibly the haem is so constrained as to remain low although five coordinate or else it may pick up a ligand from the protein, the mechanistic significance of which might be to facilitate the reduction of the ferric haem d_1.

Endogenous ligands

The structure of *Ps. aeruginosa* cytochrome c_{551} has been determined and crystallographically refined at 2.0 Å by Almassey and Dickerson (1978). The haem ligands are histidine and methionine and in view of the close spectral and magneto-optical relationship between c_{551} and the haem *c* of nitrite reductase it seems safe to conclude an identity of ligands in this case. ^{14}NO and ^{15}NO have been used as spin label probes to characterise the endogenous ligand of the high spin haem d_1 of the reduced enzyme at pH 8.0 where only this centre binds NO. Figure 2.16 shows the epr spectra (taken from Johnson *et al.*, 1980) of the ^{14}NO and ^{15}NO bound forms of the enzyme. The transition from a triplet of triplets to a doublet of triplets with a concomitant change in the larger hyperfine coupling constant from 2.4 to 3.3 mT on replacement of ^{14}NO with ^{15}NO is expected and the unchanged smaller triplet hyperfine of 0.65 mT must be assigned to another axially bound ^{14}N nucleus. Although a number of amino acids, such as histidine, arginine and lysine, or the peptide nitrogen atom, could contribute this axial ligand, the hyperfine coupling constant of 0.65 mT strongly supports imidazole from a histidyl residue for this ligand.

Exogenous ligand binding

Reduced nitrite reductase
Sodium dithionite is a versatile and widely used reducing agent in the haemoprotein field. However, its use as a reductant for nitrite reductase was precluded a number of years ago when it was shown, along with several other oxyanions of sulphur, to form a complex with the reduced enzyme (Parr *et al.*, 1974). Only those sulphur oxyanions that can form SO_3^{2-} ions (sulphite and metabisulphite) or SO_2^- (dithionite) produce the spectral shift in the α band of ferrous d_1 haem and it is interesting to speculate on the possibility that they may occupy the same site as the NO_2^- ion.

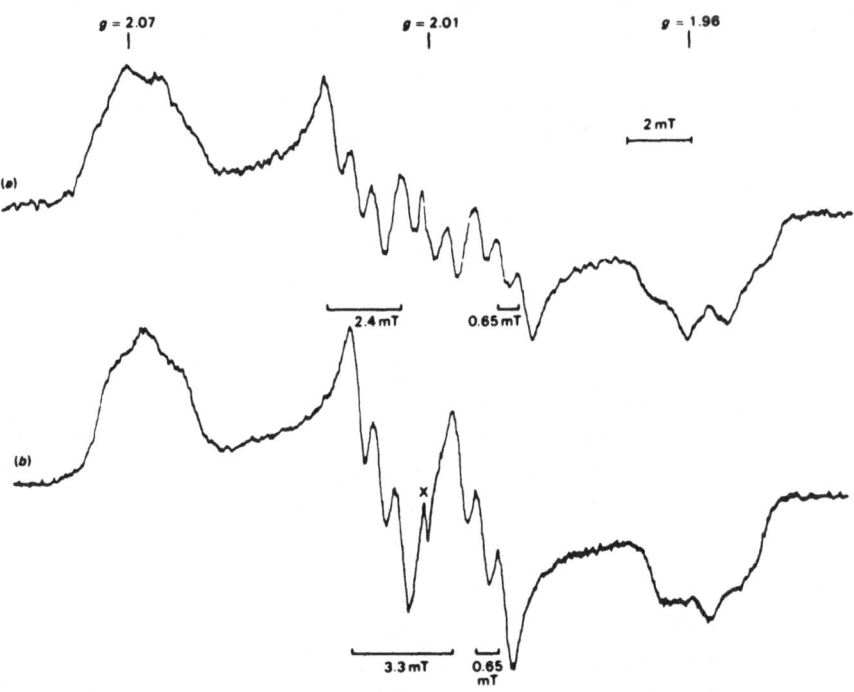

Figure 2.16 Expanded epr spectra of ^{14}NO and ^{15}NO-bound reduced *Pseudomonas* nitrite reductase at pH 8.0. (a) Shows the expanded epr spectrum of ^{14}NO-bound reduced *Pseudomonas* nitrite reductase. The protein concentration was 22 μM in the presence of 1.8 mM-KNO$_2$. (b) Shows a similar spectrum of the ^{15}NO-bound reduced enzyme in which the protein concentration was 29 μM in the presence of 1.8 mM-Na^{15}NO$_2$ (99.2 atom % ^{15}N). The epr spectra were recorded at 30 K. The microwave power was 1.25 mW, microwave frequency 9.24 GHz, field modulation amplitude 0.2 mT and receiver gain 4 × 10^5. Both samples were in 0.1 M-Tris/HCl buffer, pH 8.0, in the presence of 20 mM-ascorbate. X indicates a paramagnetic impurity in the cryostat insert. (With permission of *Biochem. J.* and author)

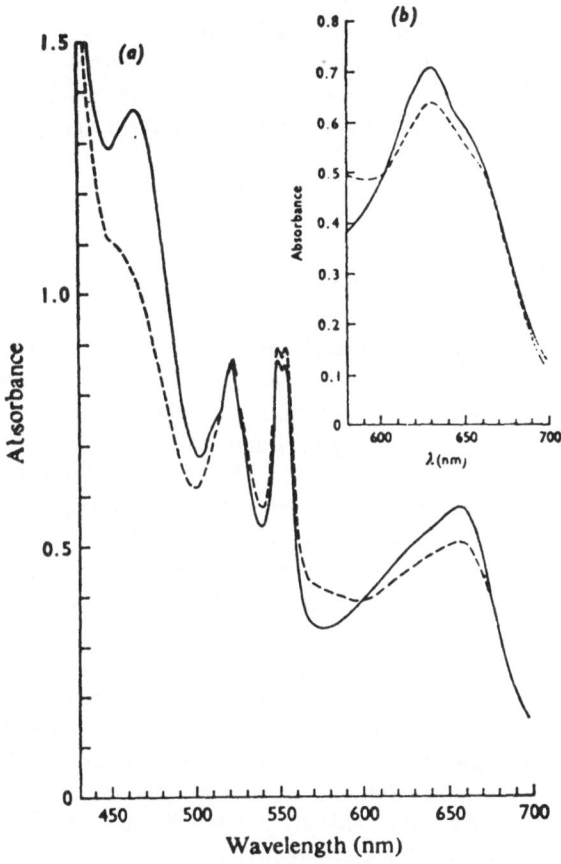

Figure 2.17 Absorption spectra of reduced *Pseudomonas* cytochrome oxidase in the absence and presence of CO. (a) —, Ascorbate-reduced *Pseudomonas* cytochrome oxidase, -----, ascorbate-reduced *Pseudomonas* cytochrome oxidase-CO complex, in 0.04 M-potassium phosphate buffer, pH 7.0. The enzyme concentration was 22 μM; CO concentration was 1 mM. Spectra were recorded at room temperature (20°C) and the light-path was 1 cm. (b) —, Dithionite-reduced *Pseudomonas* cytochrome oxidase in the red region: ----, dithionite-reduced *Pseudomonas* cytochrome oxidase-CO complex. Enzyme concentration was 26 μM, CO concentration was 1 mM. All other conditions were as above. (With permission of *Biochem. J.* and author)

The binding of CO to ascorbate reduced nitrite reductase has been studied by static titration, stopped-flow and flash-photolytic techniques (Parr *et al.*, 1975). The spectral changes on binding CO are consistent with reaction at the d_1 haem (see figure 2.17). Static titration data revealed that the binding process was non-stoichiometric with a Hill number of 1.44 indicating positive cooperativity between the d_1 haems. Furthermore, the kinetics of binding were biphasic, the faster rate exhibiting a linear dependence on CO concentration with a rate constant of 2×10^4 M^{-1}. s^{-1}, whereas the slower rate reached a pseudo first-order rate limit of 1 s^{-1}. The scheme below was devised to explain the complex binding kinetics:

$$E \xrightarrow{k_1} E \cdot CO \xrightarrow{k_2} E(CO)_2 \qquad \qquad \dots(2)$$

$$E^*CO$$

in which E^* is a form of the enzyme capable of binding only one CO molecule and k_3 represents the pseudo first-order limit of 1 s^{-1}. Binding CO to the reduced d_1 haem influences the reaction of the reduced haem c in the enzyme with oxidised azurin (Parr *et al.*, 1977) and also the oxidation of haem c by ferricyanide which is enhanced when the d_1 haem is complexed with CO (Barber *et al.*, 1978b). Thus the picture that emerges from CO binding studies is of some form of haem d_1-haem c interaction in addition to the d_1 haem cooperativity.

Cyanide binding to the d_1 haem of the reduced enzyme is rapid and complex (Barber *et al.*, 1978c). The reaction is biphasic, both phases exhibiting a linear dependence on cyanide concentration, yielding second order rates of 9.3×10^5 $M^{-1}.s^{-1}$ and 2.3×10^5 $M^{-1}.s^{-1}$, respectively with the relative proportions of the two phases varying with cyanide concentration. The binding of cyanide is not stoichiometric and the sigmoidal titration yields a value for the Hill coefficient of 2.6 with an apparent affinity constant of 5×10^{11} M^{-1}. Both haem c and d_1 bind cyanide when in the ferric state in a ligand concentration-dependent fashion, the second-order rate constant for which was approximately 0.2 M^{-1} s^{-1} at pH 7.0, increasing with increasing pH.

Nitric oxide the product of nitrite reductase activity, can bind to both haems c and d_1 in both redox states, thus NO bound forms of the enzyme may assume physiological significance in nitrite reduction. Steady state experiments in which the oxidation of azurin by nitrite under anaerobic conditions is catalysed by nitrite reductase clearly show that, at pH values near neutrality, the reaction is rapidly terminated in spite of the presence of oxidant and reductant (Silvestrini *et al.*, 1979). By lowering the pH under otherwise identical conditions or raising the nitrite concentration, the extent of the reaction can be increased, strongly suggesting that product inhibition is involved. The oxidised enzyme in the pH range 4-6 binds NO to both c and d_1 haems to give a stable product. At pH 7 and above, however, an autoreduction process resulted in the formation of the reduced ligand-bound form for the protein. Ascorbate-reduced enzyme samples between pH 4 and 7 bind NO at both haem c and d_1, although at pH 7 the rate or NO binding to haem c is very slow and at pH 8 and pH 9.0 only ferrohaem d_1 binds NO. In both oxidised and reduced nitrite reductase the haem d_1 had a much higher affinity for NO than had the c haem. Figure 2.18a shows a family of spectra produced during a titration of ascorbate-reduced enzyme with nitrite (which acts as a precursor of NO) and that the use of nitrite concentrations close to stoichiometric results in binding solely at the d_1 haem. The data in figure 2.18b is consistent with there being two binding sites per protein molecule.

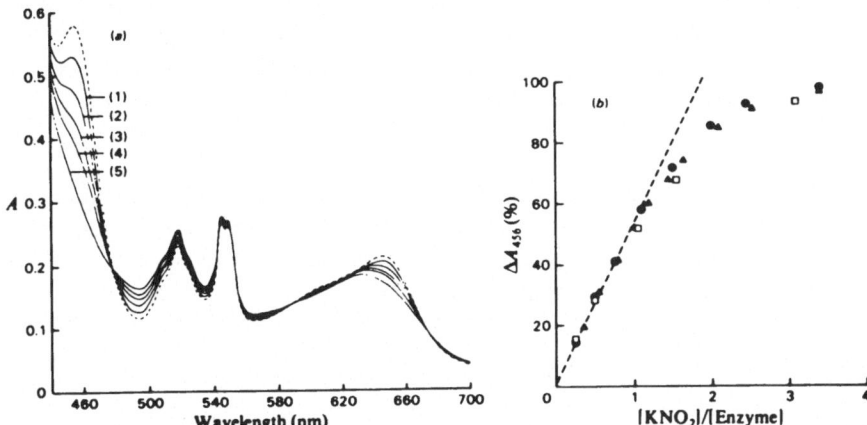

Figure 2.18 Titration of the reduced haem d_1 component of *Pseudomonas* cytochrome oxidase with nitrite. (a) ----, Spectra of 4.86 μM-reduced *Pseudomonas* cytochrome oxidase in 0.1 M-potassium phosphate buffer, pH 6.1. The enzyme was prepared anaerobically under N_2 so as to completely fill a sealed 1 cm-path-length cuvette (volume 3.7 ml) and reduced with 25 μl of 1 M-sodium ascorbate. Curves (1), (2), (3) and (4) show spectra after additions of KNO_2 to concentrations of 1.6, 3.8, 5.91 and 8.04 μM respectively. Curve (5), used as the end point, corresponds to a nitrite concentration of 27.3 μM. Some of the spectra obtained during the titration have been omitted for clarity. The temperature was 20°C. (b) The results of three experiments carried out under similar conditions to (a), at pH values of 5.1 ($\bullet$), 6.1 ($\blacktriangle$) and 7.1 ($\square$). The changes in absorbance at 456 nm were corrected for dilution and are plotted as percentages of the total against the ratio of the total concentration of nitrite added to the concentration of enzyme. (With permission of *Biochem. J.* and author)

REDOX REACTIONS

Reduction

The physiological electron donors for *P.* nitrite reductase are thought to be cytochrome c_{551} and azurin, both of which are isolated along with the enzyme during the extraction procedure. The Michaelis constants are 5.6 μM and 49 μM for c_{551} and azurin respectively at pH 7.0 and 30°C (Barber *et al.*, 1976) and the oxidised substrates act to inhibit both oxidase and nitrite reductase activities of the enzyme.

Static titrations

Ascorbate titrations by Horio *et al.* (1961) demonstrated a constant difference in half reduction potential between the *c* and d_1 haems of 70 mV, the haem d_1 being more negative. Using ferrous EDTA the redox potentials have been determined to be +288 mV for the *c* haem and +270 mV for the d_1 haem (Blatt and Pecht, 1979); however the reduction of both haems was non-stoichiometric. These findings have been interpreted in terms of negative interaction between the *c* haems and positive reaction between the haems d_1 which were of the same energy. A spectral interaction was also invoked for each pair of haems with the initial reduction causing an increase in extinction twice that of the second. Some doubt must be expressed concerning this significance of this complex interpretation, since the final reduced spectrum is different from that observed by other workers (Gudat *et al.*, 1973; Barber *et al.*, 1976; Shimada and Orii, 1978).

Kinetic studies

The spectral overlap between c_{551} and the *c* haem component of nitrite reductase has meant that most non-steady-state kinetic studies have been performed using azurin. The results of temperature jump experiments (Wilson *et al.*, 1975) designed to explore the electron exchange between azurin and the *c* haem, by isolating the d_1 haem as its CO complex, are accommodated in the following scheme:

$$A^+ + c^{3+} \rightleftharpoons X_1 \rightleftharpoons X_2 \rightleftharpoons A^{2+} + c^{2+}$$

$$\Updownarrow$$

$$(A^*)^+$$

$$\dots (3)$$

where X_1 and X_2 represent forms of a molecular complex between azurin and the haem *c* moiety of the enzyme. A and A* are two spectrally similar forms of azurin of which only A is kinetically competent, a situation consistent with work (Wilson *et al.*, 1975) which indicated two forms of azurin, only one of which can exchange electrons with c_{551}.

From rapid mixing experiments using azurin and nitrite reductase, Wharton *et al.* (1973) concluded that the initial transfer of electrons occurs between azurin and the haem *c* component of the enzyme with a molecular complex of the two proteins. Reduced azurin donates an electron to the haem *c* moiety with an apparent first-order rate constant of 29 s^{-1} while the transfer from reduced haem *c* to oxidised azurin proceeds at about 120 s^{-1}. The reaction between the copper protein and haem *c* is followed by a much slower internal reaction involving electron exchange between haem *c* and d_1. Parr *et al.* (1977) have extended these observations and have applied the results to a kinetic model originally developed from temperature jump studies (equation 3). They also observed that the addition of CO to the reduced enzyme profoundly modified the rate and profile of the ensuing reaction with oxidised azurin, which became biphasic, the fast phase accounting for over 80% of the change and tending towards a rate limit of 200 s^{-1}. Recent studies have shown that binding CO to the d_1 haem induces a large decrease (at least 80 mV) in the redox potential of the c haem component of the enzyme (Silvestrini *et al.*, 1982). These same authors report a successful attempt to observe directly and characterise the electron transfer between c_{551} and the haem c of the enzyme. Working at pH 9.1 to take advantage of a pH induced red shift in c_{551} not found in the c haem of nitrite reductase, they report a rate of 1.3×10^7 $M^{-1}.s^{-1}$ for the electron transfer from CO-bound oxidase to cytochrome c_{551}. In spite of obvious technical difficulties with the reverse process, i.e. electron transfer from reduced c_{551} to oxidised oxidase, the half time for this reaction is 0.1–0.2 s at 3.5–0.5 μM concentration compared to values of 0.4–0.6 s for electron donation from reduced azurin under comparable conditions. They conclude that azurin and cytochrome c_{551} are not equally efficient electron donors for the oxidase although the relatively small difference does not imply a sequential pathway in the electron transfer to nitrite catalysed by nitrite reductase. In addition to the physiological electron donors, *P*. nitrite reductase can accept electrons from a number of other reagents of which the most commonly employed reductants are ascorbate, hydroquinone and NADH/PMS. The Km values for hydroquinone and ascorbate at 30°C and pH 7.0 with oxygen as terminal acceptor are 30 mM and 4 mM, i.e. some three orders of magnitude higher than the protein substrates, although the simple chemical reductants have more negative redox potentials (Barber *et al.*, 1976).

Reduction of the enzyme by chromous ions (Cr^{2+}) has been followed in stopped-flow experiments (Barber *et al.*, 1977) and has proved useful in providing reduced minus oxidised difference spectra for the two haems. Reduction of the c haem was biphasic, both phases remaining constant in relative proportion and having rate constants dependent on $[Cr^{2+}]$. The dependence of haem d_1 reduction was minimal and the rate was close to the internal electron transfer from c to d_1 observed for the azurin reduction kinetics (Parr *et al.*, 1977). By raising the chromous ion concentration a clear kinetic separation of the reactions of the two chromophores was achieved enabling a reduced minus oxidised difference spectrum to be generated for each haem (see figure 2.19).

Oxidation

As mentioned earlier, *P*. nitrite reductase can accomplish the four electron reduction of oxygen to water in addition to performing its likely physiological role of catalysing the single electron reduction of nitrite to nitric oxide. A comparison of the steady-state kinetic parameters shows that the pH optimum for oxygen reduction is 5.1 whilst that for nitrite reductase activity is 6.5, with reduced c_{551} as substrate in both cases. The Km values turn out to be 28 μM for oxygen at pH 6.0 and 27°C and 53 μM at pH 6.0 and 30°C for nitrite (Yamanaka *et al.*, 1961). Kijimoto (1968) has calculated the affinity constants of nitrite and

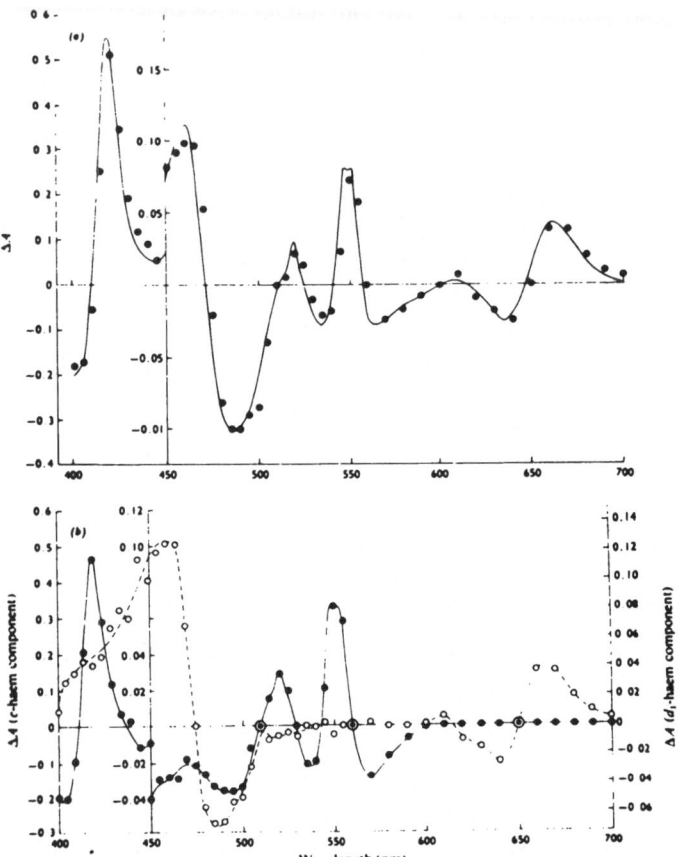

Figure 2.19 Difference spectra for the reduction of *Pseudomonas* cytochrome oxidase by Cr^{2+} ions. (a) The static difference spectrum (reduced minus oxidised) of 7.5 μM *Pseudomonas* cytochrome oxidase (—) determined by use of the Cary 118C spectrophotometer and the total kinetic difference spectrum (●), representing the change in absorbance from 3 ms, after mixing the protein with 440 μM-Cr^{2+} on the stopped-flow apparatus to the final value. The reactions were carried out in 0.1 M-sodium cacodylate, pH 7.0, at 25°C in a 2 cm-path-length cell. (b) The reduced-minus-oxidised difference spectra produced on analysis of the total kinetic difference spectrum in (a) into *c*- and d_1-haem phase. The two left-hand ordinates (400–450 nm and 450–700 nm) refer to the *c*-haem component (●); the right-hand ordinate refers to the d_1-haem (○). (With permission of *Biochem. J.* and author)

oxygen for the reduced protein as 1.25×10^6 M^{-1} and 1.67×10^4 M^{-1} respectively.

Barber *et al.* (1978b) studied the reaction between ascorbate reduced enzyme and potassium ferricyanide under nitrogen and carbon monoxide. Two ferricyanide concentration dependent processes with second-order rate constants of 9.6×10^4 M^{-1} s^{-1} and 1.5×10^4 M^{-1} s^{-1} were assigned to the direct bimolecular reactions of ferricyanide with the haem *c* and haem d_1 moieties of the enzyme respectively. Under CO the second-order rate constant for the reaction of haem c with ferricyanide was slightly enhanced at 1.3×10^5 M^{-1} s^{-1} whilst that of haem d_1 was greatly decreased being limited to the rate of CO dissociation. This latter process can be represented as follows:

$$
\begin{array}{c}
c^{2+} \\
| \\
d_1^{2+}-CO
\end{array}
\xrightarrow[1.3 \times 10^5 M^{-1}s^{-1}]{Fe(CN)_6^{3-}}
\begin{array}{c}
c^{3+} \\
| \\
d_1^{2+}-CO
\end{array}
\xrightarrow[0.03\,s^{-1}]{Slow}
\begin{array}{c}
c^{3+} \\
| \\
d_1^{2+}
\end{array}
\xrightarrow[Fast]{Fe(CN)_6^{3-}}
\begin{array}{c}
c^{3+} \\
| \\
d_1^{3+}
\end{array}
$$

whilst the process in the absence of CO proceeds through an intermediate in which both haems are formally oxidised but which differs spectrally from the 'resting' oxidised enzyme, see later.

$$
\begin{array}{c}
c^{2+} \\
| \\
d_1^{2+}
\end{array}
\xrightarrow[1.5 \times 10^4 M^{-1}s^{-1}]{\substack{9.6 \times 10^4 M^{-1}s^{-1} \\ Fe(CN)_6^{3-1}}}
\left[
\begin{array}{c}
c^{3+} \\
| \\
d_1^{3+}
\end{array}
\right]
\xrightarrow{0.17\,s^{-1}}
\begin{array}{c}
c^{3+} \\
| \\
d_1^{3+}
\end{array}
$$

Reaction with oxygen

The reaction of fully reduced enzyme with oxygen has been investigated using stopped flow techniques by both Wharton and Gibson (1976) and Greenwood *et al.* (1978) with somewhat conflicting results. Wharton and Gibson deduced that the haem d_1 was oxidised faster than the haem c, its oxidation being accurately second-order with a rate constant of 5.7×10^4 M^{-1} s^{-1} at 20°C, whilst that of the haem c was rate limited at 8 s^{-1}. From these observations they concluded that there was a sequential transfer of electrons through the enzyme to oxygen with a rate-limiting internal electron transfer from haem c to haem d_1. Greenwood *et al.*, however, found that in their experience both haems were rapidly oxidised with a bimolecular rate constant of 3.3×10^4 M^{-1} s^{-1} at 25°C that was independent of wavelength. Intermediate and slow phases were noted corresponding to monomolecular processes with rate constants of 1 s^{-1} and 0.1 s^{-1} and a scheme to provide the minimum explanation of the data was suggested to be:

$$
O_2 + \text{Fully reduced enzyme} \underset{k_{-1}}{\overset{k_{+1}}{\rightleftharpoons}} A \xrightarrow{k_{+2}} B \rightarrow C
$$

where A, B and C represent reaction intermediates.

Using the above scheme, Greenwood *et al.* (1978) were able to fit the data with values of 3×10^4 M^{-1} s^{-1}, 1.5 s^{-1} and 2.6 s^{-1} for k_{+1}, k_{-1} and k_{+2} respectively. The values of k_{+1} and k_{-1} lead to a value of $\approx 10^4$ M^{-1} for the affinity constants for oxygen binding in good agreement with the value of 1.67×10^4 M^{-1} obtained by Kijimoto (1968). Kinetic difference spectra were determined for all three phases of the reaction and these revealed that each had different characteristics. The initial fast phase involved both c and d_1 haems, the changes being consistent with c oxidation but with haem d_1 assuming a form which did not correspond to the normal oxidised state, a situation that was not restored even after the second kinetic phase, which reflected further changes in the haem d_1 component. The difference spectrum associated with the slow conversion of intermediate B into C is very reminiscent of that which was observed in the case of the slowest phase in the ferricyanide oxidation process (Barber *et al.*, 1978b). By analogy with the ferricyanide experiment, it was concluded that, in the reaction of reduced nitrite reductase with oxygen, all the expected redox changes in the enzyme occur prior to the conversion of species B into C.

Although the results of Greenwood *et al.* (1978) cannot distinguish which haem, c or d_1, is the binding site for oxygen, other ligand binding studies (Parr *et al.*, 1975; Barber *et al.*, 1978c) suggest that haem d_1 is the most likely site of attack for oxygen. If this is a correct assumption, then the fast, oxygen-dependent reaction of the haem c calls for internal electron transfer rates of at least 100 s^{-1}, i.e. greatly in excess of those observed in the anaerobic reduction of oxidised enzyme by azurin (Parr *et al.*, 1977). Ingledew and Saraste (1979) on the basis of steady-state kinetic experiments have suggested that, at high oxygen tensions, two molecules of oxygen bind, one to each haem d_1 and are reduced to peroxide with the four immediately available reducing equivalents. This step is followed by a catalase reaction. At lower oxygen tensions only a single oxygen molecule binds and is fully reduced to water.

Reaction with nitrite

No comprehensive study of the nitrite reductase activity of the enzyme has been made although Shimada and Orii (1975, 1978) have speculated about the mechanism on the basis of steady-state spectra recorded over the wavelength range 500–700 nm at 3-min intervals. The main drawback to a serious pre-steady-state investigation is that the reaction of the reduced enzyme has so far proved to be too fast to measure for conventional stopped flow equipment.

Other nitrite reductases

Liu and Peck (1981) report the isolation from *Desulphovibrio desulphuricans*, a strictly anaerobic sulphate-reducing bacteria, of a nitrite reductase capable of catalysing the six electron reduction of nitrite to ammonia. The membrane-bound enzyme has been purified to electrophoretic homogeneity and has a minimal molecular weight of 66 kD. The purified enzyme exhibits a typical c-type cytochrome spectrum with a reduced α band at 552 nm and contains six c-type haems per molecule, based on iron analysis and a comparison of the extinction coefficient of the reduced pyridine haemochromogen of the nitrite reductase with that of horse heart cytochrome c. The fully reduced enzyme prepared by either dithionite or FADH is completely autoxidisable even in the presence of cyanide. Ascorbate treatment leads to the formation of a partially reduced enzyme, suggesting, perhaps, that some of the haem groups have higher oxidation reduction potentials than others; curiously, the partially reduced enzyme is not autoxidisable or re-oxidised by nitrite or hydroxylamine. Only a fraction of the haems in the fully reduced form of the enzyme reacted with nitrite to form the haem-NO complex in contrast to the situation under NO gas, when all of the haems were found to react.

The purified enzyme shows a pH optimum in the range 8-9.5 and a high temperature for optimal activity at 57°C. The Km for nitrite was determined to be 1.14 mM and 113.5 mM for hydroxylamine. This difference in Km values seems to exclude the possibility of hydroxylamine being a free intermediate in the reduction of nitrite.

A nitrite reductase from *Achromobacter fischeri* grown under low oxygen tension has been purified by Prakash and Sadana (1972) and shown to contain two *c* haems per mole of enzyme. The absorption spectra of the oxidised and reduced forms of the enzyme as well as the nitrite or hydroxylamine reoxidised enzyme are presented in figure 2.20. The purified enzyme reduces nitrite and

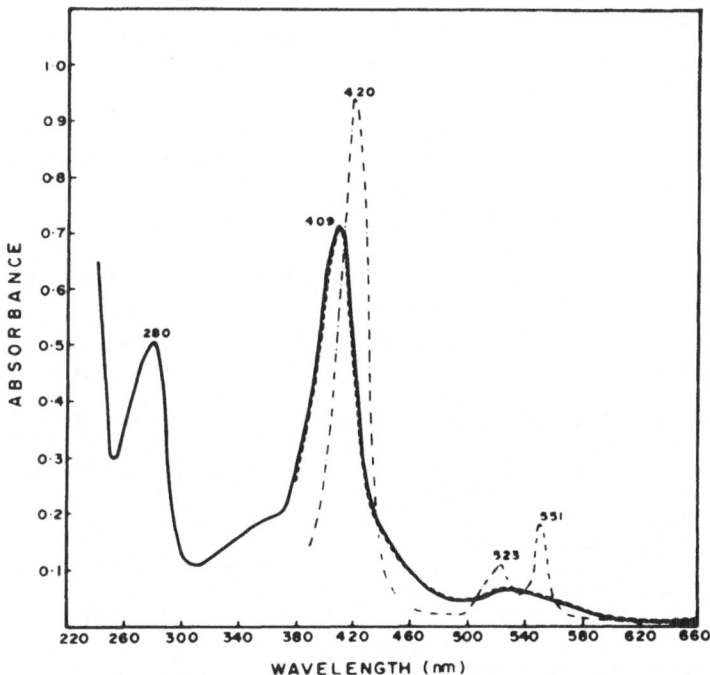

Figure 2.20 The absorption spectra of *A. fischeri* nitrite reductase, SA 1500. Spectra were measured in 0.05 M phosphate buffer, pH 7.5, enzyme protein 0.41 mg/ml; (—) oxidised; (−.−.−), reduced with dithionite; (----) NO$_2^-$ (or hydroxylamine) added to the dithionite-reduced enzyme. (With permission of *Archiv. Biochem. Biophys.*)

hydroxylamine stoichiometrically to ammonia, the respective Kms being 0.05 mM and 5-8 mM. The reduced enzyme is apparently autoxidisable and the reaction with oxygen was not affected by cyanide (10^{-2} M). In contrast, CO completely inhibited nitrite reduction and this inhibition was reversed by light in the NADH and $FADH_2$ systems; with benzyl viologen as reductant, no light reversal was observed, but this was most probably due to absorption of the effective wavelengths by the reduced dye. The differential sensitivity to cyanide and CO is curious, especially in view of the fact that endogenous ligands are probably similar to those of cytochrome *c* and thus inhibition must involve ligand displacement, a task for which cyanide might seem better equipped.

REFERENCES

Aasa, R., Ellfolk, N., Ronnberg, M. and Vanngard, T. (1981) *Biochim. Biophys. Acta*, **670**, 170

Adar, F. (1975) *Archiv. Biochem. Biophys.*, **170**, 644

Almassey, R. J. and Dickerson, R. E. (1978) *Proc. nat. Acad. Sci. U.S.*, **75**, 2674

Barber, D., Parr, S. R. and Greenwood, C. (1976) *Biochem. J.*, **157**, 431

Barber, D., Parr, S. R. and Greenwood, C. (1977) *Biochem. J.*, **163**, 629

Barber, D., Parr, S. R. and Greenwood, C. (1978a) *Biochem. J.*, **173**, 11

Barber, D., Parr, S. R. and Greenwood, C. (1978b) *Biochem. J.*, **173**, 681

Barber, D., Parr, S. R. and Greenwood, C. (1978c) *Biochem. J.*, **175**, 239

Blatt, Y. and Pecht, I. (1979) *Biochemistry*, **18**, 2917

Brunori, M., Parr, S. R., Greenwood, C. and Wilson, M. T. (1975) *Biochem. J.*, **151**, 185

Coulson, A. F. W. and Oliver, I. C. (1979) *Biochem. J.*, **181**, 159

Dickerson, R. E. and Timkovich, R. (1975) *The Enzymes*, Vol. XI (ed. Boyer, VI) p. 397

Ellfolk, N. and Soininen, R. (1970) *Acta Chem. Scand.*, **24**, 2126

Ellfolk, N. and Soininen, R. (1971) *Acta Chem. Scand.*, **25**, 1535

Flatmark, T. and Robinson, A. B. (1968) *Structure and Function of Cytochromes*, University Park Press, Manchester, p. 318

Ferguson-Miller, S., Brantigan, D. L. and Margoliash, E. (1979) *The Porphyrins*, Vol. VII (ed. Dolphin, D.), p. 149, Marcel Dekker, New York

Foote, N., Barber, D. and Greenwood, C. (1982) Unpublished

Foote, N., Thompson, A. C., Barber, D. and Greenwood, C. (1983) *Biochem. J.*, **209**, 701

Greenwood, C., Barber, D., Parr, S. R., Antonini, E., Brunori, M. and Colosimo, A. (1978) *Biochem. J.*, **173**, 11

Gudat, J. C., Singh, J. and Wharton, D. C. (1973) *Biochim. Biophys. Acta*, **292**, 376

Hill, K. E. and Wharton, D. C. (1978) *J. biol. Chem.*, **253**, 489

Hollenberg, P. F., Hager, L. P., Blumberg, W. E. and Persach, J. (1980) *J. biol. Chem.*, **241**, 5347

Horio, T. (1958a) *J. Biochem.*, **45**, 195

Horio, T. (1958b) *J. Biochem.*, **45**, 267

Horio, T., Higashi, T., Saragawa, M., Kusai, K., Nakai, M. and Okunuki, K. (1958) *Biochim. Biophys. Acta*, **29**, 297

Horio, T., Higashi, T., Yamanaka, T., Matsubara, H. and Okunuki, K. (1961a) *J. biol. Chem.*, **236**, 944

Horio, T., Kamin, M. D. and de Klerk, ? .(1961b) *J. biol. Chem.*, **236**, 2783

Ingledew, W. J. and Saraste, M. (1975) *Biochem. Soc. Trans.*, **7**, 166

Johnson, M. K., Thomson, A. J., Walsh, T. A., Barber, D. and Greenwood, C. (1980) *Biochem. J.*, **189**, 285

Keilin, D. (1966) *The History of Cell Respiration and Cytochromes*, Cambridge University Press, London and New York

Kijimoto, S. (1968) *Ann. Rep. Works*, **16**, 1, Osaka University

Kuronen, T. and Ellfolk, N. (1972) *Biochim. Biophys. Acta*, **275**, 308
Kuronen, T., Saraste, M. and Ellfolk, N. (1975) *Biochim. Biophys. Acta*, **393**, 48
Lemberg, R. and Barrett, J. (1973) *The Cytochromes*, Academic Press, New York, p. 240
Lenhoff, H. M. and Kaplan, N. O. (1953) *Nature, Lond.*, **172**, 730
Lenhoff, H. M. and Kaplan, N. O. (1956) *J. biol. Chem.*, **220**, 967
Liu, M. C. and Peck, H. D. (1981) *J. biol. Chem.*, **256**, 13159
Margoliash, E. and Schejter, A. (1966) *Adv. Prot. Chem.*, **21**, 113
Mitra, S. and Bersohn, R. (1980) *Biochemistry*, **19**, 3200
Morita, Y. and Kameda, R. (1957) *Mem. Res. Inst. Food Sci.*, **12**, 1, Kyoto University
Orii, Y., Shimada, H., Nozawa, T. and Hatano, H. (1977) *Biochem. Biophys. Res. Commun.*, **76**, 983
Parr, S. R., Wilson, M. T. and Greenwood, C. (1974) *Biochem. J.*, **139**, 273
Parr, S. R., Wilson, M. T. and Greenwood, C. (1975) *Biochem. J.*, **151**, 51
Parr, S. R., Barber, D., Greenwood, C., Phillips, B. W. and Melling, J. (1976) *Biochem. J.*, **157**, 423
Parr, S. R., Barber, D., Greenwood, C. and Brunori, M. (1977) *Biochem. J.*, **167**, 447
Paulos, T. L. and Kraut, J. (1980) *J. biol. Chem.*, **255**, 8199
Payne, W. J. (1973) *Bacteriol. Rev.*, **37**, 409
Pichinoty, F. (1964) *Biochim. Biophys. Acta*, **89**, 378
Prakash, O. M. and Sadana, J. C. (1972) *Archiv. Biochem. Biophys.*, **148**, 614
Ronnberg, M., Ellfolk, N. and Soininen, R. (1979) *Biochim. Biophys. Acta*, **578**, 392
Ronnberg, M., Osterlund, K. and Ellfolk, N. (1980) *Biochim. Biophys. Acta*, **626**, 23
Ronnberg, M., Araiso, T., Ellfolk, N. and Dunford, B. (1981a) *Archiv. Biochem. Biophys.*, **207**, 197
Ronnberg, M., Araiso, T., Ellfolk, N. and Dunford, B. (1981b) *J. biol. Chem.*, **256**, 2471
Salemme, F. R. (1977) *Ann. Rev. Biochem.*, **46**, 299
Saraste, M., Vistanen, I. and Kuronen, T. (1977) *Biochim. Biophys. Acta*, **492**, 156
Shimada, H. and Orii, Y. (1975) *FEBS Letts.*, **54**, 237
Shimada, H. and Orii, Y. (1978) *J. Biochem.*, **84**, 1553
Silvestrini, M. C., Colosimo, A., Brunori, M., Walsh, T. A., Barber, D. and Greenwood, C. (1979) *Biochem. J.*, **183**, 701
Silvestrini, M. C., Tordi, M. G., Colosimo, A., Antonini, E. and Brunori, M. (1982) *Biochem. J.*, **203**, 445
Soininen, R., Sojonen, H. and Ellfolk, N. (1970) *Acta Chem. Scand.*, **24**, 2314
Soininen, R. and Ellfolk, N. (1973a) *Acta Chem. Scand.*, **27**, 35ʹ
Soininen, R. and Ellfolk, N. (1973b) *Acta Chem. Scand.*, **27**, 2193
Springall, J. P., Stillman, M. J. and Thomson, A. J. (1976) *Biochim. Biophys. Acta*, **453**, 494
Takano, T., Dickerson, R. E., Schickman, S. A. and Meyer, T. E. (1979) *J. molec. Biol.*, **133**, 185
Thomson, A. J., Brittain, T., Greenwood, C. and Springall, J. P. (1977) *Biochem. J.*, **165**, 327
Timkovich, R. (1979) *The Porphyrins*, Vol. VII (ed. Dolphin, D.) p. 241, Marcel Dekker, New York
Verma, A. L. and Bernstein, ?. (1974) *J. Raman Spectrosc.*, **2**, 163
Vickery, L. E., Palmer, G. and Wharton, D. C. (1978) *Biochem. Biophys. Res. Commun.*, **80**, 458
Walsh, T. A., Johnson, M. K., Greenwood, C., Barber, D., Springall, J. P. and Thomson, A. J. (1979) *Biochem. J.*, **177**, 29
Walsh, T. A., Johnson, M. K., Barber, D., Thomson, A. J. and Greenwood, C. (1980) *J. inorg. Biochem.*, **14**, 15
Wharton, D. C., Gudat, J. C. and Gibson, Q. H. (1973) *Biochim. Biophys. Acta*, **292**, 611
Wharton, D. C. and Gibson, Q. H. (1976) *Biochim. Biophys. Acta*, **430**, 445
Wilson, M. T., Greenwood, C., Brunori, M. and Antonini, E. (1975) *Biochem. J.*, **145**, 449
Yamanaka, T., Ota, A. and Okunuki, K. (1960) *Biochim. Biophys. Acta*, **44**, 397
Yamanaka, T., Ota, A. and Okunuki, K. (1961) *Biochim. Biophys. Acta*, **53**, 294
Yamanaka, T., Kijimoto, S., Okunuki, K. and Kusai, K. (1963) *Nature, Lond.*, **194**, 759
Yamazaki, I., Nakajima, R., Honma, H. and Temura, M. (1968) In: *Structure and Function*

of Cytochromes (ed. Okunuki, K., Kamen, M. and Sekuzu, I.), University Park Press, Manchester, p. 552

Yonetani, T., Wilson, D. F. and Seamonds, B. (1966) *J. biol. Chem.*, **241**, 5347

3
Iron–Sulphur Proteins

A. J. Thomson

INTRODUCTION

Proteins containing iron–sulphur clusters comprise a huge class with a wide range of biological functions. They are distributed in all known living organisms including plants. In spite of this widespread occurrence they were recognised (Beinert, 1973) as a distinct group of metalloproteins only in the 1960s. It was the sharp optical absorption bands of the cytochromes that led to their early discovery in the late 19th century and, indeed, to their generic name. In contrast, the optical absorption bands of all iron–sulphur clusters are very broad, almost featureless and therefore difficult to recognise. On the other hand the electron paramagnetic resonance (epr) spectra of iron–sulphur clusters are distinctive. It was this fact that led to the discovery and subsequent detailed investigations of iron–sulphur proteins. Since then a vast literature on the subject has grown up with scientists from many different disciplines being engaged in the investigation of the structures and functions of the proteins. In addition, inorganic chemists have synthesised the metal centres, devoid of protein, in order to characterise more fully the structural and electronic features of the active sites (Lovenberg, 1973a,b, 1977; Spiro, 1982).

Iron–sulphur proteins contain iron atoms bound to sulphide ion, the so-called acid-labile sulphur, forming a core structure or cluster which is linked to the polypeptide chain by the thiolate side chains of cysteine residues. (For a concise summary of the classification and nomenclature of iron–sulphur proteins see *Eur. J. Biochem.*, **93**, 427–430, 1979.) The core structures which have been characterised crystallographically in proteins are shown in figure 3.1. They are, respectively, a one-iron [Rd], a two iron–two sulphide centre, $[2Fe-2S]^{2+}$, a four-iron–four-sulphide cluster, $[4Fe-4S]^{n+}$, and, the most recent addition to the list, a three-iron–three sulphide structure $[3Fe-3S]^{n+}$.

In the case of the one- and four-iron clusters which are structurally defined there are only thiolate (cysteinate) ligands binding the core to the protein.

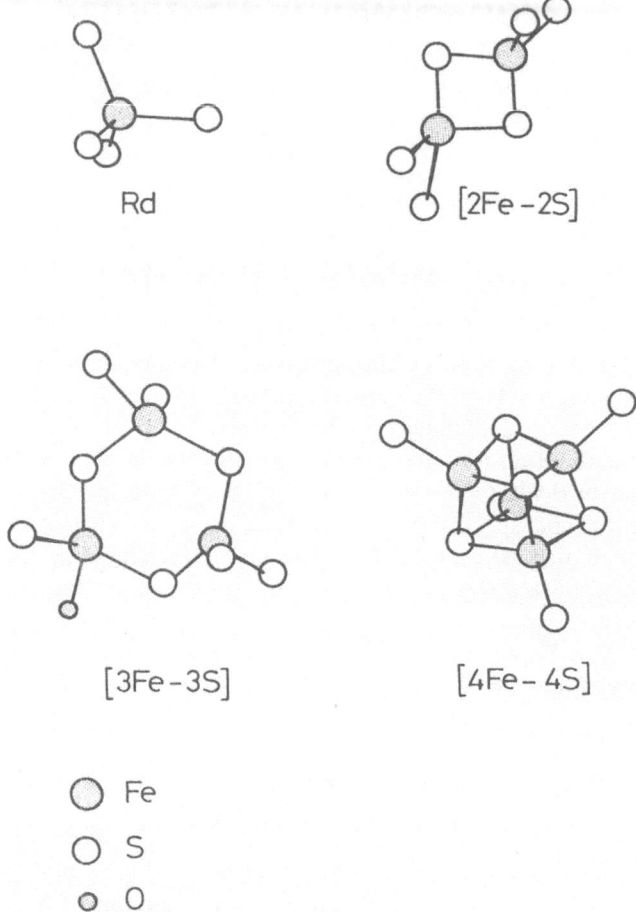

Figure 3.1 Structures of iron–sulphur clusters determined by protein crystallography. The sulphur atoms are of two types, viz. S^{2-} ion if bridging two or three iron atoms and $-CH_2-S^-$, from cysteine residues, if mono-dentate ligands to iron. In the [3Fe-3S] structure one of the terminal iron ligands is thought to be an oxygen atom, probably from water or hydroxide ion

Recently it has been shown that in the two-iron clusters of the Rieske protein from *Thermus thermophilus* there are non-cysteinyl ligands binding the core in addition to thiolate ligands. For the three-iron cluster crystallographically defined by Stout and co-workers there is one non-thiolate ligand, possibly hydroxyl ion or water but apparently not a protein side-chain.

The structural details of the three-iron cluster still remain ill-defined and controversial in some proteins. Discussion of the nature of the uncertainty is given in a later section of the chapter. The small group of proteins, called rubredoxins, [Rd], contain no acid-labile sulphur. In this case the iron atom is linked directly to four thiolate ligands from cysteine residues. These proteins are always included in the class of iron–sulphur proteins. A list of the various classes of rubredoxins and ferredoxins is given in table 3.1. This is not a comprehensive table but it does serve to show the range of protein and cluster types.

Iron–sulphur proteins have a variety of functions which can be illustrated by consideration of the proteins present in the obligate, anaerobic bacterium *Clostridium pasteurianum* (figure 3.2). This bacterium contains a water soluble, low molecular weight iron–sulphur protein, ferredoxin, which possesses two clusters of the [4Fe-4S] type. It will transfer a pair of electrons at a redox potential of -385 mV. The ferredoxin functions as a carrier of pairs of electrons between the reduction products of fermentation and the various terminal enzymes which require electrons at low reduction potentials to function. Many of these enzymes contain iron–sulphur centres, and most of them are of ill-defined structure. For examples, one of the terminal respiratory proteins is hydrogenase which catalyses the reduction of protons to hydrogen gas. This enzyme contains 12Fe and $12S^{2-}$ ions grouped into iron–sulphur clusters, probably of the $[4Fe-4S]^{n+}$ type. Another of the enzymes catalyses the reduction of sulphite to sulphide and accepts reducing equivalents from ferredoxin.

In all of these terminal enzymes iron–sulphur clusters are present. However, it is not clear what function the iron–sulphur centre, if any, performs in addition to transferring electrons. It may well be that the clusters act as electron storage centres, the capacitors of the electron transport chains of biology. In this way a number of electrons can be stored ready for rapid multi-electron reduction of substrates.

Iron–sulphur clusters can perform other enzymatic functions besides electron transfer. For the purpose of classification the iron–sulphur enzymes can be separated into simple and complex proteins, the latter containing additional prosthetic groups such as molybdenum ion, flavin or haem, see table 3.2. Many of the complex proteins are enzymes. However, it is not always clear whether the iron–sulphur cluster plays a part in the binding of substrate and its subsequent turnover or whether its role is always to store and transfer electrons. In these systems the identification of the cluster type is not always straightforward. Hence several spectroscopic methods including Mössbauer, epr, magnetic circular dichroism (mcd) and Resonance Raman (rr) spectroscopies are now being used for this task.

There are other centres as yet structurally undefined which spectroscopic measurement indicate are new types of iron–sulphur clusters. For example, in the nitrogenase enzyme system one of the components, called fraction 1 which binds dinitrogen contains three types of iron centres. One of the clusters, called a 'P' cluster, has been assigned on the basis of extrusion experiments as a $[4Fe-4S]^{\circ}$ cluster in the zero-oxidation level. When oxidised, this cluster in the protein has an unusual combination of properties that gives rise to an epr-silent but paramagnetic state with an odd number of electrons. This and the other centres in nitrogenase are fully described by Lowe, Thorneley and Smith in chapter 6.

Table 3.1 Rubredoxins and ferredoxins

Protein	Source	Type of cluster	Molecular weight $\times 10^{-3}$	E_M (mV)	References
Rubredoxin	Clostridium pasteurianum	[Rd]	6	-60	Jensen (1974)
Desulphoredoxin	Desulphovibrio gigas	2[Rd]	7.9	-35	Moura et al. (1987)
2-Fe ferredoxin	Spinach	[2Fe-2S]	10.5	-420	Hall and Rao (1977)
2-Fe ferredoxin	Spirulina platensis (alga)	[2Fe-2S]	10.5	-390	Tsukihara et al. (1978)
Adrenodoxin	Adrenal mitochondrion	[2Fe-2S]	12	-270	Estabrook et al. (1973)
Putidaredoxin	Pseudomonas putida	[2Fe-2S]	12.5	-240	Tsai et al. (1971)
HiPIP	Chromatium vinosum	$[4Fe-4S]^{2+;3+}$	9.5	+350	Carter et al. (1971)
4Fe ferredoxin	Bacillus stearothermophilus	$[4Fe-4S]^{1+;2+}$	9.12	-280	Mullinger et al. (1975)
Ferredoxin I and II	B. Polymyxa	$[4Fe-4S]^{1+;2+}$	~8	-380	Stombaugh et al. (1973)
Ferredoxin I and II	Desulphovibrio africanus	$[4Fe-4S]^{1+;2+}$		-420	Cammack et al. (1977)
Ferredoxin I	Desulphovibrio gigas	$[4Fe-4S]^{1+;2+}$	18 (trimer)	-455	Adman et al. (1976)
8Fe ferredoxin	Peptococcus aerogenes	$2[4Fe-4S]^{1+;2+}$	6		Mortenson and Nakos (1973)
	Cl. pasteurianum		6	-400	

Ferredoxin II	*Desulphovibrio gigas*	$[3Fe-xS]^{n+}$	24 (tetramer)	−130	Bruschi *et al.* (1976)
Ferredoxin	*Methanosarcina barkeri*	$[3Fe-xS]^{n+}$		negative	Moura *et al.* (1982)
Ferredoxin I	*Azotobacter vinelandii*	$[3Fe-xS]^{n+}$ $[4Fe\pm4S]^{m+}$	14	−460 mV ?	Yoch and Arnan (1972) Ghosh *et al.* (1981)
Ferredoxin	*A. chroococcum*	$[3Fe-xS]^{n+}$ $[4Fe\pm4S]^{m+}$	−	−	Yates, G. (unpublished)
Ferredoxin	*Thermus thermophilus*	$[3Fe-xS]$ $[4Fe\pm4S]^{1+,2+}$	10.5	−250 −530	Ohnishi *et al.* (1980) Fee *et al.* (1984)

Table 3.2 (Adapted from R. Cammack) Enzymes containing iron–sulphur clusters

Protein, function	Typical source	Molecular weight	Fe–S content	Other groups	References
Glutamine amidoribosyltransferase (EC 2.4.2.14)	*B. subtilis*	4×50	[4Fe–4S]		Itakura and Holmes (1979) Vollmer *et al.* 1983
2-methoxybenzoate O-demethyl monooxygenase, Fe-S component (EC 1.14.99.15)	*Pseudomonas putida*	126 trimer or tetramer	2–3[2Fe–2S]?		Bernhardt *et al.* (1978)
Aconitase (EC 4.2.1.3) Citrate → isocitrate	Beef heart	89	[3Fe–4S] inactive [4Fe–4S] active		Beinert *et al.* (1983)
Succinate dehydrogenase (complex II) (EC 1.3.99.1) Succinate + UQ → Fumarate + UQH$_2$	Mitochondria, and aerobic and photosynthetic bacteria	97 (70 + 27)	2[Fe–2S] +[4Fe–4S]$^{2+,3+}$	FAD covalently bound	Beinert *et al.* (1977) Davis *et al.* (1977) Salerno *et al.* (1979)
NADH dehydrogenase (Complex I) (EC 1.6.99.3) NADH + H + UQ → NAD$^+$ + UQH$_2$	Mitochondria, aerobic and photosynthetic bacteria	~16Fe, S	2[Fe–2S] 3[4Fe–4S]?	flavin FMN	Orme-Johnson *et al.* (1974b) Ohnishi (1975) Albracht and Subramanian (1977)
Fe–S + Molybdenum					
Nitrate Reductase (EC 1.7.99.4)	*Paracoccus denitrificans*	320	20Fe, S	1.5 Mo	Forget (1974)
	E. coli	220	2Fe-S centres	< 1 Mo	Vincent and Bray (1978)
	Klebsiella aerogenes	260	8–12Fe, 8S	< 1 Mo	Van't Riet *et al.* (1975)
Ferredoxin: CO$_2$ reductase (EC 1.2.7–)	*C. pasteurianum*	118	24Fe, 24S	1 Mo	Scherer and Thauer (1978)
Nitrogenase (EC 1.18.2.1) N$_2$ + 3Fd$_{red}$ + nATP → 2NH$_4^+$ + 3Fd$_{ox}$ + nADP + nP$_i$	*C. pasteurianum* *A. vinelandii* *K. pneumoniae*	MoFe protein 220 (2 × 50 + 2 × 60) Fe Protein	30Fe, 30S [4Fe–4S]	2 Mo	Zumft *et al.* (1974); Eady and Postgate (1974) Orme-Johnson and Davis (1977) Lowe *et al.* (1978)
Mo-Fe-S protein	*D. gigas*	120	~6[2Fe–2S]	Mo	Moura *et al.* (1978b)
Mo-Fe-S protein	*D. africanus*	112	20Fe, 20S [4Fe–4S]$^{2+,3+}$	5–6 Mo	Hatchikian *et al.* (1979)

Fe–S + Mo + Selenium

Enzyme (EC) / reaction	Source	M.W.	Fe–S	Other cofactors	References
Xanthine dehydrogenase (EC 1.2.1.37)	*C. acidi-urici*	130	4Fe-4S	Flavin	Andreesen et al. (1979)

Fe–S + Flavin + Molybdenum

Enzyme (EC) / reaction	Source	M.W.	Fe–S	Other cofactors	References
Xanthine oxidase (EC 1.2.3.2) $Xanthine + O_2 \rightarrow urate\ H_2O_2 + O_2^-$	Milk	280 dimer	4[2Fe-2S]	2 Mo 2 FAD	Bray (1975) Olson et al. (1974)
Xanthine dehydrogenase (EC 1.2.1.37)	Liver	280 dimer	4[2Fe-2S]	2 Mo 2 FAD	Cammack et al. (1976) Bray (1975)
Aldehyde oxidase (EC 1.2.3.1)	Liver	270	4 2Fe-2S?	2 Mo 2 FAD	Rajagopalan et al. (1968) Bray (1975)

Fe–S + haem

Enzyme (EC) / reaction	Source	M.W.	Fe–S	Other cofactors	References
UQ-cytochrome c reductase (Complex III) (EC 1.10.2.2)	mitochondria	250	[2Fe-2S]	Cytochromes b-562,b-566 and c_1	Orme-Johnson et al. (1974a) Rieske (1976) Trumpower and Edwards (1979)
Nitrate reductase (EC.1.7.7.1) $NO_2^- + 6Fd_{red} \rightarrow NH_4^+ + 6Fd_{ox}$	chloroplasts	60	[4Fe-4S]	Sirohaem	Aparicio et al. (1975) Cammack et al. (1978) Christner et al. (1981)
Sulphite reductase dissimilatory (EC 1.8.7.1) $SO_2^{2-} + 6Fd_{red} \rightarrow S^= + 6Fd_{ox}$	*Desulphovibrio* spp.	200 ($\alpha_2\beta_2$)	2[4Fe-4S]$^{1+;2+}$	2 Sirohaem	Lee and Peck (1971)

Fe–S + Flavin + haem

Enzyme (EC) / reaction	Source	M.W.	Fe–S	Other cofactors	References
Sulphite reductase, assimilatory (EC 1.8.7.1)	Spinach		2[4Fe-4S]?	Sirohaem	Siegel (1978)
Sulphite reductase assimilatory (EC 1.8.7.1) $SO_3^= + 3NADPH \rightarrow S^= + 3NADP^+$	*E. coli*	670	4[4Fe-4S]	4 FAD 4 FMN 4 Sirohaem	Siegel et al. (1974)
Nitrite reductase $NO_2^- + 3NADPH \rightarrow NH_4 + 3NADP^+$	*Neurospora crassa*	290	[4Fe-4S]	FAD Sirohaem	Greenbaum et al. (1978)

Fe–S + thiamin diphosphate (TDP)

Enzyme (EC) / reaction	Source	M.W.	Fe–S	Other cofactors	References
Pyruvate : ferredoxin oxidoreductase (EC 1.2.7.1) $Pyruvate + 2Fd_{ox} + CoA$ $= Acetyl\ CoA + 2Fd_{red} + CO_2$	*C. pasteurianum*	240	6Fe, 6S	TDP	Uyeda and Rabinowitz (1971)
	Halobacterium halobium	258 ($2 \times 86 + 2 \times 43$)	2[4Fe-4S]		Kerscher and Oesterhelt (1977)
	Photosynthetic bacteria				Gehring and Arnon (1972)

(*continue overleaf*)

Table 3.2 (continued)

Protein, function	Typical source	Molecular weight	Fe–S content	Other groups	References
2-oxoglutarate : ferredoxin oxidoreductase (EC 1.2.7.3)	*H. halobium*	250 (2.87 + 2 × 38)	[4Fe–4S]?	2TDP	Kersher and Oesterhelt (1977)
2-oxobutyrate : ferredoxin oxidoreductase (EC 1.2.7.2)	Photosynthetic bacteria		Fe–S?	TDP	Buchanan (1969)
Fe–S + chlorophyll Photosystem I	Chloroplasts		2[4Fe–4S]	p-700, chlorophyll a and other electron carriers	Malkin and Bearden (1978) Evans *et al.* (1974) Cammack and Evans (1975)

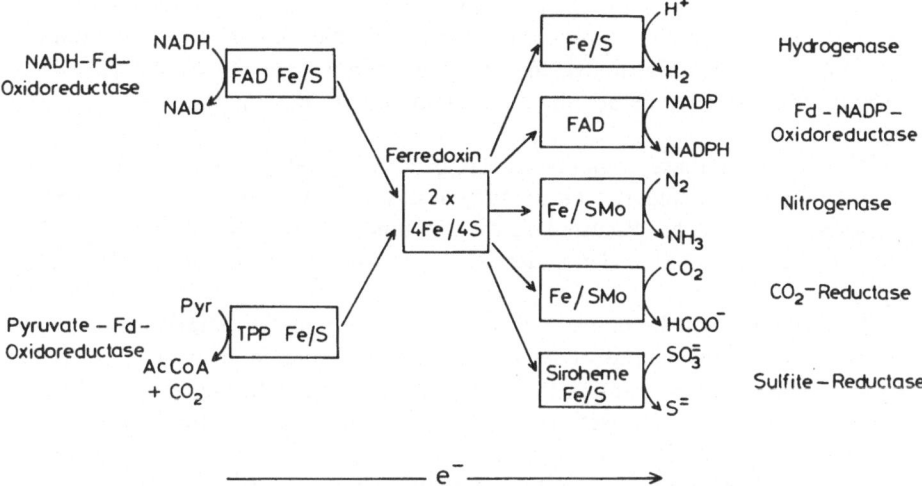

Figure 3.2 A summary of the processes in the obligate anaerobic bacterium, *Clostridium pasteurianum*, which involve iron–sulphur clusters. The soluble ferredoxin acts as an electron carrier between the reactants on the left and the terminal enzymes on the right. Flavodoxin can substitute for ferredoxin as an electron carrier. Abbreviation: Fd, ferredoxin; TPP, thiamine pyrophosphate; FAD, flavin adenine dinucleotide. (Reproduced with permission from Thauer, R. K. and Schönheit, P. In: *Iron–Sulfur Proteins* (ed. T. G. Spiro), John Wiley, New York, 1982, Chap. 8, fig. 1)

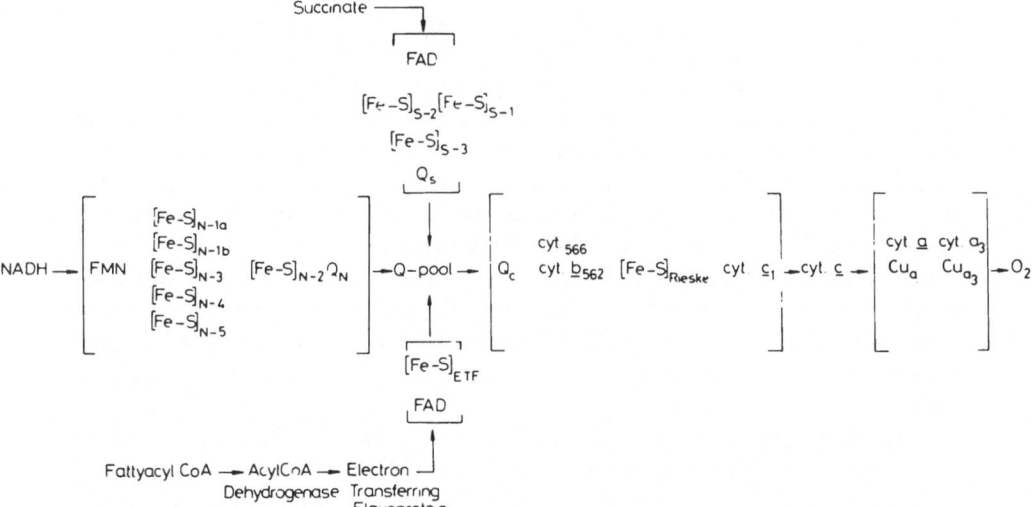

Figure 3.3 The redox components of the respiratory chain from the mitochondrial membrane. Iron–sulphur clusters associated with the NADH-ubiquinone and succinate-ubiquinone reductase portions of the chain are indicated by suffixes N-x, and S-x. Q indicates pools of ubiquinone in the respective protein compartments. (Reproduced by permission from Ohnishi, T. and Salerno, J. C. In: *Iron–Sulfur Proteins* (ed. T. G. Spiro), John Wiley, New York, 1982, Chap. 7, Scheme 1)

The electron transport chain of the mitochondrian contains many iron-sulphur clusters which are primarily involved in the dehydrogenase section of the respiratory chain although important exceptions are the Rieske [2Fe-2S] protein and the enzyme aconitase. A summary of the components is given in figure 3.3.

The study of these components has become a highly specialised aspect of iron–sulphur biochemistry. These enzymes, with the important exception of aconitase, will not be discussed further in this chapter. The interested reader should refer to recent reviews such as the chapter by Ohnishi and Salerno in Spiro (1982).

This chapter will be concerned with the clusters of the simple iron–sulphur proteins which are mainly of bacterial origin and soluble. The objective is to present a coherent account of the structural features, both molecular and electronic, of the reasonably well-defined clusters and to refer to function where it can be understood. The final sections of the chapter refer to some of the more exciting areas of iron–sulphur biochemistry which are rapidly developing at present. This is intended to give the reader an impression of the areas in which significant new results might be expected in the near future.

[2Fe-2S] CONTAINING PROTEINS

Historically these are an important sub-class of iron–sulphur proteins because it was the epr signal (Palmer and Sands, 1966; Hall *et al.*, 1966) of the ferredoxin from spinach, quite atypical of iron compounds, which suggested that a novel dimeric, spin-coupled structure might be present (Brintzinger *et al.*, 1966; Gibson *et al.*, 1966). Curiously, the structure proposed on the basis of the epr and, later, Mössbauer studies (Cammack *et al.*, 1971; Dunham *et al.*, 1971) remained unproven by direct crystallography until very recently when the first crystal structure of a protein containing a 2Fe–2S cluster was reported. This was for the ferredoxin from *Spirulina platensis* (Ogawa *et al.*, 1977; Tsukihara *et al.*, 1978; Fukuyama *et al.*, 1980). The difficulty in obtaining a structure has arisen from the reluctance of ferredoxins of this type to crystallise. However, inorganic model compounds of the type $[Fe_2S_2(SR)_4]^{2-}$ were prepared as early as 1973 and (Mayerle *et al.*, 1973, 1975) shown by x-ray crystallography to have the dimeric structure so long assumed on the evidence of spectroscopic data.

The best definition of the structure of the 2Fe-2S cluster comes from the high resolution x-ray crystallography of the model compounds $[Fe_2S_2(S_2\text{-}o\text{-}xyl)_2]^{2-}$ and $[Fe_2S_2(S\text{-}p\text{-}tolyl)_4]^{2-}$ (Mayerle *et al.*, 1973, 1975). Figure 3.4 shows the structure and some of the important bond lengths. The iron–iron distance is 2.69 Å, very close to that in the [4Fe-4S] core structure. Although the x-ray structure of *Spirulina platensis* ferredoxin is only partially refined, the dimension of the core of the protein bound cluster and the model are similar. For example, the Fe—Fe distance is 2.72 Å.

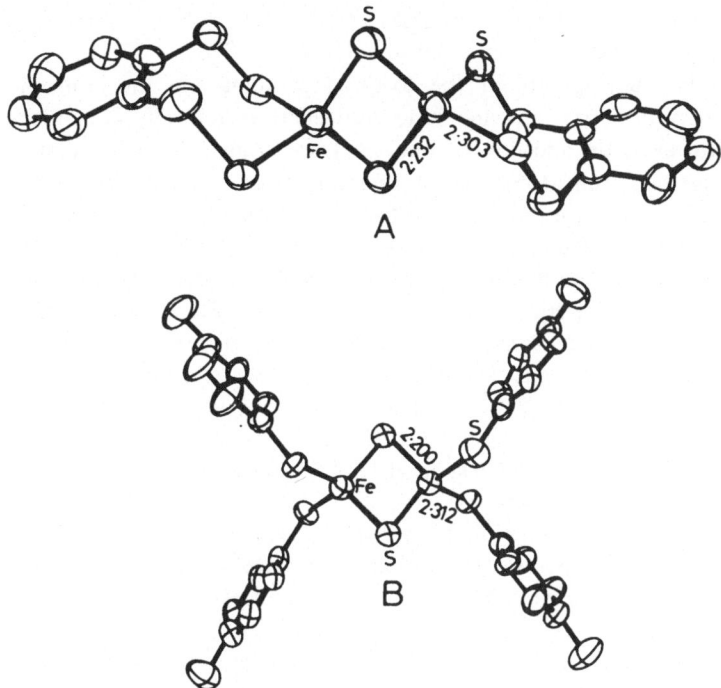

Figure 3.4 Structures of the [2Fe-2S] analogues of iron-sulphur proteins, namely, $[Fe_2 S_2 (S_2\text{-}o\text{-}xyl)_2]^{2-}$ (A) and $[Fe_2 S_2 (S\text{-}p\text{-}tolyl)_4]^{2-}$ (B). (Reprinted with permission from Mayerle *et al.* 1975, *J.Am. chem. Soc.*, **97**, 1032, American Chemical Society)

In the oxidised state the dimer contains two ferric ions in a roughly tetra-hedral environment of sulphur ligands. They are high-spin ferric ions each with a total of five unpaired d-electrons. However, the metal ions interact with one another antiferromagnetically to give a manifold of exchange-coupled ground state levels characterised by total spin quantum numbers of 0, 1, 2, 3, 4 and 5. The levels are ordered such that the S = 0 level is lowest. Hence there is no epr signal from the oxidised form of the cluster and it is diamagnetic at low tempera-tures (Palmer, 1973). This is confirmed both by Mössbauer (Cammack *et al.*, 1971) and mcd (Thomson *et al.*, 1977) spectroscopy. At room temperature thermal population of the S = 1 level may occur giving rise to some paramag-netism, most readily detectable by means of paramagnetic shifts of nmr signals of protons lying close to the metal centres (Palmer, 1973). The energy gap between the S = 0 and S = 1 level depends upon the magnitude of the antiferro-magnetic exchange which is, in turn, expected to be sensitively dependent upon metal–metal distance and possibly also upon the conformation of the ligands. Accurate measurement of this energy gap is difficult. It requires the measurement of the temperature dependence of the magnetic susceptibilty up to high temperatures. Estimates have been made of the magnitude of the antiferromagnetic coupling in a few cases and it is clear that it does change very considerably from one protein to another (Palmer, 1973).

The Mössbauer spectra of the oxidised proteins and the inorganic model show that both ferric ions are quite equivalent structurally (Cammack *et al.*, 1971). However, on addition of one electron to give the reduced state of the protein the two iron atoms became inequivalent on the Mössbauer time-scale. The isomer shifts correspond closely to ions of formal oxidation states Fe^{3+} and Fe^{2+}. Hence the added electron appears to reside entirely on one of the pair of metal atoms which are therefore to be described as a valence trapped dimer. The reduced proteins give epr spectra as shown in figure 3.7. The dimer is still to be regarded as anti-ferromagnetically coupled high-spin ferric ($S_1 = \frac{5}{2}$) and high-spin ferrous ($S_2 = 2$) ions. The coupling gives rise to a manifold of exchange coupled levels $S = \frac{1}{2}, \frac{3}{2}, \frac{5}{2}, \frac{7}{2}, \frac{9}{2}$. The lowest level has a spin $S = \frac{1}{2}$ and is the one which gives rise to the epr signal. Estimates of the energy gap between the lowest ($S = \frac{1}{2}$) and next highest level ($S = \frac{3}{2}$) have been made in a few cases using magnetic susceptibility.

The optical properties of the $[2Fe\text{-}2S]^{2+/1+}$ cores bear out the analysis of the Mössbauer and epr spectra. In the oxidised state there are broad absorption bands in the visible region. These can be assigned as charge-transfer transitions both from the sulphide ions and the thiolate ligands to the ferric ions. In the near infrared region between 700 and 1200 nm there are some weaker absorption bands arising from the d-d transitions of the high-spin ferric ion. No electronic absorption bands have been detected to longer wavelength in the oxidised state. On one-electron reduction the visible region absorption spectrum becomes weaker by a factor of about two and less well-defined. The remaining absorption intensity is probably due to ligand → Fe^{III} charge-transfer but since there is only a single ferric ion in the reduced protein compared with a pair in the oxidised state the transition intensity should drop. In the near infrared region new electronic transitions can be detected between 1500 and 2000 nm (Eaton *et al.*, 1972). These are assigned to a d-d transition of the ferrous ion. The d-d transitions of the remaining ferric ion are still detectable above 1200 nm. Hence the mixed valence nature of the reduced dimer is confirmed by optical spectroscopy.

The distinctive feature of the two-iron proteins is the very intense natural circular dichroism in the visible region, both in the oxidised and reduced states (Stephens *et al.*, 1978). The cd is largely invariant to the type of protein surrounding the cluster. Moreover the signal strength is very much larger than for any of the other iron–sulphur centres. The source of this rotational strength has not been satisfactorily explained. Presumably it must come from the small distortions impressed upon the dimer geometry by the protein to give chirality. Thus it is surprising that it is not more dependent upon protein structure. Possibly this argues for a high degree of conservation of protein structure in the local vicinity of the cluster. However, this conclusion seems to be at odds with the wide variation in mid-point potential illustrated in table 3.1. Thus the strong natural cd signal must be taken, along with the valence trapped nature of the reduced cluster, as an intrinsic property of the centre.

The mcd spectra of the reduced $[2Fe\text{-}2S]^+$ centres in several proteins have been reported (Thomson *et al.*, 1977; Johnson *et al.*, 1982). The spectra are, as expected, temperature dependent, since the ground state is paramagnetic, and are intense at low temperature. The spectra resolve into a wealth of fine structure which contrasts with the diffuse nature of the absorption spectrum. Detailed assignment and interpretation of the fine structure has not been achieved and will, undoubtedly, be difficult. However, the fact of interest is that the mcd spectra are rather different for each protein type in a way that appears to parallel other properties such as redox potential. Hence the mcd spectra of reduced spinach and *Spirulina maxima* ferredoxins are similar to one another but distinctly different from those of adrenodoxin and putidaredoxin.

A clue to the possible reasons for these differences has come recently from the work of Fee and collaborators [Fee *et al.*, 1983] on the purification and characterisation of a [2Fe-2S] containing protein from the bacterium *Thermus thermophilus*. They have called this protein the Rieske centre for the following reasons. Rieske and co-workers (Rieske *et al.*, 1964) first purified an iron–sulphur protein from the bc_1 complex of bovine mitochondria. It was subsequently shown that this protein acts within the bc_1 complex as a ubiquinol-cytochrome c_1/ubisemiquinone-cytochrome b oxidoreductase. A variety of bacterial and mitochondrial Rieske proteins have since been studied *in situ* and E_M values between $+150$ mV and $+330$ mV determined. These values are between 400 and 600 mV higher than those observed for plant-type [2Fe-2S] ferredoxins. The protein isolated from *T. thermophilus* is reducible by ascorbate so it presumably has an E_M value of about 0 V. It has been purified, analysed and its amino-acid composition determined. The data show the presence of a novel iron–sulphur centre. Spectroscopic analysis including epr and Mössbauer techniques confirm that the protein possesses two [2Fe-2S] cores. However, since there are only 4 cysteine residues in the protein each [2Fe-2S] cluster is coordinated by only two cysteine ligands. Each cluster must possess two non-cysteine ligands, presumably protein derived. Fee and his co-workers have failed to find any Raman lines characteristic of tyrosine. This appears to leave nitrogen atoms as the alternative for the ligands. Two plausible structural models are proposed.

These results provide the first example of a two-iron cluster with ligands other than sulphur and give a rationale for the high redox potential of the Rieske-type centre. A comparison is drawn between the epr, Mössbauer, optical spectra and redox properties of the *T. thermophilus* Rieske centre and the iron–sulphur clusters of the benzene dioxygenase of *Pseudomonas putida*, indeed with the NADH-dependent dioxygenases of a number of micro-organisms. It now seems likely that all of these proteins contain the same novel type of [2Fe-2S] centre.

[4Fe-4S] CLUSTERS

The structural features of the $[4Fe\text{-}4S]^{n+}$ containing clusters are well defined in several proteins and in an extensive series of model compounds (Berg and Holm, 1982). In all cases the clusters are of the cubane type, $[Fe_4S_4(SR)_4]^{n-}$ (figure 3.5).

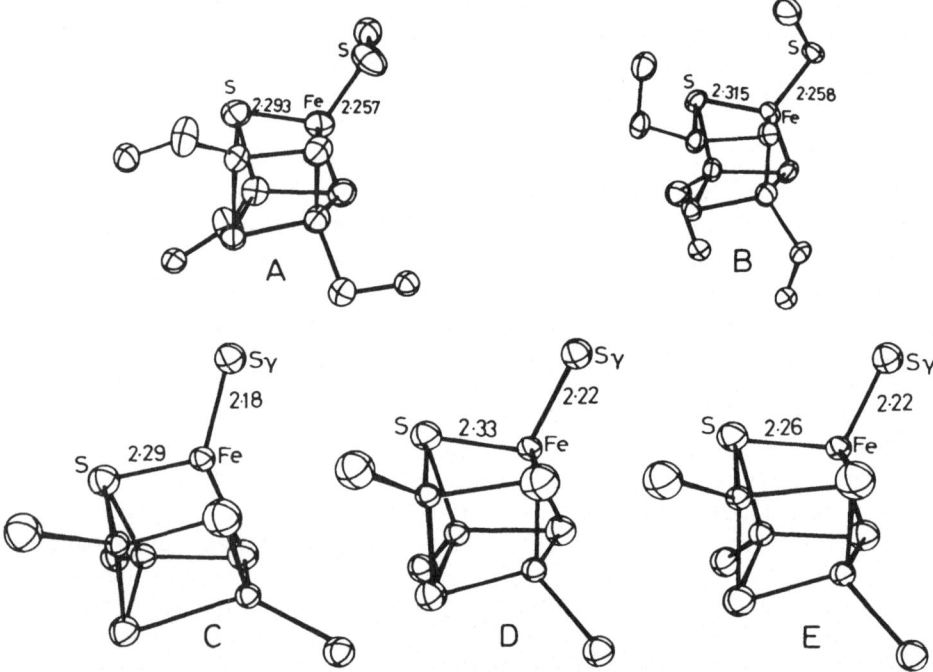

Figure 3.5 [4Fe-4S] cluster viewed from the same perspective; (A) $[Fe_4S_4(SPh)_4]^{2-}$; (B) $[Fe_4S_4(SCH_2\text{-}Ph)_4]^{2-}$; (C) *Peptococcus aerogenes* ferredoxin, oxidised (Adman *et al.*, 1973); (D) *Chromatium* HiPIP, reduced; (E) Chromatium HiPIP, oxidised (Carter *et al.*, 1974a). (Reproduced by permission from Holm, R. H. and Ibers, J. A. In: *Iron-Sulfur Proteins*, Volume III (ed. W. Lovenberg), Academic Press, London, 1977, Chap. 7, fig. 3)

It is now established that this structure can exist in one of three physiologically active oxidation levels (Carter *et al.*, 1972). This realisation brought immediate order and sense to a great deal of experimental data. The three oxidation levels can be represented in a variety of ways thus

$$[4Fe\text{-}4S]^{1+} \rightleftharpoons [4Fe\text{-}4S]^{2+} \rightleftharpoons [4Fe\text{-}4S]^{3+} \tag{1}$$

$$[Fe_4S_4(SR)_4]^{3-} \rightleftharpoons [Fe_4S_4(SR)_4]^{2-} \rightleftharpoons [Fe_4S_4(SR)_4]^{-1} \tag{2}$$

The representation used in equation (1) gives the core structures with the cysteine ligands removed. Since each of the cysteine ligands carries a formal charge of -1 the overall core charges can be arrived at. The formulae used in equation (2) represent the clusters as self-contained coordination compounds. Thus SR represents a thiolate ligand which is cysteinate ion in the case of the protein bound cluster. The clusters are mixed-valence compounds with the core $[4Fe-4S]^{2+}$ corresponding formally to $2Fe^{III} + 2Fe^{II}$ ions.

Figure 3.6 illustrates the magnetic properties and typical mid-point potentials of the clusters in proteins. There are two classes of proteins, namely, the 'high-potential iron–sulphur proteins' (HiPIP) and the ferredoxins which may contain a single or a pair of $[4Fe-4S]$ clusters. The HiPIP proteins, so-named because of their positive ($\sim +350\,mV$) mid-point potentials, occur in photosynthetic bacteria such as *Chromatium vinosum*. There is only a single cubane cluster present. The protein cluster is paramagnetic giving epr signals at $g_{\parallel} = 2.12$, $g_{\perp} = 2.04$ in the oxidised state. On one-electron reduction an epr-silent state is obtained which is diamagnetic at cryogenic temperatures. Further reduction of the cluster is possible only if the protein is partially denatured by the addition of DMSO, up to 80% by volume, followed by reduction with sodium dithionite (Cammack, 1973). The state so produced gives an axial epr spectrum with g-values of 2.05, 1.94. The mid-point potential for this step to the super-reduced state has been estimated to be $-600\,mV$ or lower. This experiment to produce a lower, epr-active oxidation state of HiPIP was a crucial one confirming the hypothesis due to Carter that the 4Fe-4S cubane cluster could exist in one of *three* oxidation levels.

A large number of ferredoxins are known which contain either a single or a pair of $[4Fe-4S]$ clusters. They are diamagnetic at low temperature in the oxidised state, although often there exists a weak epr signal at $g = 2.01$ corresponding to a low percentage of spins. The clusters undergo reduction at rather negative mid-point potentials between -300 and $-600\,mV$ to yield an epr active oxidation state, $[4Fe-4S]^{1+}$. In the case of the protein containing a single $[4Fe-4S]$ cluster the epr spectrum is a rhombic signal with g-factors typically of 2.06, 1.92, 1.88. When the protein possesses a pair of $[4Fe-4S]$ clusters the epr spectrum is complex, due to spin–spin interaction between the single unpaired electron on each cluster (figure 3.7). However, by titrating the ferredoxin with dithionite a semi-reduced state is reached in which only one of each pair of clusters is reduced. Then a simple rhombic epr signal is obtained (figure 3.7).

Attempts were made to 'super-oxidise' the $[4Fe-4S]$ clusters in ferredoxin from *Clostridium pasteurianum* by the addition of the oxidising agent, hexacyanoferrate (III) (Sweeney *et al.*, 1975). The epr properties were followed. It was observed that the signal at $g = 2.01$, invariably present in the ferredoxin as extracted increased in magnitude after treatment with hexacyanoferrate (III). It was suggested that this signal corresponded to the clusters being oxidised to the $[4Fe-4S]^{3+}$ state, that is, an oxidation level equivalent to that in oxidised HiPIP. The authors of this original paper reported that re-reduction of the pro-

Core notation	$[4Fe-4S]^{3+}$	$[4Fe-4S]^{2+}$	$[4Fe-4S]^{1+}$
Total electronic spin	$\frac{1}{2}$	0	$\frac{1}{2}$
Formal oxidation state	$3Fe(III)+1Fe(II)$	$2Fe(III)+2Fe(II)$	$1Fe(III)+3Fe(II)$
HiPIP states, EPR signals	$g = 2.12, 2.04$	EPR silent	$g = 2.04, 1.93$ 'Super-reduced'
Bacterial ferredoxins, $2[4Fe-4S]$, and $[4Fe-4S]$	Cluster degradation	EPR silent (weak, variable intensity $g = 2.01$ signal from 3Fe-xS cluster)	$g = 2.07, 1.93, 1.89$ additional features in proteins with interacting clusters

$+350\,mV$ →← (between $[4Fe-4S]^{3+}$ and $[4Fe-4S]^{2+}$)

$Fe(CN)_6{}^{3-}$ (associated with $+350\,mV$ interconversion)

$> 70\%$ DMSO → (between $[4Fe-4S]^{2+}$ and $[4Fe-4S]^{1+}$)

$-450\,mV$ (between $[4Fe-4S]^{2+}$ and $[4Fe-4S]^{1+}$)

Figure 3.6 Accessible oxidation levels of $[4Fe-4S]^{n+}$ core in proteins with a summary of their magnetic properties and methods of interconversion

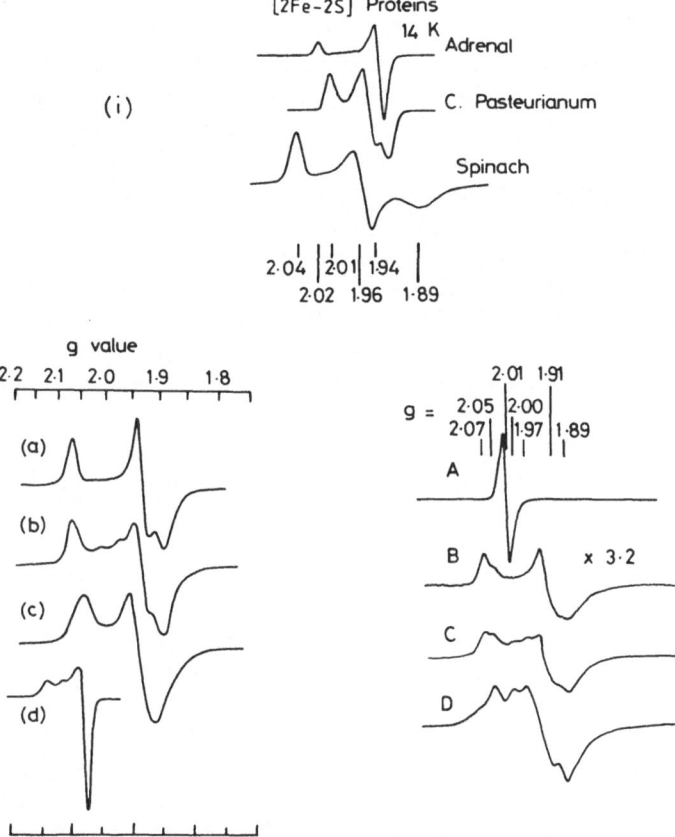

Figure 3.7 EPR spectra of [2Fe-2S] and [4Fe-4S] clusters in protein. (i) EPR spectra of iron–sulphur proteins containing the [2Fe-2S]$^+$ core. The protein source is indicated in the figure. The protein concentrations were approximately 10^{-4} M; the samples were reduced with sodium dithionite. The temperature of measurement was 14 K at a microwave power of 0.9 mW. (Reproduced with permission from Orme-Johnson, W. H. and Sands, R. H. In: *Iron-Sulfur Proteins* (ed. W. Lovenberg,), Vol. II, Academic Press, New York, 1973). (ii) EPR spectra of proteins containing [4Fe-4S] centres (a) reduced *Bacillus stearothermophilus* ferredoxin (b) partially reduced *Clostridium pasteurianum* ferredoxin (c) super-reduced *Chromatium* HiPIP in 80% DMSO and (d) oxidised Chromatium HiPIP. (Reproduced with permission from Cammack, R., Dickson, D. P. E. and Johnson, C. E. In: *Iron-Sulfur Proteins* (ed. W. Lovenberg), Vol. III, Academic Press, New York, 1977, Chap. 8, fig. 6) (iii) EPR spectra taken during a reductive titration of the 2[4Fe-4S] ferredoxin from *Clostridium acidi-urici* with sodium dithionite. The protein concentration was 8.2×10^4 M in 0.1 M Tris-HCl buffer, pH 7.5. (A) no dithionite, gain 1; (B) 0.44 electrons added per molecule, gain 3.2; (C) 1.44 electrons added per molecule, gain 1; (D) 2.0 electrons added per molecule, gain 1. (Reproduced with permission from Orme-Johnson, W. H. and Sands, R. H. In: *Iron-Sulfur Proteins* (ed. W. Lovenberg), Vol. II, Academic Press, New York, 1973)

tein after treatment with hexacyanoferrate (III) ion failed to regenerate the epr signal expected from a fully reduced $[4Fe\text{-}4S]^+$ cluster. This immediately suggests that cluster destruction or damage has occurred. Nevertheless, in spite of the cautious interpretation placed upon their own experiments by the authors this work was seized upon by others as evidence for the ability of [4Fe-4S] clusters in ferredoxins to be superoxidised to the $[4Fe\text{-}4S]^{3+}$ state provided that oxidising agents of sufficiently high potential were used. This is now known to be an erroneous view (Thomson *et al.*, 1981).

X-ray crystal structures are now available for three proteins which contain one or more four-iron clusters. They are oxidised and reduced HiPIP from *Chromatium* (Carter *et al.*, 1971, 1974a,b) containing a single cluster, the oxidised state of the eight-iron ferredoxin from *Peptococcus aerogenes* (Adman *et al.*, 1973, 1975, 1976), and the oxidised state of the seven-iron ferredoxin; Fd 1 of *Azotobacter vinelandii* (Stout *et al.*, 1980; Ghosh *et al.*, 1981). Thus a detailed comparison can be made of the cluster geometries in different protein environments. It is clear that the structure of the core [4Fe-4S] is essentially the same in each of the protein environments. Nevertheless significant but subtle changes in stereochemistry are apparent.

HiPIP is the only protein to have had an X-ray structure determination conducted on both oxidation states. There is a statistically significant difference between the mean Fe-Fe distances in the oxidised (Fe-Fe = 2.72 Å) and the reduced (Fe-Fe = 2.81 Å) cluster. There is also a slight change in overall geometry. The oxidised form of the [4Fe-4S] core contains a more nearly tetrahedral array of the 4Fe ions. The reduced cluster has a significant distortion from tetrahedral (Td) symmetry such that the tetrahedron is slightly flattened along the S_4 axis to give a tetragonal distortion. This leads to two sets of Fe-Fe distances with a range of 2.74–2.87 Å in HiPIP reduced. This cluster asymmetry is seen in the structure of the clusters of oxidised Fd, *P. aerogenes* and in many model compounds of the type $[Fe_4S_4(SR)_4]^{2-}$ (Berg and Holm, 1982). Consequently it can be assumed that a tetragonally compressed structure is the intrinsically stable structure of the $[4Fe\text{-}4S]^{2+}$ core. In a search for the electronic features of the core which underlie this distortion from regular tetrahedral geometry it has been suggested that a Jahn–Teller effect is operative (Yang *et al.*, 1975). However, it has been pointed out that this conclusion is unlikely to be valid (Thomson, 1981).

It has not yet proved possible to solve the structure of a protein containing a reduced $[4Fe\text{-}4S]^+$ core cluster. However, a very detailed study has been made of the geometries of reduced model compounds such as $[Fe_4S_4(SPh)_4]^{3-}$ and $[Fe_4S_4(SCH_2Ph)_4]^{3-}$ (Berg and Holm, 1982). No such constancy of core dimensions exists in reduced analogues. Thus the anions in $[Et_3MeN]_3[Fe_4S_4(SPh)_4]$ have an idealised D_{2d} core symmetry arising from elongation along the S_4 axis. The anions of $[Et_4N]_3[Fe_4S_4(SCH_2Ph)_4]$ show a distortion that is C_{2v} in the solid state (Berg *et al.*, 1979). Extensive spectroscopic studies of the complexes in the solid state and in solution show that the two trianions are different from one another in the solid state but are similar in solution (Laskowski *et al.*, 1978). The structures appear to be tetragonally elongated in solution but it is clear that $[Fe_4S_4(SR)_4]^{3-}$ species are much more susceptible to core distortions in different environments than are the $[Fe_4S_4(SR)_4]^{2-}$ analogues. The spin state and magnetic properties are also considerably altered by the environment. This may be a pointer to the understanding of some of the features of the unusual 'P' clusters in the enzyme nitrogenase.

[3Fe-xS] CLUSTERS

The realisation that iron–sulphur clusters containing 3Fe atoms existed came about in 1980 as a result of a combined epr and Mössbauer study of the ferredoxin, FdI, from *Azotobacter vinelandii* (Emptage *et al.*, 1980). This ferredoxin was analysed to have between 6 and 8 Fe atoms per molecular weight of 14 000 with an equivalent number of S^{2-} ions. As prepared the protein has an epr signal with $g = 2.01$ which integrates to one electron per cluster (Sweeney *et al.*, 1975). On reduction with sodium dithionite at pH 7.4 this signal is lost at a mid-point potential of -450 mV and no new signal is obtained. Treatment of protein with hexacyanoferrate (III) causes a change in the epr signal. The $g = 2.01$ line remains unaltered but new broad features appear to low and high field of this signal. Quantitation of the spins indicates that one extra spin is accounted for by the oxidation product. Moreover it appears with a mid-point potential of $+350$ mV. Thus there are two epr detectable centres whose signals disappear on reduction. Since at that time, 1975, only HIPIP clusters were known to have this character, both centres were assigned to this cluster type, namely, $[4Fe-4S]^{3+/1+}$. However this led to the uncomfortable conclusion that two clusters undergoing the same one-electron reduction process do so at mid-point potentials 750 mV apart. This work was followed by an x-ray crystallographic study (Stout, 1979). The low resolution refinement showed two clusters of very different size and so the conclusion was reached that [4Fe-4S] and [2Fe-2S] clusters were present, the former being assigned to the high potential centre and the latter to the low potential centre. This conclusion in turn raised a paradox. It required the postulate of a [2Fe-2S] cluster with an epr signal in the oxidised state, a unique situation. It was this that prompted the Mössbauer study of Münck and his co-workers (Emptage *et al.*, 1980). It was immediately clear that the cluster giving rise to the epr signal at $g = 2.01$ was novel. On reduction of this cluster the epr signal was lost but a paramagnet with an even electron spin was observed. Two types of iron were detectable in the reduced state and the integration of the Mössbauer intensity gave a ratio of 2 : 1 types of iron for the reduced state.

The unique magnetic states of this cluster immediately led to the conclusion of the presence of a novel type of Fe-S cluster and the ratio of 2 : 1 for the iron types strongly suggested a cluster with 3Fe atoms or a multiple thereof. The crystallographic structure was further refined and indeed the presence of a new cluster was seen with 3Fe and $3S^{2-}$ ions liganded to the peptide chain via 5 cysteine residues (Stout *et al.*, 1980; Ghosh *et al.*, 1981). There was a sixth cluster ligand originally thought to be the oxygen atom of a glutamic acid side chain but now suggested to be the oxygen atom either of an hydroxide or water molecule.

The sulphate reducing bacterium *Desulphovibrio gigas*, possesses two ferredoxins, I and II, with a remarkable relationship (Bruschi *et al.*, 1976). Both have the same polypetide chain (Bruschi, 1979) which possesses 6 cysteine residues but Fd I has a [4Fe-4S] cluster and Fd II a [3Fe-xS] cluster. The two clusters can be interconverted by the addition of Fe to Fd II to generate Fd I and by the oxidation of Fd I with hexacyanoferrate (III) ion to yield Fd II. Fd II is spectroscopically and analytically pure and possesses only one cluster per polypeptide chain. Therefore it has become the benchmark protein for the spectroscopic characterisation of the [3Fe-xS] cluster.

Different Fd II preparations yield, to within 5%, 3Fe atoms per protein monomer. In the oxidised state Fd II is epr active (figure 3.8), with a spectrum that can be fitted to a set of three *g*-values $g_1 = 2.02$, $g_2 = 2.00$ and $g_3 = 1.97$ respectively (Huynh *et al.*, 1980). The Mössbauer spectrum at 77 K shows a single quadrupole doublet showing that all 3Fe sites are equivalent and with an isomer shift ($F_{Fe} = 0.27$ nm/s relative to Fe metal) suggestive of a high-spin ferric ion in a tetrahedral environment of S atoms. In a weak magnetic field at 1.5 K the Mössbauer spectrum broadens into one with magnetic hyperfine structure resulting from the interaction of a slowly relaxing electron spin with the magnetic moment of the iron nucleus. This spectrum can be analysed into three subspectra associated with rather different hyperfine fields. In particular, one site is characterised by an unusually small magnetic hyperfine interaction. At first sight this conclusion conflicts with the observation of a single quadrupole

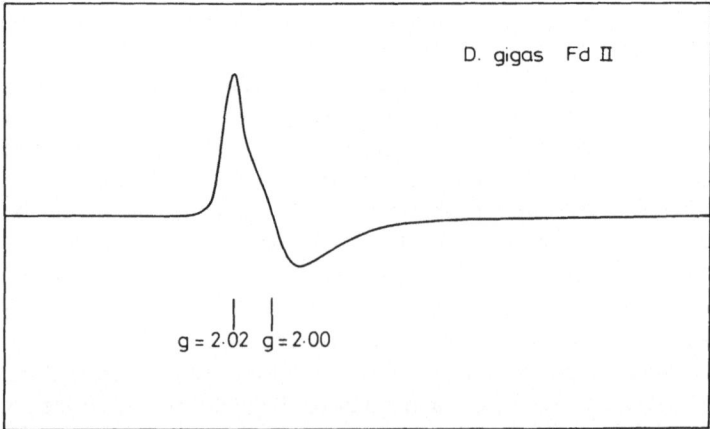

Figure 3.8 EPR spectrum of oxidised, Fd II, *Desulphovibrio gigas*, 0.180 mM. Spectrum recorded at following instrument settings; microwave power 30 μW; frequency 9.21 GHz; modulation amplitude 0.4 mT and temperature 8 K. (Reproduced with permission from Huynh *et al.*, 1980, Chap. 4, fig. 3)

doublet at higher temperatures for all the Fe atoms. However, the resolution of this apparent paradox appeared from a simple spin-coupling model of the trimer (Kent *et al.*, 1980). It turns out to be a feature of antiferromagnetic coupling of three $S = \frac{5}{2}$ ferric ions that one Fe (III) ion experiences a smaller hyperfine field than the other two even when the three exchange coupling constants are equal.

On the addition of one electron Fd II becomes epr silent but remains paramagnetic. The Mössbauer spectrum shows a pair of quadrupole doublets at 4.2 K, in zero magnetic field, in the intensity ratio of 2 : 1. The magnetic spectrum indicates an even-electron spin system (Huynh *et al.*, 1980).

Low temperature magnetic circular dichroism (mcd) spectroscopy of the oxidised and reduced forms of Fd II, *D. gigas*, confirms the assignment of the two oxidation states to paramagnets with a spin $S = \frac{1}{2}$ for the oxidised cluster (Thomson *et al.*, 1981a). It is possible to interpret the magnetisation properties of the reduced cluster as an $S = 2$ state with a predominantly axial zero-field splitting that leaves a doublet $M_S = \pm 2$ lowest in energy. Low temperature mcd spectra provide an excellent spectroscopic fingerprint to identify the 3Fe cluster in proteins. The mcd spectrum of the oxidised cluster is highly structured and quite distinct from that of the mcd spectrum of oxidised HiPIP, the only other iron–sulphur cluster known to give the epr signal in the oxidised state (Johnson *et al.*, 1981). The spectrum of the reduced cluster is much less featured but the strong temperature dependence, the steep magnetisation curve in the absence of an epr signal are characteristic features of the centre.

There are now a variety of spectroscopic signatures that can be taken to be diagnostic of the presence of a 3Fe cluster in a protein (table 3.3). A list is given in table 3.3 of the proteins which have been shown to possess a 3Fe cluster.

At first it was assumed that all 3Fe clusters had the same stoicheiometry and structure as that seen in the x-ray crystal structure of Fd I, *Azotobacter vinelandii*. However, there is now a substantial body of evidence building up which suggests that the structure seen by x-ray diffraction is not compatible with all the evidence on the structures of other 3Fe clusters (Beinert and Thomson, 1983).

The first hint was provided by mcd experiments on the product of oxidation of the 2[4Fe-4S] ferredoxin from *Clostridium pasteurianum* by hexacyanoferrate (III) (Thomson *et al.*, 1981b). It could be conclusively demonstrated that the oxidation led to irreversible change to the clusters and the production of a cluster identical with that in Fd II, *D. gigas*. In other words a three-iron cluster had been generated from the four-iron cluster. Since Prussian blue is precipitated in this reaction the loss of iron is generated by precipitation with hexacyanoferrate (III). A subsequent quantitative study of the epr properties of the resulting cluster suggested that one three-iron cluster is formed from the two [4Fe-4S] clusters (Johnson *et al.*, 1982). However, it should be recognised that all of these earlier studies were carried out in the presence of excess hexacyanoferrate (III) ion. It is not known whether treatment with hexacyanoferrate (III)

Table 3.3 Proteins which contain a 3-Fe cluster

Protein; source	Molecular weight $\times 10^{-3}$	E_M (mV)	EPR g-values	Mossbauer parameters	
				δ (isomer shift)	ΔE_Q (quadrupole splitting)
FdII, *D. gigas*					
Oxid			1.97, 2.00, 2.02	0.27 ± 0.03	0.54 ± 0.03
	18 (trimer)				
Red		−130	silent (S = 2)	0.46 ± 0.02 0.03 ± 0.02	1.47 ± 0.08 (I) 0.47 ± 0.02 (II)
Aconitase, Beef heart					
Oxid			2.01	0.27	0.71
	~80				
Red		~+100	silent	0.45 0.30	1.34 0.49
Fd, *C. pasteurianum* [Fe(CN)$_6^{3-}$ treated]					
Oxid			2.01	—	—
	?				
Red		—	silent (S = 2)	—	—
Fd, *B. stearothermophilus* [Fe(CN)$_6^{3-}$ treated]					
Oxid			2.01	0.24 ± 0.01	0.57 ± 0.02
	?	~0			
Red			silent (S = 2)	0.48 0.29	1.52 0.41
Fd, *M. barkeri*					
Oxid		—	2.02, 2.00	n.d.	n.d.
	20–22 (oligomeric)				
Red		—	silent	0.46 ± 0.04 0.30 ± 0.06	1.40 ± 0.04 0.4 ± 0.06

				0.27 ± 0.04	0.63 ± 0.05
Fd I, *A. vinelandii*					
Oxid			2.01		
Red	(7Fe)	−420	silent	0.47 0.29	1.45 (I) 0.40 (II)
Fd, *A. chroococcum*					
Oxide			2.01	—	—
Red	(7Fe)	~−400	silent	—	—
Fd, *T. thermophilus*					
Oxid	9.2		1.93, 1.98, 2.02	—	—
Red	(7Fe)	−230	silent	—	—
Fd, *T. aquaticus*					
Oxid	—			—	—
Red	(7Fe)	~−230	silent	—	—

The following proteins are suspected to have 3-Fe clusters by the presence of a 'g = 2.01' EPR signal. Fds from *Rhodospirillum rubrum*, *Corynebacterium autotrophicum*, *Mycobacterium flavum*, *Spirillum lipoferum*, *Pseudomonas ovalis*, *Mycobacterium smegmatis*; Succinate dehydrogenase and fumarate reductase; hydrogenases from *D. gigas* and *D. desulphuricans*; nitrate reductases from *Micrococcus denitrificans*, *Pseudomonas aeroginosa*, *Escherichia coli*.

References to work quoted in this table are given at the appropriate point in the text.

at lower concentration will produce one three-iron cluster from each four-iron cluster.

Subsequently it was shown that Fd I and Fd II from *D. gigas* could be reversibly interconvereted (Moura *et al.*, 1982). Addition of Fe(II) and sulphide ion to Fd II gave Fd I, the [4Fe-4S] cluster being reformed. Treatment of Fd I with a [4Fe-4S] cluster gave Fd II, a three-iron centre. Thus relatively facile 3Fe ↔ 4Fe cluster interconversion was established. This leads to the suspicion that the gross protein reorganisation required to go from the structure of a [4Fe-4S] cluster liganded by 4 cysteine residues to a 3Fe cluster liganded by 5 or 6 cysteines is unlikely to be occurring.

Further evidence was provided by EXAFS data on Fd II, *D. gigas* (Antonio *et al.*, 1982). This shows quite clear evidence for a Fe-Fe distance in both the oxidised and reduced states of 2.7 Å. This is quite at odds with the Fe-Fe distance of 4.0 Å determined by x-ray crystallography in Fd I, *Azotobacter vinelandii* (Ghosh *et al.*, 1981).

The next piece of evidence came from a study of the enzyme aconitase (Ruzicka and Beinert, 1978). This enzyme, as extracted, yields an epr signal at $g = 2.01$ which integrates to 1.0 spin per mole of enzyme. The protein is epr silent in the reduced state. In this state the enzyme is inactive. Mössbauer studies (Kent *et al.*, 1982) showed that the enzyme is a 3Fe cluster in this inactive state with properties similar to those of the centre in Fd II, *D. gigas* and Fd I, *A. vinelandii*. On addition of Fe^{2+} only the three-iron cluster re-builds into a $[4Fe-4S]^{2+}$ cluster and the enzyme recovers activity. Careful analyses of the Fe/S^{2-} ratios in the inactive form of the enzyme have given values close to 0.74 whereas analyses on the active form of the enzyme yield a ratio close to unity (Beinert *et al.*, 1983). Great care was taken to make the analysis for sulphur specific for acid labile sulphide, S^{2-}. It is known that sulphur can be stored in proteins in the zero oxidation state as sulphane or sulphur-zero ($S°$) (Petering *et al.*, 1971). This occurs in the protein rhodanese, for example, which contains a cysteine thiol which binds sulphur-zero as $-CH_2-S°-S^-$. Therefore it was concluded that the three-iron cluster in aconitase has the stoichiometry of Fe_3S_4 (Beinert *et al.*, 1983). The EXAFS spectra of the enzyme also show the presence of Fe-Fe distances of 2.8 Å, as in Fd II, *D. gigas*.

All of these observations can be nicely accommodated if the four-iron–three-iron interconversion process is represented as in figure 3.9 (Beinert and Thomson, 1983). Loss of an iron from one corner of the [4Fe-4S] cube will lead to a chair form of the three-iron cluster. If the number of binding cysteines drops from four to three a regular $[3Fe-4S]^+$ trimer with one S^{2-} ion tricapping three Fe atoms can be formed. Alternatively, if the number of cysteine thiols liganding the cluster remains at four a rather less regular structure will result. In both cases an Fe-Fe distance of 2.8 Å can be maintained, consistent with the EXAFS data on aconitase and on Fd II, *D. gigas*. The structures drawn in figure 3.9 seem most plausible representations of the centre in aconitase and Fd II, *D. gigas* although in the latter case there are no Fe/S^{2-} stoicheiometries available.

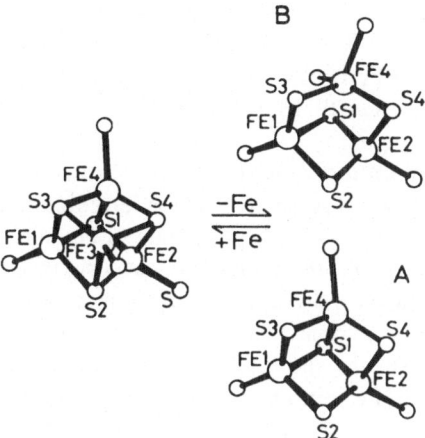

Figure 3.9 Molecular models depicting typical [4Fe-4S] cluster and the possible structures A and B resulting from the removal of one iron atom. The terminal sulphur atoms belong to protein-bound cysteine groups. Note that the core structure [3Fe-4S] is bound to three- and four- cysteine residues, in A and B, respectively. (Reproduced with permission from Beinert, H. and Thomson, A. J. 1983, Scheme I)

The three-iron cluster observed crystallographically (figure 3.1), appears to be one possible variation of the 3Fe structure with a stoicheiometry of $[3Fe-3S]^{3+}$ and six additional ligands and an Fe-Fe distance of 4.0 Å. The other variant exemplified by Fd II, *D. gigas* and inactive aconitase is $[3Fe-4S]^{+}$ with Fe-Fe distances of 2.8 Å. It has been suggested that it is the number of cysteine thiol residues available within the protein pocket to the cluster which may control the type of 3Fe cluster obtained (Beinert and Thomson, 1983). The curious fact remains that the Mössbauer spectra do not readily distinguish between the three-iron clusters of Fd I, *A. vinelandii*, Fd II, *D. gigas* and aconitase. Admittedly the Mössbauer analysis was performed on a solution of Fd I, *A. vinelandii* and not on the crystalline phase of the protein. It is possible that the solution and crystalline phases of the protein have different structures. This appears to be the only way, at present, to reconcile the differences between the interpretations.

FERREDOXINS CONTAINING 3Fe, 4Fe, 7Fe AND 8Fe PER MOLE

The ferredoxins have as their only function electron transfer. There are now examples known in which all the three types of iron–sulphur cluster can be found. Table 3.1 summarises some examples with redox potentials. There are a number of notable features of this list. First, Fd II, *D. gigas* and ferredoxin from *Methanosarcina barkeri* (Moura *et al.*, 1982) are the only two to contain a single 3Fe centre. *M. barkeri* has a sequence which shows 8 cysteines and a remarkable homology with that of the 8Fe ferredoxin from *Cl. pasteurianum*. It is quite unclear what features of the sequences determine that *M. barkeri* should have a single 3Fe cluster rather than a pair of [4Fe-4S] clusters. On reduction a small $g = 1.94$ epr signal is found, typical of a [4Fe-4S] cluster suggesting that this is the vestiges of a 4Fe centre. It is important now to discover whether this ferredoxin is an 8Fe protein that has lost part of its iron to give an artefactual 3Fe cluster.

Fd I and Fd II from *Desulphovibrio africanus* both contain a single [4Fe-4S] cluster with no $g = 2.01$ signal as isolated. These ferredoxins appear to be amongst the purest examples of single [4Fe-4S] proteins so far isolated.

The ferredoxin from *T. thermophilus* was originally thought to contain a HiPIP [4Fe-4S] cluster plus a [2Fe-2S] centre with redox potentials of -250 mV and -530 mV, respectively (Ohnishi *et al.*, 1980)). A more recent study using both Mossbauer and e.p.r. spectroscopies has confirmed that this ferredoxin contains a 3Fe centre with a redox potential of -250 mV and a [4Fe-4S] cluster which reduces at -530 mV to give the e.p.r.-detectable cluster.

A comparative study of the resonance Raman (rr) spectra of the oxidised 3Fe clusters has been reported in the following proteins: *A. vinelandii* Fd I, *T. Thermophilus* Fd, *D. gigas* Fd II and the hexacyanoferrate(III)-damaged *Cl. pasteurianum* Fd (Johnson *et al.*, 1983). The vibrational modes of the [3Fe-xS] core skeleton are resonantly enhanced and normal coordinate analysis can be made in terms of various structural models. The conclusions drawn are unambiguously in favour of a [3Fe-4S] structure corresponding to a [4Fe-4S] cubane cluster with one iron atom and once cysteine ligand missing, that is, structure A, figure 3.9. The vibrational frequencies cannot be reconciled with a nearly planar [3Fe-3S] structure observed in the X-ray structural analysis of *A. vinelandii* Fd I. The rr of a single crystal of this ferredoxin has been recorded and is identical to that of a solution of the protein showing clearly that in the oxidised state there is no change in structure on changing the phase.

EXTRUSION, RE-CONSTITUTION AND INTER-CONVERSION OF CLUSTERS

Studies of inorganic models of the iron–sulphur centres in proteins quickly led to the discovery that clusters are relatively labile. Prior to this, work by bio-chemists had established that iron–sulphur clusters could be removed from a protein in a reaction which destroyed the cluster to yield apo-proteins but that the iron–sulphur cluster could be put back into the protein by a re-constitution process using iron and sulphide (Malkin and Rabinowitz, 1966). Re-constitution can also be effected by a reaction between apo-protein and a model inorganic cluster compound. Hence it is convenient to distinguish between three types of reactions which iron–sulphur proteins may undergo that involve the loss, gain or structural change of the iron–sulphur cluster. We call these extrusion, re-constitution and inter-conversion.

Extrusion reactions

There are three methods of bringing about removal of the core cluster. Two involve the destruction of the cluster and one the removal of an intact cluster.

Removal of the iron–sulphur core in a process termed active centre extrusion (Que *et al.*, 1975) can be brought about by addition of an organic solvent such as DMSO to a level of 80% in water (pH 8.5) to unfold in part the protein. Also a thiol is added to replace the cysteine thiolate ligands of the protein; viz.

$$\text{Holoprotein} + \text{R}'\text{SH} \longrightarrow [\text{Fe}_2\text{S}_2(\text{SR}')_4]^{2-} \text{ and/or}$$

$$[\text{Fe}_4\text{S}_4(\text{SR}')_4]^{2-} + \text{apoprotein}$$

The reaction involves the transfer of the core [nFe-nS] from one ligand environment to another. This reaction has been studied in detail in order to determine the experimental conditions under which the core structure can be extruded quantitatively without undergoing change (see Berg and Holm, 1982, for a detailed discussion). The reaction has been used in an attempt to identify the clusters present in complex proteins, such as hydrogenase (Erbes *et al.*, 1975; Gillum *et al.*, 1977) and nitrogenase (Kurtz *et al.*, 1979), and mitochondrial electron transport chains, which have defied positive identification by spectroscopic methods. The methods have been calibrated by extrusion of iron–sulphur core from proteins with structurally well-defined clusters. Thus the [4Fe-4S]$^{2+}$ core is quantitatively extruded from *Clostridium pasteurianum* ferredoxin in 80% DMSO-H$_2$O, pH 8.5, with benzenethiol or with o-xylyl-α, α'-dithiol to yield the analogues $[\text{Fe}_4\text{S}_4(\text{SPh})_4]^{2-}$ or $[\text{Fe}_4\text{S}_4(\text{S}_2\text{-}o\text{-xylyl})_2]^{2-}$.

However, there can be cluster interconversion thus

$$2[\text{Fe}_2\text{S}_2(\text{SPh})_4]^{2-} \longrightarrow [\text{Fe}_4\text{S}_4(\text{SPh})_4]^{2-} + \text{PhSSPh} + 2\text{PhS}^-$$

which occurs in 80% DMSO-H$_2$O fairly rapidly below pH 7.5 (Que *et al.*, 1975). Clearly it is necessary to suppress this dimer–tetramer conversion by use of higher pH values and excess of thiol.

An alternative extrusion procedure developed by Orme-Johnson and co-workers (Orme-Johnson and Holm, 1978) involves the use of the apo-protein of a ferredoxin known to be specific for a 4Fe or 2Fe cluster as the recipient of the extruded cluster.

A number of extrusion studies have been reported and cluster identification made on the basis of the extruded product. The product core structure was in the early experiments identified by optical spectroscopy but more recently nmr techniques have been developed involving the nmr of the ^{19}F nucleus attached to the extruding ligand (Wang *et al.*, 1979).

The extrusion method has not been pursued in recent years. An experiment with aconitase led to the extrusion of a [2Fe-2S] core structure in almost quantitative yield (Kurtz *et al.*, 1979). This experiment was performed on aconitase as extracted and it was this form that was subsequently demonstrated to possess a 3Fe cluster. Since no inorganic model compounds of 3Fe clusters have yet been synthesised there are no control experiments available to define the conditions under which a [3Fe-xS] core might be extruded. Until this background work has been done it would seem wise to refrain from structural assignments of the nature of the iron–sulphur cluster in complex proteins based upon intact core extrusion experiments. However, it is of great interest that a [2Fe-2S] core was extruded from the [3Fe-4S] core of aconitase. When structural significance this might have is unclear. Core extrusion experiments have been conducted by two groups on the 7Fe protein, Fd I from *A. vinelandii*, with conflicting results. Howard *et al.* (1976) suggested that 2[4Fe-4S] clusters were present. Averill *et al.* (1978) repeated the experiments and found evidence for only one [4Fe-4S] cluster and suggested that another cluster type of non-standard form was present.

Trichloroacetic acid (TCA) is normally used to obtain a sample of the apo-protein in a process which destroys completely the cluster yielding a precipitate of apo-protein and a solution of iron and sulphur (Hong and Rabinowitz, 1967). This is the step used prior to re-constitution with, for example, the incorporation of isotopic labels such as ^{57}Fe, ^{33}S or Se in place of S. All iron–sulphur proteins are susceptible to this drastic treatment.

Another method which has been little investigated for the removal of iron–sulphur clusters is treatment with hexacyanoferrate (III) ion in the presence of urea as a denaturing agent. In a classic study the ferredoxins from spinach and *Cl. pasteurianum* (Petering *et al.*, 1971) were treated with hexacyanoferrate ion. This led to complete removal of iron, which was oxidised to Fe (III). However, some of the sulphide ion, S^{2-}, was oxidised and trapped in the protein in the oxidation state zero probably as a trisulphide, $R-S-S^{\circ}-S-R$, involving a pair of cysteine thiol residues. The proteins were treated by a denaturing agent, urea, in this study in order to give the oxidant access to the cluster.

In the absence of a denaturing agent the reactivity of the cluster towards hexacyanoferrate appears to vary enormously. It seems likely that the sensitivity towards hexacyanoferrate parallels that of the reactivity towards atmospheric oxygen. At one extreme it has been reported that a 7Fe ferredoxin from *Thermus thermophilus* containing a 3Fe and a 4Fe cluster can be incubated for 1 h in hexacyanoferrate (III) ion at room temperature without any changes occurring in the epr spectrum (Ohnishi *et al.*, 1980). This is taken to indicate that the clusters have been unaffected by the oxidant. This ferredoxin appears remarkably stable to air. At the other extreme the iron–sulphur clusters of the nitrogenase enzyme system, those of fraction 1 and the [4Fe-4S] cluster of fraction 2, are rapidly destroyed by air and are sensitive to hexacyanoferrate ion. However, treatment of fraction 1 from *Klebsiella pneumoniae* with hexacyanoferrate (III) for short periods of time can lead to oxidation which is reversible to retain full activity. There are reports that some hydrogenases are totally inactivated by exposure to air and hexacyanoferrate (III) while others are quite resistant to exposure to those reagents. The HiPIP proteins are resistant to damage by hexacyanoferrate (III) but are, of course, oxidised to the $[4Fe-4S]^{3+}$ oxidation level.

Re-constitution and interconversion

Some of the earliest work on the chemistry of iron–sulphur proteins involved re-constitution of the iron–sulphur cores into apo-protein. This involved addition of ferrous ion, sulphide and 2-mercaptothanol (Malkin and Rabinowitz, 1966). The function of the latter is to reduce any disulphide links to free thiol groups. The optimal conditions for re-constitution of clostridial ferredoxins involve the addition of sulphide and iron above the stoicheiometric quantities (Hong and Rabinowitz, 1970). In some instances re-constitution can occur with the addition only of iron. The sulphide ion remains apparently bound in the protein. For example, if spinach ferredoxin is treated with O_2 in the presence of urea the resulting apoprotein lacks iron but retains sulphide in a bound form (Petering *et al.*, 1971). It was suggested that the sulphide is bound between two cysteine thiols as sulfane sulphur in a trisulphide, RSSSR. Reduction of this group re-releases S^{2-}. This may be an important mechanism for the storage of S^{2-} and can confuse the interpretation of re-constitution experiments if S° is not assayed.

Clusters can also be reconstituted with Se^{2-} in place of sulphide ion (Tsibris *et al.*, 1968). This can be useful for the preparation of clusters for spectral assignments.

In an important paper Hong and Rabinowitz (1970) attempted to re-constitute clostridial ferredoxin with less than the amount of Fe or S^{2-} to supply a pair of [4Fe-4S] clusters. They concluded that ferredoxin molecules deficient in iron or sulphide could not be prepared. There appeared to be a cooperative uptake of iron and sulphide to generate a complete [4Fe-4S] core. This has been called the 'all-or-none' hypothesis. This is an interesting result when compared with the experiments more recently reported in which hexacyanoferrate (III) can degrade the [4Fe-4S] cluster to a 3Fe cluster. Apparently conditions have not been found in which the 3Fe cluster can be re-constituted in clostridial ferredoxin.

The 'all-or-none' nature of the binding was further tested by attempting to remove part of the iron in the protein with the chelating agents bipyridyl and mercurials (Hong and Rabinowitz, 1970). On the basis of optical spectroscopy and analysis Hong and Rabinowitz concluded that no ferredoxin could be isolated with less than 8 moles of iron and 8 moles of sulphide. However, they did mention that bipyridyl would remove only 25% of the iron in some instances. This is an intriguing number to obtain in view of present knowledge that iron can be removed to form a 3Fe cluster (Thomson *et al.*, 1982). Unfortunately Hong and Rabinowitz did not carry out epr studies of their reaction mixtures. This would have provided a sensitive method of detecting the presence of any

3Fe clusters formed during the course of degradation. Clearly this type of experiment warrants re-investigation in the light of more recent work.

In the building up of 3Fe clusters to 4Fe clusters only two systems have been investigated. The requirement for Fe II ion only to build up 4Fe clusters from 3Fe clusters seems well established for aconitase (Kent *et al.*, 1982). The conversion of the Fd II, *D. gigas*, a 3Fe cluster to Fd I, a 4Fe, has been carried out with addition of Fe and sulphide (Moura *et al.*, 1982). No attempt apparently was made to see whether conversion could be achieved by addition only of Fe.

Re-constitution of apo-protein of the ferredoxins from *D. gigas* leads to Fd I, the [4Fe-4S] cluster, in the presence of excess iron. When limiting Fe is employed, that is 3Fe/monomer of apo-protein, about 50% of a 3Fe cluster and 50% of a [4Fe-4S] cluster are produced. A clean and complete reconstitution of a single 3Fe cluster has not yet been achieved in Fd II, *D. gigas* (Moura *et al.*, 1982). Clearly though the 'all-or-none' hypothesis does not apply to the *D. gigas* protein.

No reports have been published of attempts to convert the 3Fe cluster in Fd I, *A. vinelandii*, into a 4Fe centre. However, reconstitution of the apo-protein of this ferredoxin under rigorously anaerobic conditions using Fe(II) and sulphide ion leads to an eight-iron ferredoxin containing 2[4Fe-4S] clusters (Morgan *et al.*, 1984). One of these is reducible with dithionite to give an axial e.p.r. signal with g values of 2.04, 1.92. On exposure of the protein to air it decomposes to give $\approx 10\%$ yield of the 7Fe ferredoxin. This intriguing result further increases the probability that the *in vivo* form of the protein is the 8Fe ferredoxin.

EXAMPLES OF IRON–SULPHUR PROTEINS

In this section we describe two iron–sulphur proteins that are currently undergoing intensive investigation and which promise to be especially important to our understanding of the ways in which iron–sulphur clusters can interconvert and also to reveal whether any iron–sulphur clusters can perform catalytic functions other than electron transfer.

Aconitase

Aconitase, which was known in the 1930s (Martius, 1937), is the enzyme citrate-isocitrate hydrolyase catalysing the following reaction

$$
\begin{array}{ccc}
\text{H} & & \text{H} \\
| & & | \\
\text{H} - \text{C} - \text{COO}^- & & \text{HO} - \text{C} - \text{COO}^- \\
| & \longrightarrow & | \\
{}^-\text{OOC} - \text{C} - \text{OH} & & {}^-\text{OOC} - \text{C} - \text{H} \\
| & & | \\
\text{CH}_2\text{COOH} & & \text{CH}_2\text{COOH}
\end{array}
$$

It is, therefore, present in mitochondria. It was only in 1978 (Ruzicka and Beinert) that a connection was found between the enzymic activity of aconitase and an iron–sulphur protein with an epr signal at $g = 2.01$ in its oxidised state, previously referred to as 'HiPIP', by analogy with the protein from *Chromatium vinosum* which contains a $[4Fe-4S]^{3+,2+}$ cluster. It had been recognised for some thirty years (Dickman and Cloutier, 1951; Morrison, 1954) that the enzyme aconitase was inactive as extracted from the mitochondrion and required activation by the addition of ferrous ion and reducing agents. In 1979 (Kurtz *et al.*, 1979) the cluster extrusion technique was applied to aconitase and it was shown that 90% of the iron could be extruded in the form of [2Fe-2S] clusters. However, since the enzyme as extracted has an epr signal in the *oxidised* state this suggested a [2Fe-2S] cluster with a unique set of oxidation states.

Resolution of much of this conflicting evidence came about with the discovery of the three-iron clusters and the demonstration that transformation of the $[4Fe-4S]^{2+}$ cluster into a three-iron cluster could be effected by an oxidising agent in the 8Fe ferredoxin from *Clostridium pasteurianum* (Thomson *et al.*, 1981).

Mössbauer studies of beef heart aconitase showed that the epr signal at g = 2.01 is associated with a cluster of the three-iron type (Kent *et al.*, 1982). In dithionite reduced aconitase the characteristic reduced three-iron Mössbauer spectrum in a paramagnetic state of integral spin (S = 2) was observed. On activation of the enzyme with ferrous ion the paramagnetic 3Fe cluster was transformed into a diamagnetic S = 0 form (Kent *et al.*, 1982). In the presence of dithionite and ferrous ion it is possible to obtain a reduced form of the $[4Fe-4S]^{+}$ cluster and to elicit an epr signal at g = 2.06, 1.93, 1.86 typical of a single [4Fe-4S] ferredoxin. Thus it was concluded that aconitase as prepared routinely is an enzymatically inactive three-iron cluster. Activation with iron involves the facile re-building of a [4Fe-4S] cluster which is presumed to be the catalytically active cluster form. However, it is not known which state of the cluster is present during enzyme turnover.

One curious feature of the re-building of the three- to the four-iron cluster is the fact that reduction of inactive aconitase with dithionite alone leads to development of 70% of the maximal attainable activity. Since no ferrous ion needs to be added it has been suggested that the four-iron clusters are built up at the expense of the iron of some of the three-iron clusters; a form of cluster cannibalism. This is a remarkable notion and remains to be substantiated.

In all the interconversions it is never necessary to add sulphide. This posed the question of the source of the additional S^{2-} if the cluster conversion was [3Fe-3S] → [4Fe-4S]. As more and more preparations were analysed it became apparent that the S^{2-} values were consistently higher than the Fe values. A thorough investigation shows that the ratio of Fe/S^{2-} lies in the range 0.66–0.74. Sulfane sulphur (S°) was not detected (Beinert *et al.*, 1983). This is just the ratio required for a cluster [3Fe-4S]. Confirmation was provided when it was discovered that preparations with up to 90% activity could be prepared by a rapid purification procedure. Analysis of these preparations yields Fe/S^{2-} ratios between 0.9 and 1.03, that is, close to unity, as expected for an intact [4Fe-4S] cluster.

Mössbauer and epr spectroscopy show the essential similarity of the three-iron clusters in aconitase (Kent *et al.*, 1982), in Fd II *D. gigas* (Huynh *et al.*,

1980) and in solution phase of *A. vinelandii* Fd I (Emptage *et al.*, 1980). Low temperature mcd spectroscopy further shows the similarity between the clusters of Fd II *D. gigas* and those produced by hexacyanoferrate (III) damage to the 2[4Fe-4S] and [4Fe-4S] clusters of *Cl. pasteurianum* ferredoxins (Thomson *et al.*, 1981) and *B. stearothermophilus* ferredoxin. Thus it must be presumed that all have the basic core structure of [3Fe-4S]. It has been suggested that it is the number of available cysteine residues within the protein pocket which controls the type of 3Fe cluster formed (Beinert and Thomson, 1983), that is, whether it has the [3Fe-3S] structure defined crystallographically in *A. vinelandii* Fd I or the [3Fe-4S] cluster apparently in the other proteins in the solution phase. An extended discussion of the structural possibilities has been given in a recent review (Beinert and Thomson, 1983).

One further feature of the Mössbauer study deserves mention. Since clusters in aconitase (and in Fd II, *D. gigas*) can be converted from 3Fe to 4Fe centres an opportunity is provided for the selective labelling of sites in a 4Fe cluster with either ^{56}Fe or ^{57}Fe, Mössbauer silent and active nuclei, respectively. This elegant procedure promises to give us much useful information about the nature of specific sites. In the case of aconitase activation it was shown that the added ferrous ion goes into a unique site in the formed [4Fe-4S] cluster (Kent *et al.*, 1982). The resulting clusters have a 3 : 1 site ratio. Hence the added iron has a unique site even in the resulting [4Fe-4S] cluster. On a longer time-scale the iron in this site slowly exchanges with other sites in the cluster possibly via cluster breakdown and reassembly. These findings contrast with similar labelling studies made with Fd II *D. gigas* (Moura *et al.*, 1982) in which the added Fe labels went to 3 subsites, leaving a unique Fe site unlabelled. No exchange between subsites was observed.

Hydrogenase

The enzyme hydrogenase catalyses the reversible oxidation of hydrogen as given by the equation

$$H_2 \rightleftharpoons 2H^+ + 2e^-$$

It is clear that a wide range of bacterial and algal species possess hydrogenase activity (Adams *et al.*, 1981). Hydrogen utilisation is involved in nitrogen fixation, in photoproduction of hydrogen and fermentation of biomass to methane. The consumption of hydrogen provides organisms with a supply of reductant that may be used for energy generation. This is the respiratory role of hydrogenase. Alternatively, hydrogen production results when the proton is used as a terminal electron acceptor and enables the organism to dispose of excess reducing equivalents. The *evolution* of hydrogen usually occurs under strictly anaerobic conditions whereas the *consumption* of hydrogen may take place both by aerobic and anaerobic organisms. Under anaerobic conditions most of the hydrogen produced is used to reduce CO_2 to methane by the methanogenic bacteria. The situation is complicated in the case of aerobic H_2-oxidising bacteria, some of which contain two distinct types of hydrogenase. One is a soluble enzyme which will reduce NAD directly with H_2. The other is a membrane-bound enzyme which does not interact with nicotinamide nucleotides but instead donates electrons directly into the electron transport chain that reduces oxygen. Hence during autotrophic growth, using CO_2 only as a carbon source, the soluble hydrogenase generates NADH for CO_2 fixation while the membrane-bound hydrogenase is involved in energy production (figure 3.10).

The only generalisation apparently to be made about the range of hydrogenases is that they all contain iron–sulphur clusters. In some cases other prosthetic groups such as flavin may be present and, as recently discovered, some contain nickel ion. This it is not known whether hydrogen activation always takes place at the same type of site, or indeed, whether that site is an iron–sulphur centre. Table 3.4 lists some properties of representative examples of hydrogenases from a variety of sources.

The identification of the cluster type has so far only been made by epr spectroscopy and, in some cases, by the core extrusion technique. Few Mössbauer studies have been reported. Now that purified protein can be obtained many more spectroscopic techniques will be applied to the analysis of the clusters for comparison with those centres characterised in low molecular weight ferredoxins. There are undoubtedly a number of surprises in store. We draw attention here to features that appear to be worthy of immediate study.

The 12Fe-12S hydrogenases from *Cl. pasteurianum* and *Megasphaera elsdenii* have been subjected to core extrusion and all the iron–sulphur clusters appear to be of the [4Fe-4S] type (Erbes *et al.*, 1975; Gillum *et al.*, 1977; Van Dijk *et al.*, 1980). However, at potentials of -300 mV both hydrogenases exhibit a unique rhombic epr signal at $g = 2.099$, 2.046 and 2.005 (Chen *et al.*, 1976). Because

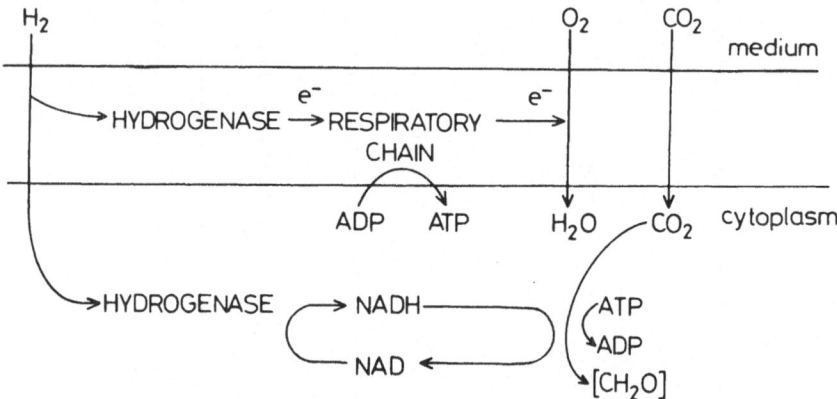

Figure 3.10 Pathways of hydrogen utilisation in the aerobic H_2-oxidising bacteria. (Taken from Adams *et al.* 1981, fig. 4)

Table 3.4 Properties of some different classes of hydrogenase

Source	Location	Fe/mol : S/mol	Molecular weight	Other groups	EPR signals	Specific[a] activity	O_2 sensitivity
Chromatium (Photosynthetic)	Membranes	4 : 4	68 000	–	oxid: $g > 2$, axial red: silent	1.9	Stable on storage Sensitive on reduction
Alcaligenes eutrophus H16 (Aerobic)	Cytoplasm	12 : 12 $2 \times$ [4Fe–4S] $2 \times$ [2Fe–2S]	205 000	2FMN	oxid: weak $g = 2.02$ red: high. temp. $g = 2.04, 1.95$ low temp. $g = 2.04, 2.00, 1.94, 1.93,$ 1.86	48	Stable
Escherichia coli (Facultative anaerobes)	Membranes	12 : 12	113 000	–	oxid: $g > 2$, axial red: complex, $g > 2$	136	Sensitive
Clostridium pasteurianum (Anaerobic)	Cytoplasm	12 : 12 $3 \times$ [4Fe–4S]	60 000	–	oxid: rhombic, $g > 2$ red: complex, $g = 2.079, 1.961,$ 1.892	4000	Very sensitive
Desulphovibrio gigas (Anaerobic)	Periplasm	12 : 12 $3 \times$ [4Fe–4S]	89 000	Ni	oxid: $g = 2.02, 2.31, 2.20, 2.0$ (Ni^{III}) red: $g = 2.17, 2.08, 2.04$	91	Sensitive

[a] Expressed as hydrogen evolution activity with dithionite-reduced methyl viologen, expressed as μmol H_2 evolved/min mg protein. This table has been adapted from Table II of Adams *et al.* (1981). References are given in that article.

all of these g-values are greater than 2 it has been supposed that these signals arise from a $[4Fe-4S]^{3+}$ cluster, that is, one of the HiPIP-type (Adams *et al.*, 1981). No true HiPIP's show signals of these g-values. They are usually almost axial and, of course, have much higher E_m values. Thus the true meaning of these signals in hydrogenase are quite uncertain and invite investigation by spectroscopic techniques other than epr.

Another fascinating problem concerns the varied oxygen sensitivity of hydrogenase. All iron–sulphur proteins are oxygen-sensitive to a greater or lesser degree. This seems to arise from oxidation of acid-labile sulphur and the cysteine residues. This leads to a gradual disintegration of the clusters. However, there appears to be a rather more specific interaction between hydrogenase and oxygen. Adams *et al.* (1981) have distinguished between reversible inactivation and irreversible inactivation. In the former case there are hydrogenases from a number of organisms, including *Alcaligines eutrophus* (Schink and Schlegel, 1979; Schlegel and Schneider, 1976) and *Chromatium* (Krasna, 1979) which are not active in the H_2-oxidation array as isolated. Several procedures serve to re-activate these enzymes including incubation under H_2, treatment with a reductant or simply removal of O_2 by evacuation. It may be that O_2 is occupying the site normally accessed by H_2. Although reduction of a protein-bound group may also be required.

Irreversible inactivation generally occurs when hydrogenase in the catalytically active state or reduced form is exposed to oxygen. Even for hydrogenase such as that from *Desulphovibrio vulgaris* (Van der Westen *et al.*, 1980), which is oxygen insensitive, as isolated 50% of activity was lost within 5 min after the enzyme was reduced. The hydrogenases from *Cl. pasteurianum* and *M. elsdenii* are irreversibly inactivated as prepared. Clearly the active centre of all hydrogenases is irreversibly damaged by oxygen. But some cells have evolved methods of protecting their hydrogenase from damage whilst it is not in the functioning state. Thus oxygen sensitivity correlates with the physiological role of the hydrogenase rather than with the oxygen sensitivity of the cell from which it is extruded. The recent discovery of cluster interconversion involving oxidation of a $[4Fe-4S]$ cluster to a $[3Fe-4S]$ cluster suggests one possible mechanism for the protection and reversible inactivation of a catalytically active centre. The presence of a $g = 2.01$ epr signal has been observed in, for example, the hydrogenase from *D. vulgaris* (Van der Westen *et al.*, 1980) and may be indicative of this process occurring.

The most exciting discovery of the last few years in the field of hydrogenase biochemsitry has been the realisation that nickel ion is present in the enzymes from a range of bacteria that employ hydrogen as a source of energy under anaerobic conditions. Table 3.5 lists the hydrogenases in which nickel has been identified both by analytical and spectroscopic (epr) methods. In addition to the enzymes given in the table others containing nickel have been isolated from *Alcaligenes eutrophus* (Friedrich *et al.*, 1982) and *Chromatium vinosum* (Albracht *et al.*, 1982). The activity of uptake hydrogenases have been shown to

Table 3.5 Properties of hydrogenases containing nickel ion

Species of bacteria	Overall reaction catalysed	Strain on genus	Properties of hydrogenase			References
			Molecular weight	Metal content per mole and likely cluster grouping	EPR characteristics	
Desulphovibrio	$4H_2 + SO_4^{2-} \rightarrow S^{2-} + 4H_2O$	*D. gigas*	89 000	0.9–1.0 Ni; 11–12 Fe; 2 × [4Fe-4S]; 1 × [3Fe-xS]?	Ni^{III} 2.31, 2.23, 2.02 [2.19, 2.16, 2.02 after H_2 reduction]; Fe-S g = 2.02 oxid.	Cammack *et al.* (1982); Moura *et al.* (1982); LeGall *et al.* (1982)
		D. desulphuricans I	77 600	0.6 Ni; 7.8 Fe; 6.8 S^{2-}		Krüger *et al.* (1982)
		II	75 500	0.6 Ni; 10.9 Fe	Ni^{III} 2.3, 2.2, 2.0; Fe-S g = 2.02 oxid. g = 1.94 red	
				11.45	Weak Ni g = 2.28 signal after H_2 reduction	
Methanobacterium	$4H_2 + CO_2 \rightarrow CH_4 + 2H_2O$	*M. bryantii*		Ni	Ni^{III} 2.30, 2.23, 2.02	Lancaster (1980)
		M. thermoautotrophicum (Marburg strain)		Ni; No Fe-S centres detected by EPR	Ni^{III}	Albracht *et al.* (1982)
		M. thermoautotrophicum (Strain ΔH) 2 types – F_{420} reducing	170 000	2.5–3.1 Ni; 33–43 Fe; 24–30 S; 2.3 FAD	Ni^{III} 2.309, 2.237, 2.017 [g = 2.196, 2.140, after H_2 reduction] Complex Fe-S signals in reduced state	Kojima *et al.* (1983)
		Methylviologen reducing	?	Ni; Fe	Ni^{III} 2.309, 2.237, 2.017 Complex Fe-S signals in reduced state	
Vibrio succinogenes	H_2 + fumarate → succinate	–	minimum 100 000	Ni; Fe 11–20 mol/mol Ni; S 8 mol/mol Ni		Unden *et al.* (1982)

be a function of the nickel content of the growth medium in *Azotobacter chroococcum* (Partridge and Yates, 1982) and *Rhodopseudomonas capsulata* (Takakuwa and Wall, 1981). It has therefore been tentatively concluded that uptake hydrogenases are in general nickel enzymes.

The search for nickel was stimulated by observations that the metal is required nutritionally for certain bacteria, notably the methanogens (Schonheit *et al.*, 1979). An investigation of the epr spectra of both soluble and membrane-bound fractions of *M. bryantii* showed signals at g = 2.30, 2.23 and 2.02 (Lancaster, 1980) which were assigned to Ni(III), a low-spin d^7 metal ion with one unpaired electron. The assignment of this type of epr signal to nickel ion has now been confirmed in four instances, namely, *D. gigas* (Moura *et al.*, 1982), *D. desulphuricans* (Krüger, 1982), *M. thermoautotrophicum* (Marburg strain) (Albracht *et al.*, 1982) and *M. thermoautotrophicum* (strain ΔH) (Kojima *et al.*, 1983) by growth of the bacteria on a medium isotopically enriched in ^{61}Ni which possesses a nuclear spin of 3/2. This gives rise to a well resolved hyperfine splitting of 4 lines on at least one of the three components of the rhombic epr signal. This is totally convincing evidence that the epr signal first attributed to nickel by Lancaster (1980) was correctly assigned. The signal must arise from the trivalent state of nickel since this has the odd number of electrons (S = $\frac{1}{2}$) required for detection by epr spectroscopy.

The mid-point potential for the reduction process Ni(III) $\rightarrow$ Ni(II) in the hydrogenase from *D. gigas* has been measured by epr spectroscopy (Cammack *et al.*, 1982). The potential is pH dependent with a value of -145 mV at pH 7.2. The pH dependence shows that one proton is taken up during reduction. The potential is higher than the H^+/H_2 couple (which is -420 mV at pH 7.0) but much lower than the redox potentials of typical inorganic Ni(III) complexes. Margerum and co-worker (Bossu and Margerum, 1977) made a series of nickel (III) complexes of amino acids by electrochemical oxidation. The potentials required were $\sim +800$ mV although the g-values obtained were not dissimilar to those observed in hydrogenases. Thus it is clear that the protein must supply a set of ligands and a suitable coordination geometry to lower the Ni(III)/Ni(II) potential by about 1 V. One puzzling common feature of the nickel (III) epr signal in hydrogenases is that it integrates only to about 50% of the nickel present detectable analytically. Thus much more needs to be learnt about the nature of the nickel ion coordination site and its involvement in catalysis before its role in hydrogenases can be appreciated.

CONCLUSIONS

There has been a re-vitalisation of interest in the biochemistry and inorganic chemistry of iron-sulphur clusters within the last three years. This has been caused by the discovery of the 3Fe clusters and by the attempts to define the structural properties of this cluster. The second finding of interest is the discovery that 3Fe and 4Fe clusters can be interconverted with relative ease in some cases. This suggests that such interconversions might well occur under physiological conditions, even if only as intermediates of cluster formation, degradation or rearrangement. It may be significant that evidence for the presence of 3Fe clusters is found in aconitase when *in vitro* and indications are that such clusters are present in some hydrogenases. It has always been puzzling that the four-iron cluster appears to have catalytic properties but yet is coordinatively saturated. There seems to be no site on the cluster where a substrate molecule such as H_2, H^+ or citrate might bind. However, the fragility

In a recent X-ray investigation of the inorganic model compound $[Fe_4 S_4 (SC_6 H_4\text{-}o\text{-}OH)_4]^{2-}$ one of the iron atoms is five-coordinate with the o-$HOC_6 H_5 S^-$ ligand forming a chelate ring to the iron (Johnson *et al.*, 1983). The Fe-OH bond length at 2.318 Å is long and represents a secondary, weaker interaction. The possibility of expansion of the iron coordination number to five within a cluster clearly points to a mechanism for the functional role in providing a site for substrate binding. We have therefore seen several properties of the iron–sulphur cluster which may be utilised in enzyme function. Cluster *lability* may have a role in control of function in anaerobic versus aerobic environments, cluster *stability* is required for electron storage and transport and, finally, expansion of the coordination number of iron can provide for substrate binding and activation.

REFERENCES

Adams, M. W. W., Mortenson, L. E. and Chen, J.-S. (1981) *Biochim. Biophys. Acta*, 594, 105

Adman, E. T., Sieker, L. C. and Jensen, L. H. (1973) *J. biol. Chem.*, 248, 3987

Adman, E. T., Sieker, L. C. and Jensen, L. H. (1976) *J. biol. Chem.*, 25, 3801

Adman, E. T., Watenpaugh, K. D. and Jensen, L. H. (1975) *Proc. nat. Acad. Sci., U.S.A.*, 72, 4854

Albracht, S. P., Graft, E. G. and Thauer, R. K. (1982) *FEBS Lett.*, 140, 311–313

Albracht, S. P. J. and Subramanian, J. (1977) *Biochim. Biophys. Acta*, 462, 36

Andreesen, J. R., Wagner, R., Imhoff, D. and Dürre, P. (1979) *Hoppe-Seylers Z. Physiol. Chem.*, 360, 1122

Antonio, M. R., Averill, B. A., Moura, I., Moura, J. J. G., Orme-Johnson, W. H., Teo, B. K. and Xavier, A. V. (1982) *J. biol. Chem.*, 257, 6646

Aparicio, P. J., Knaff, D. B. and Malkin, R. (1975) *Archiv. Biochem. Biophys.*, 169, 102–107

Averill, B. A., Bale, J. R. and Orme-Johnson, W. H. (1978) *J. Am. chem. Soc.*, 100, 3034

Beinert, H. (1973) In: *Iron–Sulphur Proteins*, Vol. I, (ed. W. Lovenberg), Academic Press, New York

Beinert, H., Ackrell, B. A. C., Vinogradov, A. D., Kearney, E. B. and Singer, T. P. (1977) *Arch. Biochem. Biophys.*, 182, 95–106

Beinert, H., Scott, R. A., Emptage, M. H., Dreyer, J.-L., Hahn, J. E., Hodgson, K. O. and Thomson, A. J. (1983) *Proc. nat. Acad. Sci., U.S.A.*, 80, 393–396

Beinert, H. and Thomson, A. J. (1983) *Arch. Biophys. Biochem.*, 222, 333–361

Berg, J. M., Hodgson, K. O. and Holm, R. H. (1979) *J. Am. chem. Soc.*, 101, 4586

Berg, J. M. and Holm, R. H. (1982) In: *Iron–Sulphur Proteins*, Vol. 4, Metal Ions in Biology Series (ed. Spiro, T. G.), Wiley, New York

Bernhardt, F.-H., Heymann, E. and Traylor, P. S. (1978) *Eur. J. Biochem.*, 92, 209–223

Bossu, F. P. and Margerum, D. W. (1977) *Inorg. Chem.*, 16, 1210

Bray, R. C. (1975) Molybdenum Iron-Sulphur Flavin Hydroxylases and Related Enzymes. In: *The Enzymes*, 3rd edn, Vol. 12, (ed. Boyer, P. D.), pp. 300–420

Brintzinger, H., Palmer, G. and Sands, R. H. (1966) *Proc. nat. Acad. Sci., U.S.A.*, 55, 397

Bruschi, M., Hatchikian, C. E., LeGall, J., Moura, J. J. G. and Xavier, A. V. (1976) *Biochim. Biophys. Acta*, **449**, 275

Bruschi, M. (1979) *Biochem. Biophys. Res. Commun.*, **91**, 623

Buchanan, B. B. (1969) *J. biol. Chem.*, **244**, 4218–4223

Cammack, R., Barber, M. and Bray, R. C. (1976) *Biochem. J.*, **157**, 469–478

Cammack, R. and Evans, M. C. W. (1975) *Biochem. Biophys. Res. Commun.*, **67**, 544–549

Cammack, R., Hucklesby, D. P. and Hewitt, E. J. (1978) *Biochem. J.*, **171**, 519

Cammack, R., Patil, D., Aguirre, R. and Hatchikian, C. E. (1982) *FEBS Lett.*, **142**, 289–292

Cammack, R., Rao, K. K., Hall, D. O. and Johnson, C. E. (1971) *Biochem. J.*, **125**, 849

Cammack, R., Rao, K. K., Hall, D. O., Moura, J. J. G., Xavier, A. V., Bruschi, M., Legall, J., Deville, A. and Gayda, J. P. (1977) *Biochem. Biophys. Acta*, **490**, 311

Cammack, R. C. (1973) *Biochem. Biophys. Res. Comm.*, **54**, 548

Carter, C. W., Jr., Freer, S. T., Xuong, Ng H., Alden, R. A. and Kraut, J. (1971) *Cold Spring Harbor Symp. Quant. Biol.*, **36**, 381

Carter, C. W., Jr., Kraut, J., Freer, S. T., Alden, R. A., Sieker, L. C., Adman, E. T. and Jensen, L. H. (1972) *Proc. nat. Acad. Sci., U.S.A.*, **69**, 3526

Carter, C. W., Jr., Kraut, J., Freer, S. T., Xuong, Ng. H., Alden, R. A. and Bartsch, R. G. (1974a) *J. biol. Chem.*, **249**, 4212

Carter, C. W., Jr., Kraut, J., Freer, S. T. and Alden, R. A. (1974b) *J. biol. Chem.*, **249**, 6339

Chen, J-S., Mortensen, L. E. and Palmer, G. (1976) In: *Iron and Copper Proteins*, (ed. Yasonobu, K. K., Mower, H. F., and Hayaishi, O.), Plenum Press, New York, pp. 68–82

Christner, J. A., Münck, E., Janick, P. A. and Siegel, L. M. (1981) *J. biol. Chem.*, **254**, 2098–2101

Davis, K. A., Hatefi, Y., Crawford, I. P. and Baltscheffsky, H. (1977) *Archiv. Biochem. Biophys.*, **180**, 459–464

Dickman, S. R. and Cloutier, A. A. (1951) *J. biol. Chem.*, **188**, 379

Dunham, W. R., Bearden, A. J., Salmeen, I. T., Palmer, G., Sands, R. H., Orme-Johnson, W. H. and Beinert, H. (1971) *Biochim. Biophys. Acta*, **253**, 134

Eady, R. R. and Postgate, J. R. (1974) *Nature, Lond.*, **249**, 805–809

Eaton, W., Palmer, G., Fee, J. A., Kimura, J. and Lovenberg, W. (1972) *Proc. nat. Acad. Sci., U.S.A.*, **68**, 3015

Emptage, M. H., Kent, T. A., Huynh, B. H., Rawlings, J., Orme-Johnson, W. H. and Münck, E. (1980) *J. biol. Chem.*, **255**, 1793

Erbes, D. L., Burris, R. H. and Orme-Johnson, W. H. (1975) *Proc. nat. Acad. Sci., U.S.A.*, **72**, 4795

Estabrook, R. W., Suzuki, K., Mason, J. E., Baron, J., Taylor, W. E., Simpson, E. R., Purvis, J. and McCarthy, J. (1973) In: *Iron–Sulphur Proteins, Vol. I* (ed. W. Lovenberg), Academic Press, New York

Evans, M. C. W., Reeves, S. G. and Cammack, R. (1974) *FEBS Lett.*, **49**, 111–114

Fee, J. A., Findling, K. L., Yoshida, T., Hille, R., Tarr, G. E., Hearshen, D. O., Durham, W. R., Day, E. P., Kent, T. A. and Münck, E. (1984) *J. biol. Chem.*, **259**, 124–133

Forget, P. (1974) *Eur. J. Biochem.*, **42**, 325–332

Friedrich, C. G., Schneider, K. and Friedrich, B. (1982). *J. Bacteriol.*, **152**, 42–48

Fukuyama, K., Hase, T., Matsumoto, S., Tsukihara, T., Katsube, Y., Tanaka, N., Kakudo, M., Wada, K. and Matsubara, H. (1980) *Nature, Lond.*, **286**, 522

Gehring, U. and Arnon, D. I. (1972) *J. biol. Chem.*, **247**, 6963–6969

Ghosh, D., Furey, W. F., O'Donnell, S. and Stout, C. D. (1981) *J. biol. Chem.*, **256**, 4185

Gibson, J. F., Hall, D. O., Thornley, J. H. M. and Whatley, F. R. (1966) *Proc. nat. Acad. Sci., U.S.A.*, **56**, 987

Gillum, W. O., Mortenson, L. E., Chen, J.-S. and Holm, R. H. (1977) *J. Am. chem. Soc.*, **99**, 584

Graf, E.-G. and Thauer, R. K. (1981) *FEBS Lett.*, **136**, 165–169

Greenbaum, P., Prodouz, K. N. and Garrett, R. H. (1978) *Biochim. Biophys. Acta*, **526**, 52–64

Hall, D. O., Gibson, J. F. and Whatley, F. R. (1966) *Biochim. Biophys. Res. Commun.*, **23**, 81

Hall, D. O. and Rao, K. K. (1977) Ferredoxin. In: *Encyclopedia of Plant Physiology* (ed. A. Trebst and M. Avron), Vol. 5, Springer-Verlag, Berlin, Heidelberg, pp. 206–216

Hatchikian, E. C., Bruschi, M., Bonicel, J. and Couchoud, P. (1979) *Biochim. Biophys. Res. Commun.*, 86, 725-734

Hong, J.-S. and Rabinowitz, J. C. (1967) *Biochem. Biophys. Res. Commun.*, 29, 246

Hong, J-S. and Rabinowitz, J. C. (1970) *J. biol. Chem.*, 245, 6574

Howard, J. B., Lorsbach, T. and Que, L. (1976) *Biochem. Biophys. Res. Commun.*, 70, 582

Huynh, B. H., Moura, J. J. G., Moura, I., Kent, T. A., LeGall, J., Xavier, A. V. and Münck, E. (1980) *J. biol. Chem.*, 255, 3242

Itakura, I. and Holmes, E. W. (1979) *J. biol. Chem.*, 254, 333-338

Jensen, L. H. (1974) *Ann. Rev. Biochem.*, 43, 461

Johnson, M. K., Thomson, A. J., Robinson, A. R., Rao, K. K. and Hall, D. O. (1981) *Biochim. Biophys. Acta*, 667, 433

Johnson, M. K., Spiro, T. G. and Mortenson, L. E. (1982) *J. biol. Chem.*, 257, 2447

Johnson, M. K., Czernuszewicz, R. S., Spiro, T. G., Fee, J. A. and Sweeney, W. V. (1983) *J. Am. chem. Soc.*, 105, 6671-6678

Johnson, R. E., Papaefthymiou, G. C., Frankel, R. B. and Holm, R. H. (1983). *J. Amer. chem. Soc.*, 105, 7280-7287

Kent, T. A., Huynh, B. H. and Münch, E. (1980) *Proc. nat. Acad. Sci., U.S.A.*, 77, 6574

Kent, T. A., Dreyer, J.-L., Kennedy, M. C., Huynh, B. H., Emptage, M. H., Beinert, H. and Münck, E. (1982) *Proc. nat. Acad. Sci., U.S.A.*, 79, 1096

Kerscher, L., Oesterhelt, D., Cammack, R. and Hall, D. O. (1976) *Eur. J. Biochem.*, 71, 101-107

Kerscher, L. and Oesterhelt, D. (1977) *FEBS Lett.*, 83, 197-201

Kojima, N., Fox, J. A., Hausinger, R. P., Daniels, L., Orme-Johnson, W. O. and Walsh, C. (1983) *Proc. nat. Acad. Sci., U.S.A.*, 80, 378-382

Krasna, A. J. (1979) *Enzyme Microb. Technol.*, 1, 165

Kruger, H.-J., Huynh, B. H., Ljungdahl, D. O., Xavier, A. V., DerVartanian, D. V., Moura, I., Peck, H. D., Jr., Teixera, M., Moura, J. J. G. and LeGall, J. (1982) *J. biol. Chem.*, 257, 14620-14623

Kurtz, D. M., Holm, R. H., Ruzicka, F. J., Beinert, H., Coles, C. J., and Singer, J. R. (1979) *J. biol. Chem.*, 254, 4967

Kurtz, D. M., Jr., McMillan, R. S., Burgess, N. K., Mortenson, L. E. and Holm, R. H. (1979) *Proc. nat. Acad. Sci., U.S.A.*, 76, 4986

Lancaster, R. J., Jr. (1980) *FEBS Lett.*, 115, 285-288

Laskowski, E., Frankel, R. B., Gillum, W. O., Papaefthymiou, G. C., Renaud, J., Ibers, J. A. and Holm, R. H. (1978) *J. Am. chem. Soc.*, 100, 5322

Lee, J. P. and Peck, H. (1971) *Biochem. Biophys. Res. Commun.*, 45, 583-586

Legall, J., Ljungdahl, D. O., Moura, I., Peck, H. D., Jr, Xavier, A. V., Moura, J. J. G., Teixera, M., Huynh, B. H. and DerVartanian, D. V. (1982) *Biochem. Biophys. Res. Commun.*, 106, 610

Lovenberg, W. (Ed.) (1973a) *Iron-Sulfur Proteins*, Vol. I, Academic Press, New York

Lovenberg, W. (Ed.) (1973b) *Iron-Sulfur Proteins*, Vol. II, Academic Press New York

Lovenberg, W. (Ed.) (1977) *Iron-Sulfur Proteins*, Vol. III, Academic Press, New York

Lowe, D. J. and Bray, R. C. (1978) *Biochem. J.*, 169, 471-479

Lowe, D. J., Eady, R. R. and Thorneley, R. N. F. (1978) *Biochem. J.*, 173, 277

Malkin, R. and Aparicio, D. J. (1975) *Biochem. Biophys. Res. Commun.*, 63, 1157

Malkin, R. and Bearden, A. J. (1978) *Biochim. Biophys. Acta*, 505, 147-181

Malkin, R. and Rabinowitz, J. C. (1966) *Biochim. Biophys. Res. Commun.*, 23, 822

Martius, C. (1937) *Z. Physiol. Chem.*, 247, 104

Mayerle, J. J., Frankel, R. B., Holm, R. H., Ibers, J. A., Phillips, W. D. and Weiher, J. F. (1973) *Proc. nat. Acad. Sci., U.S.A.*, 70, 2429

Mayerle, J. J., Denmark, S. E., DePamphilis, B. V., Ibers, J. A. and Holm, R. H. (1975) *J. Am. chem. Soc.*, 97, 1032

Morgan, T. V., Stephens, P. J., Burgess, B. R. and Stout, C. D. (1984). *FEBS Lett.*, 167, 137-141

Morrison, J. R. (1954) *Biochem. J.*, 58, 685

Mortenson, L. E. and Nakos, G. (1973) In: *Iron-Sulphur Proteins, Vol. I* (ed. W. Lovenberg), Academic Press, New York

Moura, J. J. G., Moura, I., Huynh, B. H., Kruger, J.-H., Teixeira, M., DuVarney, R. C., DerVartanian, D. V., Xavier, A. V., Peck, H. D., Jr. and LeGall, J. (1982a) *Biochem. Biophys. Res. Commun.*, 108, 1388-1393

Moura, I., Moura, J. J. G., Huynh, B-H., Santos, H., LeGall, J. and Xavier, A. V. (1982b) *Eur. J. Biochem.*, 126, 95-98

Moura, J. J. G., Xavier, A. V., Cammack, R., Hall, D. O., Bruschi, M. and LeGall, J. (1978) *Biochem. J.*, 173, 419-425

Moura, J. J. G., Moura, I., Kent, T. A., Lipscomb, J. D., Huynh, B. H., LeGall, J., Xavier, A. V. and Münck, E. (1982c) *J. biol. Chem.*, 257, 6259

Moura, I., Moura, J. J. G., Huynh, B. H., Santos, H., LeGall, J. and Xavier, A. V. (1982) *Eur. J. Biochem.*, 126, 95

Mullinger, R. N., Cammack, R., Rao, K. K., Hall, D. O., Dickson, D. P. E., Johnson, C. E., Rush, J. D. and Simopoulos, A. (1975) *Biochem. J.*, 151, 75-83

Ogawa, K., Tsukihara, T., Tahara, H., Katsube, Y., Matsu-ura, Y., Tanaka, N., Kabudo, M., Wada, K. and Matsubara, H. (1977) *J. Biochem.*, 81, 529

Ohnishi, T. (1975) *Biochem. Biophysica Acta*, 387, 475

Ohnishi, T., Blum, H., Sato, S., Nakazawa, K., Hon-nami, K. and Oshima, J. (1980) *J. biol. Chem.*, 255, 345-348

Olson, J. S., Ballou, D. P., Palmer, G. and Massey, V. (1974) *J. biol. Chem.*, 249, 4363-4382

Orme-Johnson, W. H. and Davis, L. C. (1977) Current topics and problems in the enzymology of nitrogenase. In; Lovenberg (1977) Vol. 3, pp. 15-60

Orme-Johnson, N. R., Hansen, R. E. and Beinert, H. (1974a) *J. biol. Chem.*, 249, 1928-1939

Orme-Johnson, N. R., Hansen, R. E. and Beinert, H. (1974b) *J. biol. Chem.*, 249, 1922-1927

Orme-Johnson, W. H. and Holm, R. H. (1978) *Meth. Enzymology*, 53, 268

Palmer, G. and Sands, R. H. (1966) *J. biol. Chem.*, 241, 252

Palmer, G. (1973) In: *Iron-Sulfur Proteins*, Vol. II (ed. W. Lovenberg), Academic Press, New York

Partridge, C. D. P. and Yates, M. G. (1982). *Biochem. J.*, 204, 339

Petering, D., Fee, J. A. and Palmer, G. (1971) *J. biol. Chem.*, 246, 643

Que, L., Jr., Holm, R. H. and Mortenson, L. E. (1975) *J. Am. chem. Soc.*, 97, 463

Rajagopalan, K. V., Handler, P., Palmer, G. and Beinert, H. (1968) *J. biol. Chem.*, 243, 3784-3796

Rieske, J.-S. (1976) *Biochim. Biophys. Acta*, 456, 195-247

Rieske, J. S., MacClennan, D. H. and Collman, R. (1964) *Biochem. Biophys. Res. Commun.*, 15, 338

Ruzicka, F. J. and Beinert, H. (1978) *J. biol. Chem.*, 253, 2514

Salerno, J. C., Lim, J., King, T. E., Blum, H. and Ohnishi, T. (1979) *J. biol. Chem.*, 254, 4828-4835

Sato, S., Nakazawa, K., Hon-nami, K. and Oshima, T. (1981) *Biochim. Biophys. Acta*, 668, 277-289

Scherer, P. A. and Thauer, R. K. (1978) *Eur. J. Biochem.*, 85, 125-135

Schink, B. and Schlegel, H. G. (1979) *Biochim. Biophys. Acta*, 567, 315

Schlegel, H. G. and Schneider, K. (1976) *Biochim. Biophys. Acta*, 452, 66

Schönheit, P., Moll, J. and Thanel, R. K. (1979) *Arch. Microbiol.*, 123, 105

Siegel, L. M. (1978) Structure and function of siroheme and the siroheme enzymes. In: *Mechanisms of Oxidizing Enzymes* (ed. Singer, T. P. and Ondarza, R. N.) Elsevier, Amsterdam, pp. 201-214

Siegel, L. M., Davis, D. S. and Kamin, H. (1974) *J. Biol. Chem.*, 249, 1572

Spiro, T. G. (Ed.) (1982) *Iron-Sulfur Proteins*, Vol. 4, Metal Ions in Biology series, Wiley, New York

Stephens, P. J., Thomson, A. J., Dunn, J. B. R., Keiderling, T. A., Rawlings, J., Rao, K. K. and Hall, D. O. (1978) *Biochemistry*, 17, 4770

Stombaugh, N. A., Burris, R. H. and Orme-Johnson, W. H. (1973) *J. biol. Chem.*, 248, 7951-7956

Stout, C. D. (1979) *Nature, Lond.*, 279, 83

Stout, C. D., Ghosh, D., Pattabhi, V. and Robbins, A. H. (1980) *J. biol. Chem.*, 255, 1797

Sweeney, W. V., Bearden, A. J. and Rabinowitz, J. C. (1974) *Biochem. Biophys. Res. Commun.*, 59, 188

A. J. Thomson

Sweeney, W. V., Rabinowitz, J. C. and Yoch, D. C. (1975) *J. biol. Chem.*, **250**, 7842
Takakuwa, S. and Wall, J. D. (1981) *FEMS Microbiol. Lett.*, **12**, 359
Thomson, A. J., Cammack, R., Hall, D. O., Rao, K. K., Briat, B., Rivoal, J.-C. and Badoz, J. (1977) *Biochim. Biophys. Acta*, **493**, 132
Thomson, A. J., Robinson, A. E., Johnson, M. K., Moura, J. J. G., Moura, I., Xavier, A. V. and LeGall, J. (1981a) *Biochim. Biophys. Acta*, **630**, 93
Thomson, A. J., Robinson, A. E., Johnson, M. K., Cammack, R., Rao, K. K. and Hall, D. O. (1981b) *Biochim. Biophys. Acta*, **637**, 423
Thomson, A. J. (1981) *J. chem. Soc.*, Dalton Trans, 1180
Trumpower, B. L. and Edwards, C. A. (1979) *FEBS Lett.*, **100**, 13–16
Tsai, R. L., Gunsalus, I. C. and Dus, K. (1971) *Biochem. Biophys. Res. Commun.*, **45**, 1300–1306
Tsibris, J. C. M., Namtvedt, M. J. and Gunsalus, I. C. (1968) *Biochem. Biophys. Res. Commun.*, **30**, 323
Tsukihara, T., Fukuyama, K., Tahara, H., Katsube, Y., Matsu-ura, Y., Taneka, N., Kakudo, M., Wada, K. and Matsubara, H. (1978) *J. Biochem.*, **84**, 1645–1647
Unden, G., Bocker, R., Knecht, J. and Kruger, A. (1982) *FEBS Lett.*, **145**, 230–234
Uyeda, K. and Rabinowitz, J. C. (1971) *J. biol. Chem.*, **246**, 3111–3119
Van der Westen, H. M., Mayhew, S. G. and Veeger, C. (1980) *FEMS Microbiol Lett.*, **7**, 35
Van Dijk, C., Mayhew, S. G., Grande, H. J. and Veeger, C. (1980) *Eur. J. Biochem.*, **102**, 317
Van't Riet, J., Van El, J. H., Wever, R., Van Gelder, B. F. and Planta, R. J. (1975) *Biochim. Biophys. Acta*, **405**, 306–317
Vincent, S. P. and Bray, R. C. (1978) *Biochem. J.*, **171**, 639–647
Vollmer, S. J., Switzer, R. L. and Debrunner, P. G. (1983). *J. Biol. Chem.*, **258**, 14284–14293
Wang, G. B., Kurtz, D. M. Jr., Holm, R. H., Mortenson, L. E. and Upchurch, R. G. (1979) *J. Am. chem. Soc.*, **101**, 3078
Yang, C. Y., Johnson, K. H., Holm, R. H. and Norman, J. G., Jr. (1975) *J. Am. chem. Soc.*, **97**, 6596
Yoch, D. C. and Arman, D. I. (1972). *J. biol. Chem.*, **247**, 4514
Zumft, W. G., Mortenson, L. E. and Palmer, G. (1974) *Eur. J. Biochem.*, **46**, 525–535

4

Superoxide Dismutases

A. E. G. Cass

1 INTRODUCTION

1.1 Aerobiosis

An aerobic lifestyle offers substantial bioenergetic advantages to those organisms able to use dioxygen as a terminal electron acceptor. These advantages can be readily appreciated both from a comparison of the thermodynamics of dioxygen with other terminal oxidants (table 4.1) and from the commonplace observation that the majority of present day species are aerobes.

Unfortunately, as with many desirable situations there is a concomitant hazard and this is the toxicity of dioxygen. Complete reduction of O_2 to H_2O accounts for approximately 90% of the total dioxygen consumed; the remainder is reduced to the superoxide ion, hydrogen peroxide, and the hydroxyl radical.[1] It is these partially reduced oxygen species that are suspected of mediating the toxicity of dioxygen. Biologically the response to the unavoidable generation of these intermediate reduction products has been to elaborate enzymes that remove them from the cellular environment.

Catalase was one of the first enzymes recognised as such and it disproportionates hydrogen peroxide:

$$2H_2O_2 \rightarrow 2H_2O + O_2 \tag{1}$$

Superoxide dismutase similarly disproportionates the superoxide ion:

$$2O_2^- \xrightarrow{2H^+} H_2O_2 + O_2 \tag{2}$$

The tremendous reactivity of the hydroxyl radical precludes any specific mechanism for its removal. Instead the strategy for handling this radical appears to be two-fold; firstly a likely source of the hydroxyl radical is a catalysed reaction of the superoxide ion and hydrogen peroxide:

$$O_2^- + H_2O_2 + H^+ \rightarrow OH^\bullet + H_2O + O_2 \tag{3}$$

121

Table 4.1   Redox potentials of some commonly occurring terminal electron acceptors (see Ref. 144)

Couple	E'_0
O_2/H_2O	+0.820 V
NO_3^-/NO_2^-	+0.420 V
SO_4^{2-}/S^{2-}	−0.201 V

and so by keeping the concentrations of the reactants low reaction (3) is suppressed. Secondly the presence of relatively high concentrations of good reductants in the cell, coupled with re-reduction pathways, aids in intercepting the hydroxyl radical before a critical cellular target is hit. The classical antioxidants such as ascorbate and glutathione fall into this class.

1.2   The superoxide ion

One electron reduction of dioxygen yields the superoxide radical anion; the chemical and biochemical properties of this molecule have been repeatedly reviewed in the past[1-6] and for our purposes a brief summary of the pertinent properties of the ion will be provided here and the above references should be consulted for further information.

Figure 4.1 is an oxidation state diagram for oxygen and illustrates most of the basic data on the redox properties of the superoxide ion. In particular it can be seen that O_2^- is a fairly strong reductant (E'_0 O_2^-/O_2 = −0.33 V) and a moderate oxidant (E'_0 H_2O_2/O_2^- = 0.87 V). More importantly any reaction that involves the reduction of the superoxide ion will of necessity also involve protons and thus the thermodynamics and kinetics of the reaction will be dependent upon pH. This is particularly so for the dismutation reaction, and the kinetic studies of Bielski and Allen[7] using pulse radiolysis showed the expected pH response and the second-order rate constant for the reaction is calculated to be 4×10^5 M^{-1} s^{-1}.

Such a high rate of spontaneous dismutation coupled with the low reactivity of the superoxide ion with many cellular components *in vitro* has led some to suggest that the superoxide ion is not an important intermediate in dioxygen toxicity and thus that the enzymes known as superoxide dismutases do not have the catalysis of reaction (2) as their physiological function. Both sides of this debate have been vigorously pursued in the literature[8-12] and the present author will not recap the arguments, *pro* and *con*, here. Instead this review will concentrate on the chemical and physical enzymology of those proteins known as superoxide dismutases.[13]

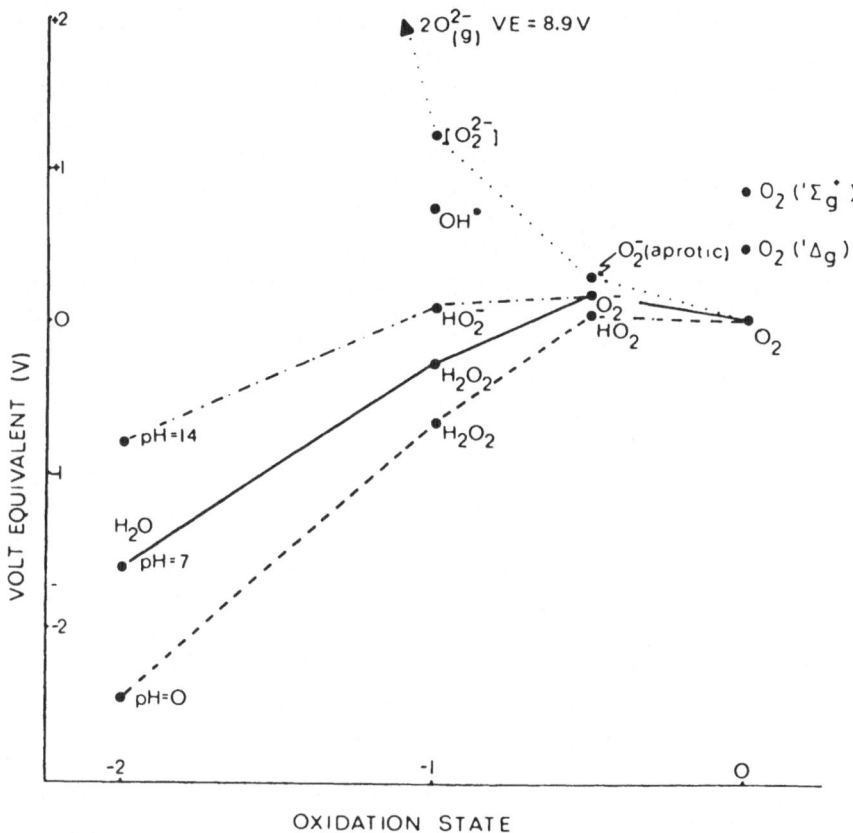

Figure 4.1 An oxidation state diagram for dioxygen. The volt equivalent is the product of the standard redox potential (E_0) and the oxidation state. This means that the gradient of the line joining any two species is equal to the standard redox potential

2 A FIRST LOOK AT THE ENZYMES

2.1 Metal content

Superoxide dismutases are metalloenzymes and contain either copper and zinc, or manganese or iron at the active site. The cuprozinc proteins had been isolated and characterised long before their enzymatic activity was discovered.[14] They are present in high concentrations in the erythrocytes, liver and brain of higher organisms and in the absence of any discernible catalytic function they were relegated to storage proteins and generically labelled cupreins.[15] Recognition of their enzyme status came by a tortuous and indirect route as has been elegantly described by their discoverers.[16] Once an enzymatic activity had been recognised it became possible to assay and isolate the enzyme from a wide variety of eukaryotic sources; and all of the cuprozinc superoxide dismutases described to date appear to share very similar properties, these are collected in table 4.2.

When the enzymes from prokaryotic sources were isolated a quite different pattern emerged. These enzymes do not contain copper and zinc but instead contain manganese[17] or iron[18] as their catalytic centres. Furthermore none of the other properties appear to be at all similar as can be seen from the comparisons of table 4.2. However there are further ramifications to this tale, for although the first superoxide dismutases isolated from eukaryotes were cuprozinc proteins the enzyme activity from mitochondrial rather than cytoplasmic sources was found to be due to a mangano-enzyme.[19,20] Mitochondrial superoxide dismutase thus has more similarities to the prokaryotic enzyme than to the cytoplasmic enzyme from the same organism.

At this stage there appears to be a relatively clear cut distinction between the cuprozinc enzymes from eukaryotic cytoplasm, the mangano-enzymes from bacteria and mitochondria and the iron enzymes from bacteria. A question now arises as to whether there is any distinction between the situations in which the iron and manganese superoxide dismutases are found in prokaryotes; the answer depends upon both the organism and the environmental conditions.

In the case of *Escherichia coli*, a facultative anaerobe, which produces both an iron and a manganese superoxide dismutase the former is constitutive whilst the latter is induced in the presence of dioxygen.[21] In addition a hybrid enzyme is also produced during aerobic growth. This latter enzyme appears to contain a sub-unit of the manganese and a sub-unit of the iron proteins; the metal present is iron.[22] In this particular organism its natural habitat can experience large fluctuations in oxygen tension and the synthesis of the different metal forms is triggered accordingly.

Table 4.2   Some properties of superoxide dismutases

Source	Molecular weight (Number of Sub-units)	Metal content	E'_0
Bovine erythrocytes	31 200 (2)*	2Cu, 2Zn	+0.28 V
Yeast cytoplasm	31 000 (2)	2Cu, 2Zn	n.d.
Yeast mitochondria	96 000 (4)	4Mn	n.d.
E. coli	40 000 (2)	2Mn	+0.31 V
E. coli	40 000 (2)	2Fe	+0.25 V
B. stearothermophilus	45 000 (2)	1Mn	+0.26 V
P. leiognathi	40 000 (2)	1Fe	n.d.
P. leiognathi	32 000 (2)	1Cu, 2Zn	n.d.

n.d. = not determined.
*Numbers in parentheses represent sub-unit numbers.

As a contrast to this the aerotolerant anaerobe *Propionibacterium shermanii* synthesises its superoxide dismutase in a form that is dependent on the availability of the metal ions (Mn or Fe) in the growth medium. When the environment is replete with iron then an iron containing superoxide dismutase is synthesised, whilst in an iron free medium the organism elaborates a manganese containing enzyme. Although the stoicheiometry of the two forms is different (1.8 Mn and 1 Fe per sub-unit) the molecular weights, amino acid compositions and N-terminal sequences are identical. This is quite unlike the *E. coli* enzymes where the two forms although similar have distinct properties. Furthermore the *P. shermanii* enzymes when converted to the apo (metal free) state are unique in being reconstitutable by either manganese or iron to yield an active enzyme irrespective of the metal ion initially present.[23]

A somewhat different example of the nutritional response of an organism is superoxide dismutase production by the fungus *Dactylium dendroides*. This is a eukaryote and so produces a cuprozinc cytoplasmic enzyme and a mitochondrial manganese one. During growth in a low copper medium the synthesis of the cytoplasmic form is suppressed and there is a corresponding increase in the synthesis of the manganese enzyme.[24]

2.2 Amino acid sequences and species comparisons

The metal ion content of the superoxide dismutases appears to reveal a relatively well defined phylogenetic pattern and it is of interest to determine whether this pattern also arises at the level of the polypeptide chain.

Four eukaryotic cuprozinc superoxide dismutases have had their primary structures determined; bovine[25] and human erythrocyte,[26,27] equine liver[28] and yeast.[29,30] These four sequences clearly reveal that a high degree of similarity exists between the four enzymes, the yeast and bovine superoxide dismutases have a 55% homology, a value comparable to the cytochrome *c*'s from the same species. A high degree of conservation in the primary structure places the cuprozinc superoxide dismutasers amongst the most slowly evolving proteins known.

Amino acid sequence data, although important in providing a point by point comparison of the primary structures, are not the only way of studying the relationship between the enzymes. A complementary technique is to use high resolution nmr. The nmr method lacks the detail of sequence determination, but it is rapid, convenient and provides structural information in three dimensions rather than the single dimension of a linear sequence. In this respect comparative ^{1}H nmr studies of bovine, human, swordfish and yeast cuprozinc superoxide dismutases have concentrated on the active site homologies.[31]

Sequences for three of these enzymes are also known (*vide supra*) and the ligand residues are all conserved. Figure 4.2 shows the ^{1}H nmr spectra of the ligand resonances of all four proteins and illustrates the close similarity of the enzymes; as the chemical shift values are sensitive to the environment of a

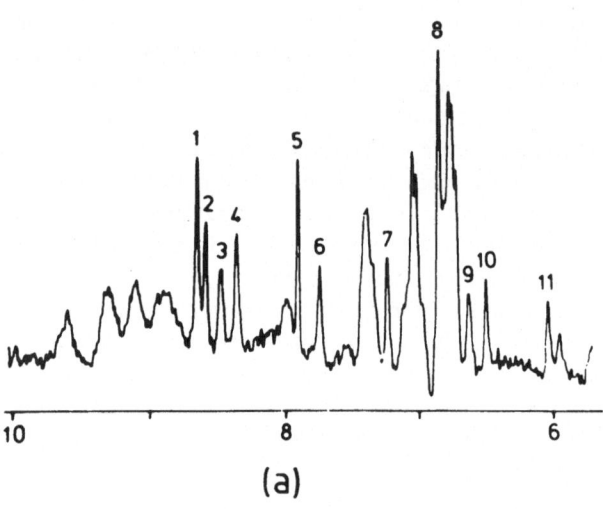

(a)

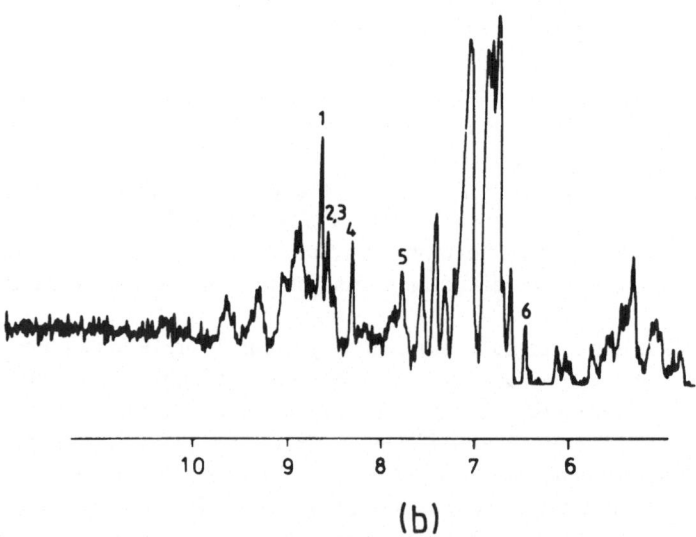

(b)

Figure 4.2 ^{1}H nmr spectra of the reduced form of copper–zinc superoxide dismutases from (a) bovine erythrocytes, (b) human erythrocytes, (c) swordfish liver and (d) yeast. The numbering refers to the assigned histidine resonances

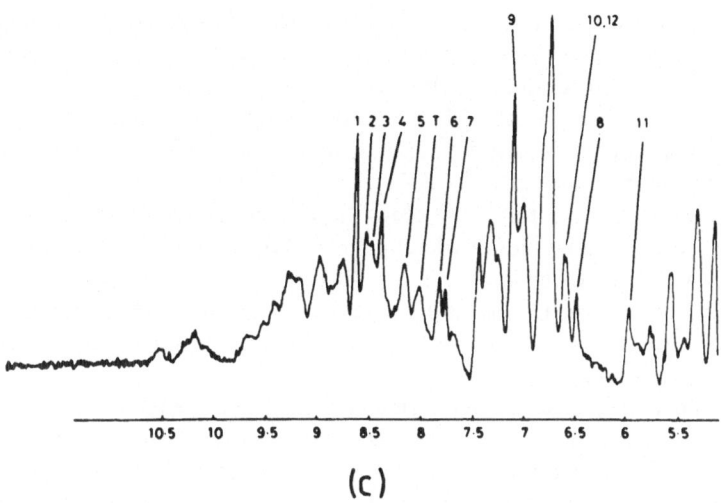

(c)

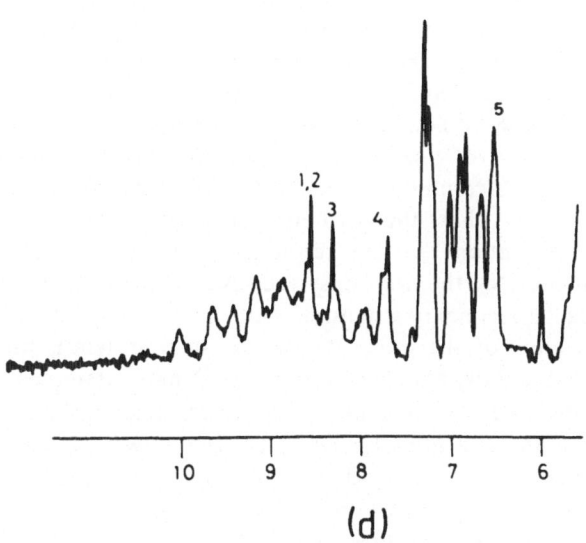

(d)

Figure 4.2 (continued)

particular nucleus then the active site structures must be very similar. A further interesting comparison can be made for the bovine and human enzymes, both have a histidine residue at position 41 that is not part of the active site. In the bovine enzyme it was shown that the C-2 proton of this residue exchanged very rapidly with solvent deuterons[32] and that the rapid exchange was due to the environment of this residue. The human enzyme not only has the same chemical shift for this proton but also shows a comparable rate of exchange, implying that the environment around histidine 41 that accounts for the rapid exchange is conserved in both proteins.[33]

Nmr studies of structural homology have been extended by comparing the spectra of the two isozymes of wheat germ cuprozinc superoxide dismutase with the spectra of the bovine liver enzyme. In this work the attention was focussed on the protons bound to the imidazole nitrogen atoms of the metal ligands and here again a close correspondence for the chemical shift values for the enzymes from different sources was found.[34]

Nmr and amino acid sequence analyses therefore concur in assigning a high degree of conservation to the structures of the cuprozinc superoxide dismutases.

Amongst the prokaryotic superoxide dismutases for which sequences are available no similarity to the eukaryotic cuprozinc enzymes can be discerned. The prokaryotic enzymes do show considerable homology amongst themselves, and a number of N-terminal sequences have been determined.[35] These, although only providing limited data, show that the manganese and iron enzymes are similar not only within a particular metal ion containing group but also between the two groups. Furthermore there appears to be considerable homology between the manganese enzymes from prokaryotes and those from mitochondria. Confirmation of the sequence homologies is provided by the three published complete sequences; those of the enzymes from *Bacillus stearothermophilus*,[36] *E. coli*[37] and *S. cerevisiae* mitochondria.[38] The bacterial enzymes have approximately 60% homology whilst the yeast enzyme is about 40% homologous to the bacterial ones; predictions of the secondary structures reveal considerable similarities with most of the non-conserved residues occurring in regions other than those of regular secondary structure.

The determination of the primary structure of a protein is still a relatively laborious task and recently Martin and Fridovich[39] have attempted to use amino acid composition data to elucidate evolutionary relationships amongst the superoxide dismutases and in particular to try and answer the curious question of bacteriocuprein. As discussed above the cuprozinc superoxide dismutases appear to occur solely in eukaryotes; however when the enzymes from the luminescent bacterium *Photobacterium leiognathi* were isolated in addition to a typical bacterial iron containing enzyme[40] a cuprozinc superoxide dismutase was also found to be present.[41] This protein was called bacteriocuprein and appeared to share many of the properties of the eukaryotic variants (table 4.2), a second example has since been isolated.[42] It was suggested that the similarity of the *Ph. leiognathi* enzyme to the eukaryotic form may have arisen as the result of a natural gene transfer from the pony fish, on which the bacterium is symbiotic.

To test this hypothesis Martin and Fridovich isolated superoxide dismutases from a number of teleost fishes and determined their amino acid compositions. They then applied a form of discriminant analysis to these data plus all of the published amino acid compositions of cuprozinc, manganese and iron enzymes and were able to demonstrate that these three types of enzyme distinguished by metal ion content also fell into three corresponding groups on the basis of amino acid composition (figure 4.3a). Iron superoxide dismutase from *Ph. leiognathi* was in the iron containing group whilst bacteriocuprein was in the cuprozinc group. The cuprozinc enzymes could be further subdivided in three classes comprising the enzymes from (i) fish, (ii) plants and fungi and (iii) birds and mammals (figure 4.3b); bacteriocuprein was included in the first group and this was argued to support the concept of a natural gene transfer.

A second example of a bacteriocuprein has recently been isolated, from *Caulobacter crescentus CB15*;[42] in contrast to *Ph. leiognathi* this organism is free living so the acquisition of a cuprozinc superoxide dismutase must have occurred by a mechanism other than gene transfer. Like the *Photobacterium* superoxide dismutase, that from *Caulobacter* possesses many similarities to the eukaryotic varieties, however compositional analysis of the trytic peptides suggests that the *Caulobacter* enzyme contains amino acid substitutions at sites strictly conserved amongst the eukaryotic enzymes. Obviously a true appreciation of the relationship between bacteriocupreins and eukaryotic superoxide dismutases will have to await complete amino acid and/or DNA sequences for representatives of both groups.

2.3 Three dimensional structures

Several of the superoxide dismutases have been the subject of x-ray crystallographic analysis but the first and still the best resolved structure is that of the bovine erythrocyte enzyme. The first structure was produced in 1974[43] and was at a resolution of 2.8 Å; recently an improved structure has been published and the subsequent discussion in this chapter is based on the 1982 model.[44] Refinement of the structure to 2.0 Å has confirmed the presence of the salient features seen in the earlier electron density map and has revealed a number of other details.

The overall molecular architecture of the sub-unit of bovine superoxide dismutase is an eight stranded β-barrel (figure 4.4). Contacts between the sub-units are extensive and there is an approximately two-fold crystallographic axis between the sub-units with some 9% of each sub-unit involved in dimer formation; the forces holding the two halves of the molecule together involve both hydrophobic and hydrophilic contributions. In addition to the regular β-sheet secondary structure there are 18 tight turns in each sub-unit as well as three long loops of non-regular secondary structure that connect strands 4 and 7, 7 and 8, and 6 and 5. A further structural feature, noted in the earlier crystallographic

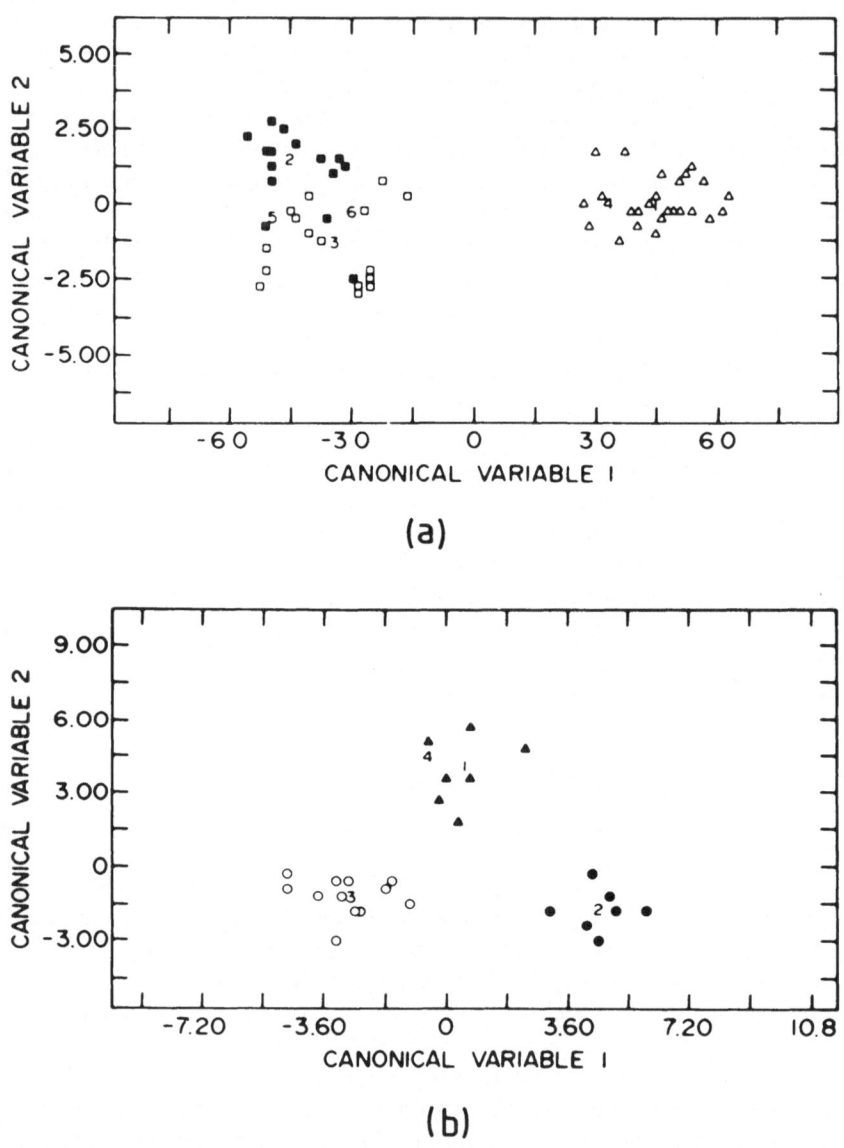

Figure 4.3(a) Discriminant analysis of the amino acid composition data for all superoxide dismutases. ΔCu/Zn enzymes; ■Fe enzymes; □Mn enzymes. (b) Shows the same analysis applied to the Cu/Zn group and reveals a further subdivision into ▲ fish, ■ plant and fungi and 0 mammals and birds. The details of the methodology and the nature of the Canonical Variables are described in ref. 39. (Reproduced from ref. 39 with permission)

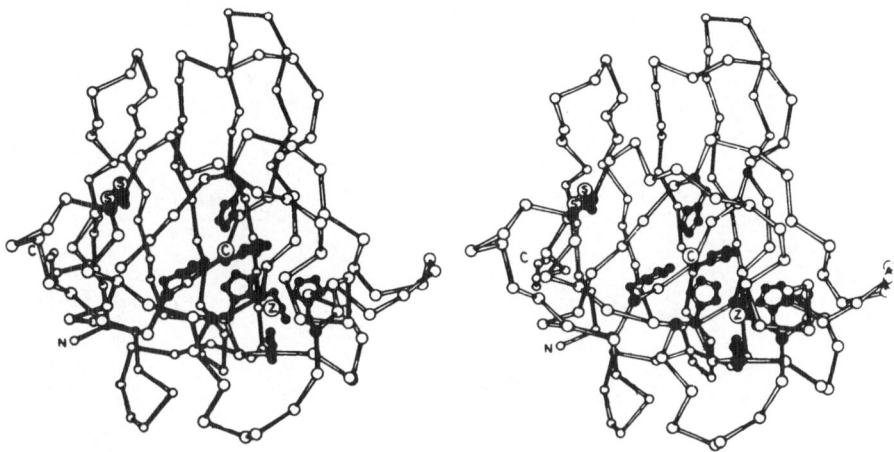

Figure 4.4 Stereo diagram of the backbone fold of bovine cuprozinc superoxide dismutase at 1.8 Å resolution. The α-carbon atoms are open circles and the metal ligand and cysteine side chain carbon atoms are the filled circles. C = Copper Atom, Z = Zinc Atom, S – S = Disulphide Bond. (Reproduced from ref. 44 with permission)

analysis, is the presence of an approximate, non-crystallographic two-fold axis passing through each sub-unit and 2.1 Å from the copper ion.[45] Although this axis divides the sub-unit into two related domains there is no similarity in amino acid sequence between the two subdomains. It was speculated that this feature reflected an early gene duplication.

The geometries of the two metal ions can be seen from the latest refinement to be a tetrahedrally distorted square plane for the copper ion and a trigonally distorted tetrahedron for the zinc ion. The two metal ions are separated by 6.3 Å with the imidazole ring of histidine 61 interposed between and coplanar with them; this residue, coordinated to both metals in the oxidised enzyme is therefore present as an imidazolate ion. A detailed view of the metal ion binding region is shown in figure 4.5 and is further discussed in section 3.2.

One of the most striking features of the structure of the cuprozinc superoxide dismutases is that they are dimeric molecules. Early work showed that the two sub-units were held together non-covalently, although the interaction was sufficiently strong to require reduction of the intrasub-unit disulphide bond and subsequent unfolding of the individual sub-units to separate them.[46] Often the presence of multiple sub-units in a protein is related to control of activity via some interaction between the active sites. A number of studies have appeared to try and elucidate the significance of the sub-unit structure of the enzyme; some of these relate to metal ion incorporation and will be discussed later in section 3.1.

A particularly important aspect of the sub-unit structure is whether the isolated monomers possess any activity and in order to try and answer this question enzymes have been sought in which the sub-unit interaction is weakened

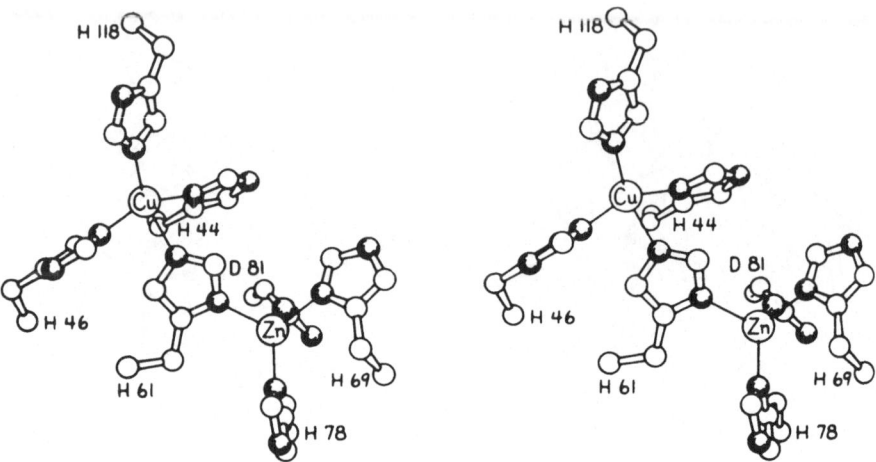

Figure 4.5 Stereo diagram of the metal binding site in bovine cuprozinc superoxide dismutase at 1.8 A resolution. (Reproduced from ref. 44 with permission)

such that monomers may be produced without the need for harsh and hence inactivating conditions. Isozyme I from wheat germ is readily dissociated into monomers by detergent treatment and under these conditions has a residual activity of 0.2%.[47] Removal of the detergent results in a biphasic recovery of activity and reassociation of the monomers. 8M urea has been reported to dissociate the enzyme from swordfish liver[48] and from bovine erythrocytes and yeast[49] into fully active monomers although in this case it is possible that the high urea concentration caused the dimers to compact, rather than dissociate, and thus give an apparently lower molecular weight on gel filtration columns. Later sedimentation equilibrium analysis under the same conditions were consistent with the enzymes still being present as dimers.[50] Hybrid proteins between sub-units from different species[49] or native and chemically modified inactive sub-units can be produced in the presence of high urea concentrations suggesting some labilisation of the dimer structure. The enzymes formed with one active and one inactive sub-unit have half of the activity of the native form consistent with each half acting independently.[50]

Extensive succinylation of the lysine residues of the bovine enzyme substantially weaken the sub-unit interactions, although the activity is also much reduced (see section 4.3).[51] Bovine superoxide dismutase has also been coupled to an insoluble support and subsequently succinylated whereupon approximately half the protein is released into solution. It is unfortunate that these authors did not determine the activity of the residual, immobilised and presumably monomeric enzyme remaining on the support as this would have indicated whether the monomers were active.

In conclusion it is apparent that the individual active sites in the dimer are non-interacting and have identical catalytic properties. It is less certain, though likely, that the isolated monomers possess at least some activity.

Structural information on the iron and manganese containing superoxide dismutases is much more meagre; although crystallisation of the enzyme from *B. stearothermophilus* was reported some years ago[52] no further information has appeared. More recently crystals of the manganese containing enzyme from the extreme thermophile *Thermus thermophilus* have been grown and are reported to diffract to 1.4 Å.[53] The space group and unit cell dimensions have been determined and the crystals contain one molecule per asymmetric unit; interestingly they are reported to be very resistant to radiation damage.

Crystal structures for iron superoxide dismutases from *E. coli* and *Pseudomonas ovalis*[54] have been reported and the two molecules appear to be rather similar. Although the resolution is not high enough to identify the metal ligands, especially in the absence of a complete sequence, the fold of the peptide backbone can be discerned. As might be expected from the other molecular properties there is no resemblance between these structures and that of the bovine enzyme.

3 THE CATALYTIC CENTRES

3.1 Reconstitution and metal substitution

All of the different metal ion containing superoxide dismutases can be converted to their metal free or apo forms and subsequently reconstituted with either the native, or other metal ions. Often much useful information may be obtained from such metal ion substitutions although one must always bear in mind that the properties of the substituted protein may not necessarily reflect those of the native protein, or even in extreme cases that the metal ions may not occupy the same binding site. Metal replacement studies in the manganese and iron enzymes are relatively straightforward in principle, although not necessarily so experimentally. In the cuprozinc dismutases with the two metal binding sites a large number of permutations are possible.

Preparation of the apo-protein generally involves treatment of the holoenzyme with a metal ion chelator under partially or totally denaturing conditions followed by transfer to a simple buffer. The actual conditions used are almost as numerous as the number of apo-proteins prepared.

The manganese superoxide dismutase from *E. coli* was the first of the bacterial enzymes to be isolated but owing to the lability of the apo-enzyme and the tight binding of the metal proved to be difficult to reconstitute reversibly. Ose and Fridovich[55] prepared the apo-enzyme but found that it was only stable in the presence of the denaturant and chelator used in its preparation. However, under these conditions it could be reversibly reconstituted with manganese, with subsequent recovery of function or with cobalt, nickel or zinc but without regaining enzyme activity.

A stable apo-manganese superoxide dismutase was prepared from the thermophile *B. stearothermophilus*[56] and this too could be restored to full activity with manganese, as well as forming inactive derivatives with several other metal ions.

Iron superoxide dismutase from *Ps. ovalis* can also be reversibly reconstituted either with the native metal ion or with cadmium, chromium or iron.[57] Again only the native metal will restore enzymatic activity. The relationship between the iron and manganese enzymes is interesting, they appear to have many features in common and will bind the complementary metal though not with restoration of function. This is more surprising when we consider that *E. coli* has a hybrid enzyme (section 2.1), obviously the active sites are similar enough to bind both metal ions but are very specific in expressing their activity. The only example of functional equivalence of the two metals is in the enzyme from *P. shermanii* where the site seems more catholic, for nutritional reasons, and pays a high price in lowered specific activity.[23]

The cuprozinc superoxide dismutases can be converted to the apo-protein by treatment with EDTA at pH 3.8 though care must be taken to remove all of the bound EDTA otherwise artefacts may appear in the subsequent reconstitution.[58] Recently a kinetic study of metal ion removal from the yeast enzyme has been reported, the chelators used were 1,10-phenanthroline, citrate and dipicolinic acid.[59] These chelating agents do not behave identically; whereas the former one removes both metal ions together the latter two produce a differential loss of copper and zinc. Total removal of the metals occurred only at pH values less than 3.5 and was biphasic with the fast phase corresponding to the loss of 50% of both the copper and the activity. The authors suggest that the low pH alone results in the removal of the metals from one sub-unit whilst chelating agents are required to remove the metal ion from the second sub-unit. This work is consistent with the observations of Valentine and coworkers[60] who found selective loss of zinc when the native enzyme was dialysed against phosphate buffer at pH 5 as phosphate is an effective coordinating ion for zinc.

Reconstitution of cuprozinc superoxide dismutase can involve replacement of either metal ion or both with a variety of others and the derivatives so formed are shown in table 4.3. The most recent studies of metal ion binding have focussed on determining the binding constants for copper under a variety of

Table 4.3   Metal derivatives of bovine cuprozinc superoxide dismutase

Copper site	Zinc site					
	Cu	Zn	Co	Cd	Hg	E
Cu	68	137	58	138	138	138, 60
Zn	–	66, 139	–	–	–	–
Co	–	–	140	–	–	–
Ag	141	–	141	–	–	–
E	–	66, 139	142	–	143	–

E = metal site unoccupied.
– = derivative not prepared.

conditions. Work by Valentine *et al.*[61] showed that in the copper-only enzyme at low pH each metal ion occupied the native copper site of individual sub-units. As the pH was raised the copper ions migrated to produce apo sub-units and sub-units with copper in both its native site and in the adjacent zinc site forming a binuclear complex. This behaviour has since been quantified by equilibrium dialysis measurements, and at pH 4 the dissociation constants for copper in the copper site and in the zinc site are 1 nM and approximately 1 mM respectively.[62] These studies were subsequently extended over a range of pH values and the apparent dissociation constants as a function of pH for copper in the two sites do approach one another at alkaline values as shown in figure 4.6.[63]

In a semiquantitative study of copper binding to the zinc-only protein or copper removal from the holo-protein it was found that in analysing the distribution of products the latter was a simple statistical process whilst the former

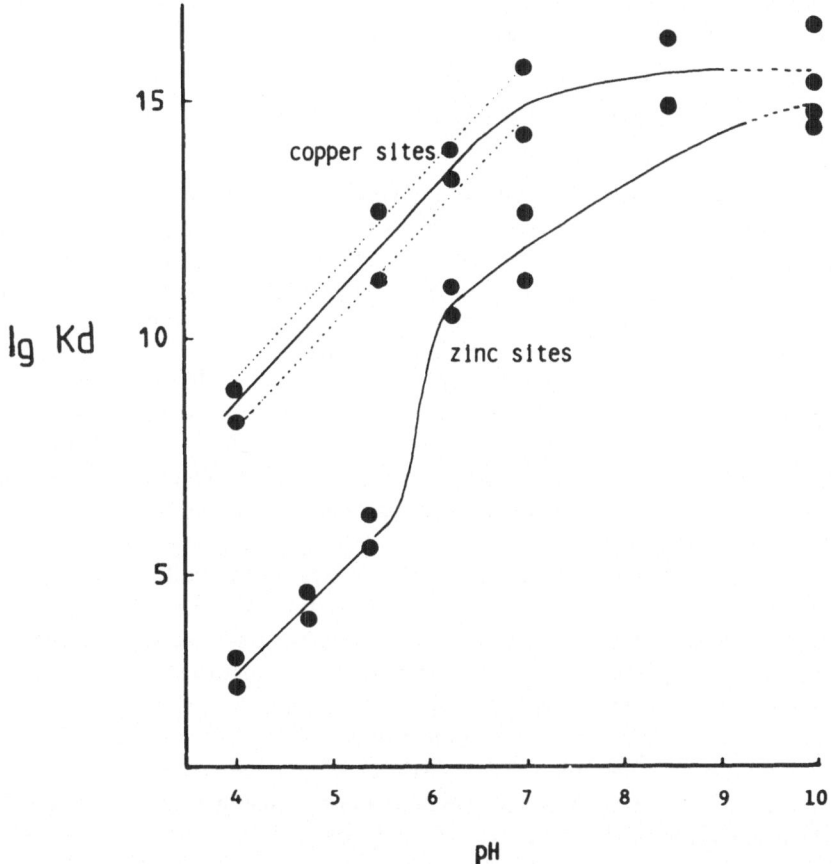

Figure 4.6 Dependence of the binding constant for copper to bovine superoxide dismutase as a function of pH. (Reproduced from ref. 63 with permission)

was not.[64] Copper binding to the zinc-only protein showed a preferential incorporation into molecules already containing a copper ion, that is binding of one copper ion lowers the activation energy for the binding of the second copper ion. A second paper studied the activities of partially reconstituted samples and the authors concluded that the molecules with only one copper ion had twice the specific activity of the two copper species, consistent with an anti-cooperative interaction between the sites.[65]

Zinc binding to the apo-protein has been less well characterised owing to the lack of suitable spectroscopic or enzyme activity markers for incorporation. A proton nmr study showed that binding of zinc to its native site was at least an order of magnitude tighter than binding to the copper site and that it preformed the latter. This paper pointed out that the apparent inconsistency between the nmr data and previous circular dichroism studies was due to the fact that the latter technique measures the regular secondary structure of the sheet which is the same in both holo- and apo-proteins. Apparently metal ion binding only causes structural changes in the active site loop.[66]

3.2 Structure and spectra of the active sites

When the structures of the metal binding sites in proteins are discussed two distinct but not exclusive meanings are involved. One is geometric and refers to the dispositions of the atoms in space; this geometric description is obtained from fitting the x-ray diffraction data to an electron density map. A second use of the term 'active site structure' refers to the electronic states of the metal ion that are deduced from various spectroscopic techniques, from these measurements the natures of the metal–ligand bonds are determined. It is implicit in deriving a geometric description from an electronic one that the relationships found in small molecules between electronic and structural parameters can also be applied to macromolecules. In the discussion that follows we shall first describe the structure of the active site as revealed from the electron density map and then consider how the various spectroscopic data can be interpreted in the light of this structure.

Figure 4.5 illustrates the nature of the binuclear metal binding site of bovine cuprozinc superoxide dismutase at 2 Å resolution. As briefly mentioned in section 2.3 the overall nature of the region may be described as a tetragonally distorted square planar copper site and a trigonally distorted tetrahedral zinc site with the two metal ions 6.3 Å apart. The imidazole ring of histidine 61 is between and coplanar with the two metal ions and thus appears to be bridging them as an imidazolate anion. These features were apparent in the earlier 2.8 Å structure, however what has been revealed upon refinement is how closely the metal ligands interact with the rest of the protein.[44] Orientation of both main chains and side chains of the ligands appear to be stabilised by hydrogen bonding; all but two of the possible hydrogen bonding atoms in the seven ligands form links to other parts of the protein. A further connection between the copper and zinc sites is provided by the carboxylate group of aspartate 122 which hydrogen bonds to both histidine 44 (a copper ligand) and histidine 69 (a zinc ligand). The whole metal binding region in this protein therefore seems to be a highly organised structure.

A water molecule is bound to the copper ion as evidenced by a peak in the electron density map approximately 3 Å from the copper in an axial direction. More water molecules are present in a hydrogen bonded network in a channel leading from the surface of the molecule to the copper ion. In contrast the zinc ion is inaccessible to the solvent.

The spectroscopic data available can be largely reconciled with the structures of the copper and zinc sites described above. The earliest epr data were interpreted in terms of an axial copper(II) ion with some rhombic distortion[67] and the observations of magnetic coupling between the native copper(II) ion and a copper(II)[68] or cobalt(II)[58,69,70] ion substituted for zinc suggested that the two sites were closely linked. A detailed analysis of the single crystal epr spectra has recently been published[71] and computer simulations have yielded the directions of the g and A tensors relative to the crystal axes and hence the copper ligands. As suggested by the powder spectra the g_z direction is nearly perpendicular to the plane of the copper ligands. Further information on the magnetic properties of the copper(II) ion is provided by electron spin echo (ese) and electron-nuclear double resonance (endor) spectroscopies. Both of these variations on the epr experiment are sensitive to the chemical nature of the ligand field and provide complementary information; ese detects nuclei that are weakly coupled to the paramagnetic centre whilst endor detects strongly coupled nuclei.

In ese spectroscopy it is the non-coordinated imidazole nitrogen nuclei of the histidine residues that modulate the decay of a pulsed epr signal. The spectra of the holo, cadmium substituted and zinc free forms of the enzyme were obtained and although no detailed analysis was presented there were clear differences beween spectra in which the zinc site was occupied and those in which it was empty. The former were considerably more complex and this was suggested to be consistent with a bridging imidazolate between the two metal ions.[72] Endor spectroscopy shows the coupling to the coordinated nitrogen atoms and here again the spectrum of the native enzyme was not easily interpreted. Analysis of the spectra suggested that the copper was bound by inequivalent nitrogen atoms and considerable changes in the hyperfine interactions were seen when anions were added (section 4.4).[73]

Further spectroscopic evidence for a bridging imidazolate comes from x-ray absorption spectroscopy of the oxidised and reduced forms.[74] This method also reveals no change in the zinc site whilst the copper changes appreciably.

One set of spectroscopic results that seems, as yet, to conflict with idea of a bridging imidazolate between the two metal ions is that obtained from perturbed angular correlation (pac) spectroscopy on the yeast enzyme substituted with cadmium.[75] Briefly this technique measures the angular distribution of γ-rays emitted from a metal ion incorporated into the protein, in this case the metal is cadmium in the zinc site. Pac spectra are sensitive to motions on the timescale of μs and to the charges on the ligands. Both the cadmium only and the cadmium–copper(II) derivatives have essentially identical spectra suggesting that histidine 63 (corresponding to histidine 61 in the bovine enzyme) is in the same environment in both forms, i.e. not bridging. In the cadmium–copper(II) protein the pac spectra can only be interpreted by invoking two inter-converting forms whilst the cadmium–copper(I) derivative seems to exist as a single form. The interpretation of the pac results is further complicated by the results of ^{133}Cd nmr studies which gave quite different results. The earliest investigation revealed substantial chemical shift differences for the cadmium nucleus in the reduced and cadmium only derivatives,[76] however these results are probably artefactual as no precautions were taken to remove bound EDTA used in preparing the apo-enzyme.[77] Recently it has been shown that there is little difference between these two derivatives, with chemical shifts of 320 and 310 ppm respectively.[78]

One point which should be borne in mind when comparing the conclusions of the different types of spectroscopy is that each method senses the structure on a characteristic time-scale. A feature that appears constant over a period of milliseconds may in fact be an average of two structures inter-converting in microseconds.

Spectroscopic studies of the other metallo-forms of the enzyme are less extensive than for the cuprozinc one. In the case of the mangano-enzyme from *E. coli* no manganese epr spectrum could be detected until the enzyme had been denatured, whereupon the six line signal of the hexaquo manganese(II) ion appeared.[17] Evidence for manganese(III) in the oxidised form of the enzyme was provided by magnetic susceptibility measurements over a cryogenic temperature range.[79] The iron superoxide dismutases do give an epr spectrum and although different sources give signals differing in detail the overall appearance is of high spin iron(III) in a rhombic site. In the case of the enzyme from *Azotobacter vinelandii* very good agreement is obtained between the experimental and calculated g values.[80]

Many of the spectroscopic techniques used to study metalloproteins naturally focus on the metal ion or its first coordination sphere, however one particular method that has the ability to observe a much greater number of atoms in the molecule is nmr. The type of problems therefore approachable by this method are complementary to those addressed by the metal centred spectroscopies. In a high resolution proton nmr study of the bovine enzyme we were able to demonstrate that the nmr data were consistent with the crystallographic data, i.e. that the structures in the crystal and in solution were the same within the limitations of both methods.[81]

Spectroscopic methods are particularly powerful when combined with chemical modification of the enzyme and one type of minimal modification is the replacement of a hydrogen atom by deuterium or tritium. This reaction has been used to good effect in determining the metal ligands in yeast cuprozinc superoxide dismutase. Experiments using high resolution ^{1}H nmr showed that the exchange of the histidine C2 protons by solvent deuterons depended upon whether or not the histidine residue was coordinated to a metal ion; if it was then the exchange was very slow.[82]

Subsequently tritium exchange and analysis of the tritiated protein showed that the copper and zinc ligands were as for the bovine enzyme; however the incorporation into the shared ligand (histidine 63) seemed to show that its coordination properties were different from the other ligands.[83]

A review of the physicochemical properties of imidazolate bridged copper complexes and a comparison with the bovine cuprozinc superoxide dismutase has recently been published.[84]

4 KINETICS AND MECHANISM

4.1 Rate studies

As we discussed in the introduction the superoxide ion is not stable towards dismutation in aqueous media, and this provides a number of obstacles to the study of the kinetics of superoxide dismutase catalysis. Competitive methods can be used,[85,86] however the application of transient techniques is much more satisfactory and pulse radiolysis has proven the method of choice in this case. Irradiation of oxygenated aqueous solutions with electrons results in the generation of the superoxide ion, either from direct electron capture by the dioxygen molecule or by the reactions of the breakdown products of the water. The decay of the superoxide ion either in the presence or absence of the enzyme can be followed optically.

In the case of the cuprozinc superoxide dismutase the mechanism of substrate turnover appears to be a simple ping-pong reaction with the copper(II) ion being reduced and then re-oxidised by the substrate:[87]

$$\text{E. Cu(II)} + O_2{}^- \longrightarrow \text{E. Cu(I)} + O_2 \tag{4}$$

$$\text{E. Cu(I)} + O_2{}^- + 2H^+ \longrightarrow \text{E. Cu(II)} + H_2O_2 \tag{5}$$

Rate constants for the two steps are approximately 10^9 M^{-1} s^{-1} and the reaction is at least partially diffusion controlled as increasing the viscosity of the medium lowers the rate, and the activation energy is characteristic of a diffusion controlled reaction.[88] Two other interesting observations arose from the pulse radiolysis studies; first the rate of the enzyme catalysed reaction is independent of pH over a wide range (5-9) in contrast to the spontaneous reaction, and secondly no saturation of the enzyme could be observed even at the highest substrate levels available.

The simple ping-pong scheme that describes the cuprozinc superoxide dismutases does not appear to apply to the bacterial manganese enzyme; pulse radiolysis studies have been performed with both the *E. coli*[89] and *B. stearo-*

thermophilus[90] forms. Superoxide dismutase from the former organism shows a first order decay of the substrate when the ratio of superoxide to enzyme is less than 0.1, that is at low ratios the rate law is similar to the cuprozinc forms. At high ratios the decay is again first order but the rate constants are different and the authors interpret this as indicating more than two redox forms of the metal ion are involved in parallel catalytic cycles.[89]

B. stearothermophilus manganese superoxide dismutase also shows two types of kinetic pattern dependent on the substrate to enzyme ratio. A first-order decay is also observed for this enzyme at low substrate levels but as the substrate concentration is increased the decay of the superoxide ion approaches zero order. In addition to measuring the superoxide concentration, changes in the optical spectra of the manganese in both the native and hydrogen peroxide reduced forms of the enzyme were monitored and the authors suggested the following scheme:[90]

$$EA + O_2^- \longrightarrow EB + O_2 \tag{6}$$

$$EB + O_2^- + 2H^+ \longrightarrow EA + H_2O_2 \tag{7}$$

$$EB + O_2^- \longrightarrow EC \tag{8}$$

$$EC \longrightarrow EA \tag{9}$$

Rate constants were calculated based on the above scheme and it is the fact that equation (9) becomes rate limiting at high substrate levels that accounts for the zero order term.

In a subsequent paper the same authors studied the effect of temperature and pH on the kinetics and found that the fast (first-order) and slow (zero-order) reactions responded differently to these variables.[91] An alternative scheme to explain these kinetics has also appeared.[92]

The iron superoxide dismutases show different kinetic patterns both with respect to the enzymes described above and also depending on the source of the enzyme. *Ph. leiognathi* iron superoxide dismutase has the same simple ping-pong behaviour of the cuprozinc form[93] whilst the *E. coli* enzyme is very different. Pulse radiolysis and stopped flow methods have been used to study this enzyme and the kinetics quite clearly reveal the characteristic saturation behaviour of the Michaelis–Meten equation.[94] The Michaelis constant is very dependent on pH above pH 9 whilst k_{cat} is essentially independent of pH; it appears that a group with a pK_a of 8.8 is controlling substrate binding, a result consistent with previous steady state assays.

Two other transient techniques apart from pulse radiolysis have been used to investigate superoxide dismutase kinetics; stopped flow methods can be employed with crown ether solutions of potassium superoxide in aprotic organic solvents. The design of an apparatus to use the stopped flow method has been described, although relatively high pH media must be used to prevent too rapid a spontaneous dismutation reaction.[95] Superoxide ions can also be generated electrochemically in aqueous solutions under suitable conditions and a polarographic assay for the enzyme based upon this has been developed[96] although here again a working pH of greater than 9 must be used. Initially kinetic data were obtained at a dropping mercury electrode by using the method of kinetic currents, a more recent method applies the rotating ring disc technique.[97] Rate constants for the bovine cuprozinc enzyme were in agreement with the pulse radiolysis results, however the polarographic method showed evidence for saturation of the enzyme, with a Michaelis constant of 0.4 mM.[98] Such a high value for Km explains why no saturation behaviour was seen with the lower substrate concentrations generated in the pulse radiolysis work.

A summary of the results of the kinetic studies on the various superoxide dismutases is collected in table 4.4.

Table 4.4 Results of kinetic analyses of superoxide dismutases

Source	Method	Kinetics	pH Dependent	Ref.
Bovine erythrocytes	Pulse radiolysis	1st order	No	87
Bovine erythrocytes	Polarography	M–M	N.d.	96
E. coli (Mn)	Pulse radiolysis	Biphasic	Yes	89
B. stearothermophilus	Pulse radiolysis	Biphasic	Yes	90
Ph. leiognathi	Pulse radiolysis	1st order	Yes	93
E. coli (Fe)	Stopped flow	M–M	Yes	94
E. coli (Fe)	Pulse radiolysis	M–M	Yes	94

M–M = Michaels–Menten (saturation) kinetics.
N.d. = not determined.

Although the use of transient kinetics provides the fundamental rate constants for the reactions of the enzymes it gives little insight into the intimate details of the mechanism. In order to understand these a variety of additional physical and chemical measurements must be made. The following sections consider reaction with products, substrate encounter, ligand exchange, redox changes and proton transfers and concludes with an overall mechanism.

4.2 Reaction with hydrogen peroxide

In common with all enzymes superoxide dismutase should catalyse the reverse of reaction (2), that is the comproportionation of dioxygen and hydrogen peroxide although the strong driving force in the direction of dismutation requires efficient trapping of the superoxide ion. Using tetranitromethane (TNM) Hodgson and Fridovich[99] showed that the reverse reaction did occur in the presence of the enzyme:

$$O_2 + H_2O_2 \rightarrow 2H^+ + 2O_2^- \tag{10}$$

$$O_2^- + TNM \rightarrow TNM^- + O_2 \tag{11}$$

As might be expected from the mechanism of the dismutation reaction and the principle of microscopic reversibility the reaction proceeds by reduction of the copper by hydrogen peroxide and then reoxidation by dioxygen.

In the presence of an excess of hydrogen peroxide and at alkaline pH values the cuprozinc enzyme is irreversibly inactivated.[100] The nature of the inactivation is quite specific although mechanistically complex.[101] It seems that the copper(I) ion reacts with hydrogen peroxide to yield a 'Fenton type reagent' represented as a hydroxyl radical coordinated to copper(II), this species is then believed to attack one or more of the ligand histidine residues with subsequent loss of activity and alteration of the optical and epr properties. Although it is not known which of the histidine residues react the observation that in the copper(II)–cobalt(II) derivative treated under the same conditions there is no loss of the magnetic coupling between the centres suggests that histidine 61 is not one.

The iron enzyme from *Ph. leiognathi* is also reduced and then inactivated by hydrogen peroxide.[93] Manganese superoxide dismutase from *B. stearothermophilus* like the cuprozinc and iron forms is reduced but it is not inactivated.[92]

4.3 Chemical modification and electrostatic effects

In our discussion of the structure of the bovine superoxide dismutase we remarked on the dimeric nature of the enzyme (section 2.3) and the reasons for this. One interesting suggestion has been put forward by Koppenol.[102] He points out that although the rate of reaction of the substrate with the copper centre is no faster than with the simple aquo copper(II) ion the enzyme is in fact much larger; this in turn means that in contrast to the aquo ion many of the collisions of the two reactants would be unsuitable for electron transfer. To compensate for this steric effect he suggests that the distribution of charged groups on the surface of the molecule acts to guide the substrate into the active site. Experimental support for this hypothesis has come from studies on the effect of chemically modifying the lysine residues in the enzyme, and from the effect of ionic strength on the activity.[103] In a commonly used assay, where superoxide dismutase activity is competed against superoxide dependent cytochrome *c* reduction there was no effect of either the ionic strength or the dielectric constant of the reaction medium. As it is known that the lysine residues around the haem edge in cytochrome *c* control its activity with anionic reductants the lack of effect of either the polarity or the ionic strength on the activity of superoxide dismutase implies a parallel and compensatory effect. The use of an alternative assay, however, shows a distinct ionic strength dependence with activity decreasing with increasing salt concentration.[104] This effect is reversed when the lysine residues are modified by acetylation or succinylation. All of these observations point to a general effect of the protein charge in controlling its reactivity with substrate.

In addition to the general effects of charge on activity there is also a specific residue at the active site that appears to be critical for activity. Arginine 141 has its guanidino group 5 Å from the copper ion and pointing into the active site cleft. Interestingly this residue was identified as the hexachloroiridate ion binding site in one of the heavy atom derivatives used in the x-ray crystallographic analysis.[43] Subsequently it was shown that reaction of the enzyme with butane-2,3-dione or phenylglyoxal resulted in loss of activity concomitant with the modification of arginine 141.[105] Although only approximately 90% of the activity was lost in the case of the bovine enzyme, greater than 99% inactivation of the yeast enzyme could be achieved with suitable reaction conditions.[106] Initially (^{14}C)-phenylglyoxal was used to identify the reactive residue but later work showed that a chromophoric derivative, 4-hydroxy-3-nitrophenylglyoxal also reacted with a single arginine residue, 143, in the yeast enzyme.[107] It was suggested on the basis of the yeast work that the residual activity in the bovine enzyme was due to incomplete reaction but the insensitivity of this residual activity to ionic strength argues for a different interpretation and implies that the fully modified bovine enzyme has partial but altered catalytic competence.

4.4 Ligand exchange and inhibition

As the substrate of the superoxide dismutases is an anion, and as the active sites contain a metal ion then ligand exchange studies should provide insight into substrate binding. In this respect the paramagnetism of the metal ions involved in catalysis has provided a powerful handle on ligand binding.

One ligand that might be expected to be present in the proximity of the metal centre is the water molecule; the unpaired electron(s) on a transition metal ion are very effective in relaxing the nuclear spins of the water molecules bound nearby and the measurement of water relaxation times can provide values for exchange rates, numbers of exchanging nuclei and their distances from the metal.[108] Most water relaxation studies have monitored the proton signals because of their high sensitivity; more recently valuable complementary data have become available from oxygen (as [17]O) relaxation measurements.

Villafranca *et al.*[109] measured the temperature and frequency dependence for solutions of both the manganese and iron enzymes from *E. coli* and fitted the resulting relaxation times (longitudinal and transverse) to obtain residence times and numbers of protons in the first coordination sphere. The manganese (III) enzyme had a single water molecule bound with a residence time of 5–20 μs, whilst the iron (III) enzyme had apparently a bound hydroxyl ion with a residence time of 5–80 μs. Although very little of the relaxation observed for the manganese enzyme was attributed to outer sphere molecules, in the iron enzyme an additional contribution of a second water molecule 5–5.5 Å from the metal was suggested.

This conclusion was later modified in the light of a room temperature determination of the electron spin relaxation time. An estimate of this suggested that in the iron enzyme all of the water relaxation was due to outer sphere protons and that there was not an exchangeable water molecule bound to the iron atom.[110]

The relatively straightforward interpretation of water proton relaxation data used in the above analyses and later in this section for the cuprozinc enzyme has recently been questioned by Koenig and Brown.[111] Although not specifically concerned with superoxide dismutases they point out a number of instances where the analysis of water relaxation by metalloproteins is not possible by using the simple theory developed for aquo ions. In particular they suggest that a water molecule hydrogen bonded to an ion coordinated to a metal centre can be relaxed as efficiently as, or more so than, a directly bonded water molecule. This mechanism, 'the fluoromet mechanism' may be of relevance to the high pH relaxation results for bovine cuprozinc superoxide dismutase (*vide infra*).

Anion binding to the iron enzyme is a two-site reaction with apparently one on the metal ion and the second nearby. Azide and fluoride ions both inhibit the enzyme and cause changes in the optical and epr spectra; however the spectroscopically determined binding constants and the kinetically measured inhibition constants are quite different.[112] Furthermore perchlorate, though not binding to the iron or causing changes in the spectrum is still an effective competitive inhibitor (K_i = 20 mM). Both enzyme activity and anion binding show very similar responses to pH; the spectroscopically determined dissociation constant for azide binding has a pK_a of 8.6, the Michaelis constant is dependent on a group with a pK_a of 8.8 whilst the change in the epr spectrum in the alkaline region is reflected in a pK_a of 9.3. Fee *et al.*[113] have combined these various observations into a scheme for both anion binding and substrate turnover that is illustrated in figure 4.7.

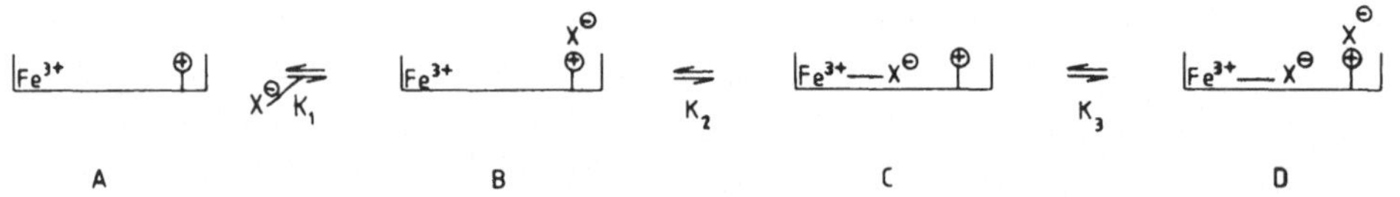

Figure 4.7 Scheme for the binding of anions to iron superoxide dismutase from *E. coli* based on kinetic and spectroscopic data. (Reproduced from ref. 113 with permission)

Bovine cuprozinc superoxide dismutase has been studied by several groups with respect to water proton relaxation rates.[114–116] The conclusions are in broad agreement at neutral pH although they differ in their interpretation of the results obtained at alkaline pH values. Around neutrality the copper(II) ion has a single bound water molecule exchanging with an off rate of *c*. 10^6 s^{-1}, this water molecule occupies an axial position, consistent with the crystal structure.

As the pH is increased to values above 9 the relaxation rate of the water protons is dramatically enhanced and this has been interpreted in two quite different ways. Terenzi *et al.*[116] considered that the increase in relaxation rate was due to deprotonation of the water molecule with subsequent shortening of the metal–oxygen bond and faster relaxation of the residual proton. Boden *et al.*[115] on the other hand felt that the bond length calculated by the Italian group was too short and suggested instead substitution of one of the equatorial histidine ligands by a hydroxide so increasing to three the number of protons in the inner sphere. Support for an equatorially coordinated water molecule/hydroxide ion comes from oxygen-17 relaxation measurements. As the oxygen nucleus is directly coordinated to the copper ion the dominant relaxation mechanism is a contact effect, and as the unpaired spin is largely in the equatorial plane an axially bound oxygen atom is likely to be little affected. Consistent with this there is little paramagnetic enhancement at low pH. As the pH is increased the relaxation rate rises, as expected for a further, equatorially bound, water or hydroxide ligand.[117]

Whatever the exact nature of the enhanced relaxation rate seen at high pH it correlates with a loss of enzyme activity. In kinetic studies the hydroxide ion has been shown to be a competitive inhibitor of the enzyme.[118]

The structure of the copper site and the presence of a coordinated water molecule suggests that anions should also bind to the metal ion. Neither water nor other ligands bind to the zinc. Although many anions interact with the copper centre there is increasing evidence that they may not all do so in the same fashion. Species such as cyanide or azide bind strongly, cause large changes in the optical and epr spectra, abolish the enhanced water relaxation and inhibit enzyme activity.

The nature of anion binding to the enzyme is still uncertain, the magnetic data are generally consistent with equatorial coordination especially with the observation that for suitably isotopically labelled anions superhyperfine structure is seen in the epr spectrum.[67] Different behaviour with respect to water relaxation appears to occur with different anions, cyanide completely inhibits the paramagnetic contribution to the water relaxation[115] and thiocyanate hardly affects it at all even when by other criteria the protein–ligand complex is 90% formed.[119] Two interpretations of these results have been suggested, in one the exogenous ligand displaces the water molecule and the site rearranges to place the incoming ligand in the equatorial plane;[120] in the second the incoming ligand displaces one of the endogenous ligands and the axial water molecule is then labilised depending on the ligand field strength of the anion.[121] Although it is not yet clear which if either of these interpretations is correct the problem bears some resemblance to fluoride binding to galactose oxidase. In this case there are two anion binding sites on the copper(II) ion, one equatorial that is 'seen' by epr and one axial 'seen' by ^{19}F relaxation studies. The binding of cyanide to this enzyme displaces the equatorial fluoride ion but also causes considerable changes in the relaxation of the axial fluoride ion.[122]

Thiocyanate binding to the antiferromagnetically coupled copper(II)-copper(II) derivative destroys the magnetic interaction between the two metal centres, implying displacement of the bridging imidazolate anion.[123] Endor studies on the native enzyme at high pH, where a hydroxide ion is thought to be bound equatorially, are also consistent with the breaking of the imidazolate bridge.[73]

Fluoride ion binding to the oxidised form of the enzyme has been monitored by ^{19}F relaxation measurements and in addition to deriving the kinetic and thermodynamic binding parameters[124] can be used to assay the enzyme levels in tissues as the relaxation is proportional to the enzyme concentration.[125] ^{19}F relaxation measurements have also been employed to determine the amount of the copper(II) form present during steady state enzyme turnover.[126]

The oxidised form of superoxide dismutase provides a very powerful probe of anation reactions by nature of its unpaired electron. However anion binding to the reduced form is much less easily measured. Inhibition of activity has suggested that a number of anions may bind[118] although this approach suffers from two drawbacks; firstly inhibition may be due to binding to either or both redox states. Secondly for weakly coordinated anions, the concentrations employed may be inhibitory due to ionic strength effects. Chloride binding to the reduced form of the enzyme has been confirmed by ^{35}Cl nmr studies.[127] The chlorine nucleus has a spin of 3/2 and so in the presence of an electric field gradient will be efficiently relaxed through a nuclear quadrupole mechanism. In solution the hydrated ion is surrounded by a relatively uniform electric field, however upon coordination to the copper(I) ion the field gradient is increased and the relaxation is enhanced. Inhibition of this effect by cyanide ion is consistent with the latter ion also binding to the reduced form.

Anion binding to the copper(I) derivative has also been probed by observing the chemical shifts of the histidine C2 protons of the residues coordinated to the metal ion.[128] In a mixture of chloride and perchlorate at constant ionic strength progressive shifts were observed with increasing chloride concentration. Only the resonances from the copper ligands were affected. However as figure 4.8 shows the perturbation does not appear to be a simple function of either the nature of the ion or the ionic strength and we have speculated that the copper ion is responsive to the environment of the whole active site in addition to changes in its first coordination sphere. Specifically we suggested that an important role may be played by the ordered water molecules in the active site cleft, an idea that we shall return to at the end of this chapter.

4.5 Redox potentials

The formal potentials for the O_2/O_2^- and O_2^-/H_2O_2 couples at pH 7 are -0.33 V and 0.87 V respectively (section 1.2), so if the proposed ping-pong mechanism is valid then the copper (II)/(I) potential must lie between these

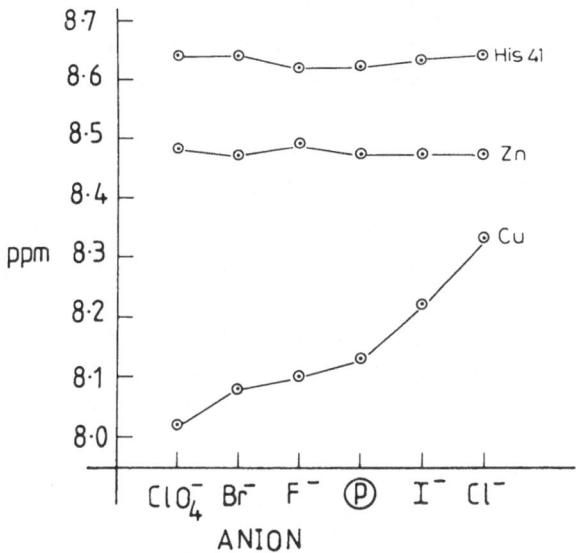

Figure 4.8 Effect of anions on the chemical shifts of resonances in bovine cuprozinc super-oxide dismutase. The resonances of histidine 41 (non-ligand), a zinc ligand (Zn) and a copper ligand (Cu) are shown

two values. Similar limits also apply to the manganese (III)/(II) and iron (III)/(II) potentials in those enzymes.

Fee and Dicorleto[129] first determined the redox potential of the cuprozinc enzyme by titration with ferri/ferrocyanide and obtained a value of 0.42 V at pH 7; one of the drawbacks of this method is that the titrant may bind to the protein and subsequently perturb the potential. A later, potentiometric determination revised the value to 0.28 V.[130] In both studies the redox potential was pH dependent, changing by approximately 50 mV per pH unit, as expected if a single proton was being taken up with each electron; and this was ascribed to protonation of the bridging imidazolate when the copper was reduced.

In this potentiometric study the two redox centres did not appear to be equivalent, but rather the reduction occurred by two distinct steps implying that the redox potential of one copper ion is sensitive to the redox state of the other. In addition the presence of hexachloroiridate, known from the crystal structure to bind close to arginine 141, renders one of the copper ions irreducible by the mediator. As the chemical modification studies have revealed the importance of this residue for the enzyme activity (section 4.3), this poses interesting consequences for the mechanism.

The manganese ions of the superoxide dismutases from *B. stearothermophilus* and *E. coli* have midpoint potentials of 0.26 and 0.31 V and also show a pH dependence in their potentials although the nature of the group(s) being protonated in these enzymes is unknown.[130] Iron containing superoxide dismutases from three different bacteria have had their midpoint potentials determined by the use of mediators and the three values are very similar (0.25 V).[131]

4.6 Proton Transfers

Superoxide dismutases are amongst the fastest enzymes known, with turnover numbers of approximately 10^6 s^{-1} and second-order rate constants in excess of 10^8 M^{-1} s^{-1}. These rapid reactions are comparable to those of another very fast enzyme, carbonic anhydrase, and like this latter molecule the source of the protons involved in the catalysis is an interesting question. It seems clear that the bulk solvent is not involved with the cuprozinc enzymes as the observed rate constants are independent of pH in the range 5–9, and as the reduction of the superoxide ion must be accompanied by the transfer of at least one proton to the developing peroxide anion then the protons must come from the protein. Alternatively the apparent pH at the active site must be much lower than in the bulk solution, say around 4–5 irrespective of the external pH. An alteration in local pH can be achieved by creating a suitable electrostatic potential at the active site, and may be important for cytochrome P450cam[132] and for the hydroperoxidases.[133] There is no experimental evidence for this being so for the superoxide dismutases, indeed the binding of hydroxide ions to the copper at alkaline pH (section 4.4) argues against this.

In the case of the manganese enzymes there is a pH dependence of activity and for the iron enzyme from *E. coli* the same is true although in the latter case the source of the pH versus activity profile can be determined thanks to the observed saturation kinetics. Above pH values of about 9 the Michaelis constant rapidly increases whereas k_{cat} is essentially unaffected. This is very similar to the dependence of azide binding upon pH (section 4.4), a group with a pK_a around 9 appears to be controlling anion binding, whether substrate or inhibitor, whilst the independence of k_{cat} on pH implies that in this enzyme also the proton donor is not the bulk solvent.[113]

The nature of the proton donor in the cuprozinc superoxide dismutases has not been unequivocally assigned although the two most mooted choices are arginine 141 and histidine 61. It is the imidazole ring of the latter that bridges the two metal ions in the oxidised state, whilst in the reduced state it is protonated and coordinated solely to the zinc ion. This change upon reduction is not unexpected when the properties of the +2 and +1 oxidation states of copper are compared. Furthermore transient kinetic studies on the copper–cobalt derivative using pulse radiolysis are consistent with protonation of the bridge during turnover.[134] The role of arginine 141 as a proton donor is supported by the observation that when this residue is chemically modified then activity is lost (section 4.3) although this is more likely due to an effect on superoxide binding.

4.7 An alternative mechanism

The most widely quoted mechanism of action for the cuprozinc superoxide dismutases involves inner sphere electron transfer to or from the metal ion and protonation/deprotonation of the imidazole ring of histidine 61.[13] However the observation that in the derivative where the zinc ion is replaced by copper addition of thiocyanate breaks the bridge without affecting activity suggests that the bridge is not necessary for activity.[123] We proposed an alternative mechanism some years ago,[128] based largely on the effects of anions on the proton nmr spectrum of the reduced enzyme. In this section we will develop a more detailed discussion of our alternative mechanism.

The superoxide ion is a relatively strongly hydrated species, having an enthalpy of hydration between that of the fluoride and chloride ions (see ref. 2). An inner sphere reaction of O_2^- with a metal ion therefore implies that one or more water molecules are removed from the hydration sphere of both reactants, a step that could add considerably to the activation energy. This is analogous to the fluoride ion where the highly nucleophilic character of F^- is masked by the tightly bound hydration shell. Experimental evidence for the hydrated nature of O_2^- is provided both by the more negative reduction potential for dioxygen in aprotic media and by the epr studies that show the ion to be hydrated in aprotic media unless considerable care is taken to remove all the moisture.[135]

We suggest that the binding of the substrate in the active site occurs through hydrogen bonds both to the axial water molecule on the copper ion and to guanidino group of arginine 141 as shown in figure 4.9. This arrangement leads to a good orientation of the in-plane orbital on the copper which contains its unpaired electron and the molecular orbital on the superoxide ion which contains its unpaired electron. Subsequently outer sphere electron transfer occurs to yield dioxygen and the reduced enzyme.

In the reoxidation reaction the same pattern of binding occurs and there is again an outer sphere electron transfer reaction; however this time a simultaneous proton transfer from the coordinated water molecule also takes place as the nascent peroxide ion develops. After the hydrogen peroxide has dissociated the reprotonation of the copper(II)-hydroxide takes place via a series of hydrogen bond rearrangements in the ordered water structure in the cleft leading to the active site, which finally results in a hydroxide ion being ejected at the protein surface. This mechanism does not need to involve translation of water molecules in the protein but rather a readjustment of bond lengths and nuclear positions and so will be extremely fast; because of the similarity of this step to proton conduction in ice we have called this a Grotthus mechanism; the complete cycle is shown in figure 4.9.

Although as described the mechanism refers to the cuprozinc enzyme, for which we have the most complete information, it may also be broadly the same for the other metallo-forms. Indeed as discussed for the iron enzyme from *E. coli* there is evidence for an outer sphere anion binding site that behaves in a

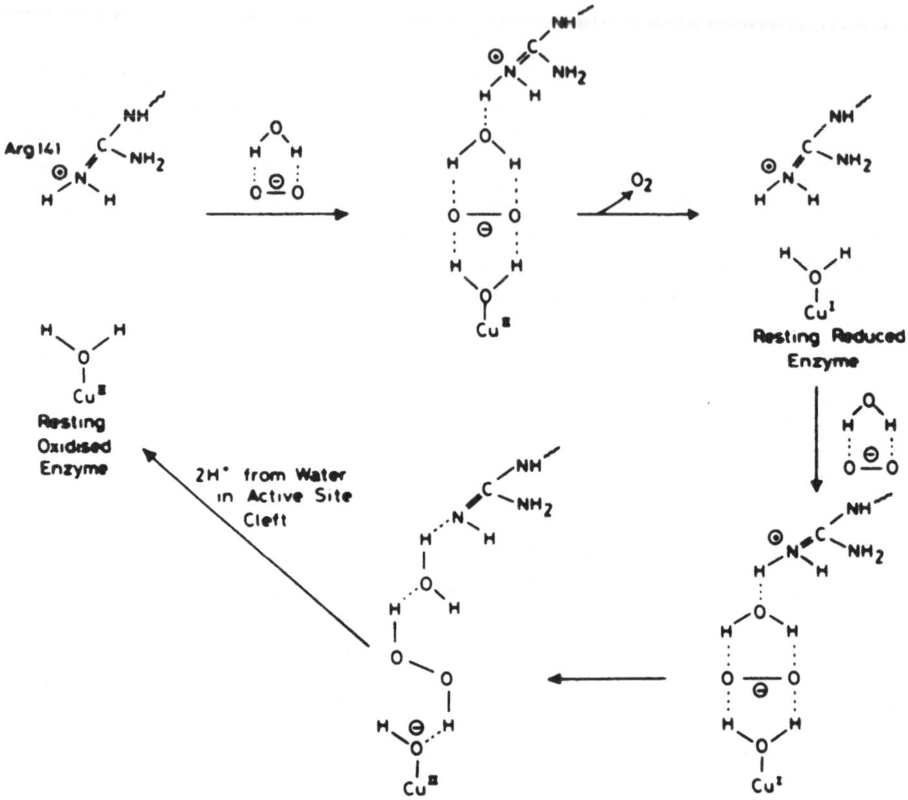

Figure 4.9 A proposed mechanism for cuprozinc superoxide dismutase involving Arg 141 and outersphere electron transfer to and from the copper

similar fashion to the substrate binding site; unlike the cuprozinc enzyme however occupancy of this site is controlled by a group with a pK_a of around 9 and this may be a lysine residue functionally analogous to arginine 141.

Our description of the Grotthus mechanism as outlined above, and illustrated in figure 4.9 takes no account of the breaking and protonation of the copper-imidazolate bond when the metal is reduced. This reaction is now well established and has had an electron transfer role assigned to it, we wish to suggest another interpretation. As described earlier the role of the zinc ion appears to be one of organising the copper binding site and possibly facilitating copper incorporation; the introduction of a shared ligand obviously provides an intimate association of the two metal ions. A consequence of the presence of this shared metal ligand is that when the copper(II) is reduced the copper(I) is destabilised both by the presence of the charged ligand and by the high coordination number. Protonation of the bridging imidazolate ion would solve both these problems. We therefore suggest that the bridging imidazolate ion functions to link the two metal binding sites during metal incorporation and the protonation and breaking of the copper nitrogen bond is simply a consequence of the different chemistries of copper(I) and copper(II).

5 CONCLUSIONS, SUPEROXIDE DISMUTASE?

This review of the superoxide dismutases has concentrated primarily on their molecular properties, very little has been said of their physiology. The inter-relationships of oxygen toxicity, free radical biology and the superoxide dismutases have been repeatedly reviewed and as intimated at the beginning of this chapter the debate over the roles of the superoxide dismutases *in vivo* has by no means abated. However before concluding this chapter it is pertinent to try and address one question concerning the superoxide dismutase activity of these proteins.

The rate constants for the enzymatic dismutation of O_2^- are very high and yet the aquo copper(II) ion is an effective catalyst at pH 7[136] thus why bother with an enzyme? There are several possible answers to this and they are not connected with catalysis *per se*, perhaps the most striking reasons relate to control. The levels of free transition metal ions *in vivo* must be kept low, largely because of their ability to catalyse a variety of potentially deleterious oxidations. Relatively low concentrations of metal ions such as iron and copper can catalyse lipid peroxidation or ascorbate autoxidation and thus the organisms must keep their concentrations depressed. Additionally there are occasions where the presence of enhanced superoxide levels is important to the organism, the respiratory burst during phagocytosis is one example. In both cases the spatial control of enzyme activity is more conveniently effected with macromolecules rather than with small metal complexes.

Finally control in a different sense is needed to cope with nutritional changes; we discussed earlier how a fungus switched between different metallo-forms in response to its environmental copper levels, in higher organisms during periods of copper deficiency the levels of superoxide dismutase activity are amongst the last to be affected. Apparently the enzyme is important enough to warrant maintenance in an active state at the expense of other copper proteins. This kind of selectivity would be hard to achieve if the simple metal ion or a small complex was the catalyst.

In conclusion we can say that although the molecular properties of the enzymes discussed in this review do not prove that superoxide dismutation is their physiological function they nonetheless appear to be singularly well suited to this task.

ACKNOWLEDGMENTS

I wish to thank Drs Allen Hill and Joe Bannister for introducing me to the super-oxide dismutases and for many enlightening discussions on their chemistry. This review was written whilst the author was a British Petroleum Junior Research Fellow in Enzyme Studies at St Hugh's College, Oxford and I thank both organisations for their support.

REFERENCES

1. Hill, H. A. O. (1979) In: *Oxygen Free Radicals and Tissue Damage*, Ciba Foundation Symposium 65, Excerpta Medica, p. 5
2. Fee, J. A. and Valentine, J. S. (1977) In: *Superoxide and Superoxide Dismutases* (ed. Michelson, A. M., McCord, J. M. and Fridovich, I.), Academic Press, New York, p. 19
3. Valentine, J. S. and Sawyer, D. T. (1981) *Acc. Chem. Res.*, **14**, 393
4. Fridovich, I. (1972) *Acc. Chem. Res.*, **5**, 321
5. Lee-Ruff, E. (1977) *Chem. Soc. Rev.*, **6**, 195
6. Bors, W., Saran, M., Lengfelder, E., Spottl, R. and Michel, C. (1974) *Current Topics Rad. Res. Quart.*, **9**, 247
7. Bielski, B. H. J. and Allen, A. O. (1977) *J. Phys. Chem.*, **81**, 1048
8. Fee, J. A. (1981) In: *Oxygen and Oxy Radicals in Chemistry and Biology* (ed. Rodgers, M. J. A. and Powers, E. L.), Academic Press, New York, p. 205
9. Fee, J. A. (1980) In: *Metal Ion Activation of Dioxygen* (ed. Spiro, T. G.), Wiley, New York, p. 209
10. Fridovich, I. (1981) In: Ref. 8, p. 197
11. Halliwell, B. (1982) *Trends Biochem. Sci.*, **7**, 270
12. Fee, J. A. (1982) *Trends Biochem. Sci.*, **7**, 84
13. Valentine, J. S. and Pantoliano, M. W. (1982) In: *Copper Proteins* (ed. Spiro, T. G.), Wiley, New York, p. 291
14. Mann, T. and Keilin, D. (1938) *Proc. Roy. Soc. (Lond.)*, **B126**, 303
15. Carrico, R. J. and Deutsch, H. F. (1969) *J. biol. Chem.*, **244**, 6087
16. McCord, J. M. and Fridovich, I. (1977) In: Ref. 2, p. 1
17. Keele, B. B., McCord, J. M. and Fridovich, I. (1970) *J. biol. Chem.*, **245**, 6176
18. Yost, F. J. and Fridovich, I. (1973) *J. biol. Chem.*, **248**, 4905
19. Ravindranath, S. D. and Fridovich, I. (1975) *J. biol. Chem.*, **250**, 6107
20. Weisiger, R. A. and Fridovich, I. (1973) *J. biol. Chem.*, **248**, 4793
21. Hassan, H. M. and Fridovich, I. (1978) *Life Sci. Res. Rep.*, **13**, 179
22. Dougherty, H. W., Sadowski, S. J. and Baker, E. E. (1978) *J. biol. Chem.*, **253**, 5220
23. Meier, B., Barra, D., Bossa, F., Calabrese, L. and Rotilio, G. (1982) *J. biol. Chem.*, **257**, 13977
24. Shatzman, A. R. and Kosman, D. J. (1978) *Biochim. Biophys. Acta*, **544**, 163
25. Steinman, H. M., Naik, V. R., Abernethy, J. L. and Hill, R. L. (1974) *J. biol. Chem.*, **249**, 7326
26. Jabusch, J. R., Farb, D. L., Kerschensteiner, D. A. and Deutsch, H. F. (1980) *Biochemistry (USA)*, **19**, 2310
27. Barra, D., Martini, F., Bannister, J. V., Schinina, M. H., Rotilio, G., Bannister, W. H. and Bossa, F. (1980) *FEBS Lett.*, **120**, 53
28. Lerch, K. and Ammer, D. (1981) *J. biol. Chem.*, **256**, 11545
29. Johansen, J. T., Overballe-Petersen, C., Martin, B., Hasemann, V. and Svendsen, I. (1979) *Carlsberg Res. Commun.*, **44**, 201
30. Steinman, H. M. (1980) *J. biol. Chem.*, **255**, 6758
31. Bannister, J. V., Bannister, W. H., Cass, A. E. G., Hill, H. A. O. and Johansen, J. T. (1980) In: *Chemical and Biochemical Aspects of Superoxide Dismutases* (ed. Bannister, J. V. and Hill, H. A. O.), Elsevier, Amsterdam, p. 284
32. Cass, A. E. G., Hill, H. A. O., Smith, B. E., Bannister, J. V. and Bannister, W. H. (1977) *Biochem. J.*, **165**, 587
33. Hill, H. A. O., Lee, W.-K., Bannister, J. V. and Bannister, W. H. (1980) *Biochem. J.*, **185**, 245
34. Burger, A. R., Lippard, S. J., Pantoliano, M. W. and Valentine, J. S. (1980) *Biochemistry* (USA), **19**, 4139

35. Muno, D., Isobe, T., Okuyama, T., Ichihara, K., Noda, Y., Kusunose, E. and Kusunose, M. (1981) *Biochem. Int.*, **2**, 33
36. Brock, C. J. and Walker, J. E. (1980) *Biochemistry (USA)*, **19**, 2873
37. Steinman, H. M. (1978) *J. biol. Chem.*, **253**, 8708
38. Ditlow, C., Johansen, J. T., Martin, B. M. and Svendsen, I. (1982) *Carlsberg Res. Commun.*, **47**, 163
39. Martin, J. P. and Fridovich, I. (1981) *J. biol. Chem.*, **256**, 6080
40. Puget, K. and Michelson, A. M. (1974) *Biochimie*, **56**, 1255
41. Puget, K. and Michelson, A. M. (1974) *Biochem. Biophys. Res. Commun.*, **58**, 830
42. Steinman, H. M. (1982) *J. biol. Chem.*, **257**, 10283
43. Richardson, J. S., Thomas, K. A., Rubin, B. H. and Richardson, D. C. (1975) *Proc. nat. Acad. Sci. (USA)*, **72**, 1349
44. Tanier, J. A., Getzoff, E. D., Beem, K. M., Richardson, J. S. and Richardson, D. C. (1982) *J. molec. Biol.*, **160**, 181
45. McLachlan, A. D. (1980) *Nature, Lond.*, **285**, 267
46. Abernethy, J. L., Steinman, H. M. and Hill, R. L. (1974) *J. biol. Chem.*, **249**, 7339
47. Rigo, A., Marmocchi, F., Cocco, D., Viglino, P. and Rotilio, G. (1978) *Biochemistry (USA)*, **17**, 534
48. Bannister, J. V., Anastasi, A. and Bannister, W. H. (1978) *Biochem. Biophys. Res. Commun.*, **81**, 469
49. Marmocchi, F., Vernardi, G., Bossa, F., Rigo, A. and Rotilio, G. (1978) *FEBS Lett.*, **94**, 109
50. Malinowski, D. P. and Fridovich, I. (1979) *Biochemistry (USA)*, **18**, 237
51. Marmocchi, F., Mavelli, I., Rigo, A., Stevanato, R., Bossa, F. and Rotilio, G. (1982) *Biochemistry (USA)*, **21**, 2853
52. Smit, J. D. G., Pulver-Sladek, J. and Jansonius, J. N. (1977) *J. molec. Biol.*, **112**, 491
53. Stallings, W. C., Pattridge, K. A., Powers, T. B., Fee, J. A. and Ludwig, M. L. (1981) *J. biol. Chem.*, **256**, 5857
54. Stallings, W. C., Powers, T. B., Pattridge, K. A., Fee, J. A., Ludwig, M. L., Ponzi, D. R. and Petsko, G. A. (1982) *Int. Union Biochem. Meeting*, Perth, Book of Abstracts (POS 004-141), p. 319
55. Ose, D. E. and Fridovich, I. (1979) *Arch. Biochem. Biophys.*, **194**, 360
56. Brock, C. J., Harris, J. I. and Sato, S. (1976) *J. molec. Biol.*, **107**, 175
57. Yamakura, F. and Suzuki, K. (1976) *Biochem. Biophys. Res. Commun.*, **72**, 1108
58. Fee, J. A. (1973) *J. biol. Chem.*, **248**, 4229
59. Dunbar, J. C., Johansen, J. T. and Uchida, T. (1982) *Carlsberg Res. Commun.*, **47**, 81
60. Pantoliano, M. W., McDonnell, P. J. and Valentine, J. S. (1979) *J. Am. chem. Soc.*, **101**, 6454
61. Valentine, J. S., Pantoliano, M. W., McDonnell, P. J., Burger, A. R. and Lippard, S. J. (1979) *Proc. nat. Acad. Sci. (USA)*, **76**, 4245
62. Hirose, J., Iwatzuka, K. and Kidani, Y. (1981) *Biochem. Biophys. Res. Commun.*, **98**, 58
63. Hirose, J., Ohhira, T., Hirata, H. and Kidani, Y. (1982) *Archiv. Biochem. Biophys.*, **218**, 179
64. Rigo, A., Viglino, P., Calabrese, L., Cocco, D. and Rotilio, G. (1977) *Biochem. J.*, **161**, 27
65. Rigo, A., Terenzi, M., Viglino, P., Cocco, D. and Rotilio, G. (1977) *Biochem. J.*, **161**, 31
66. Cass, A. E. G., Hill, H. A. O., Bannister, J. V. and Bannister, W. H. (1979) *Biochem. J.*, **177**, 477
67. Rotilio, G., Finazzi-Agro, A., Calabrese, L., Bossa, F., Guerrieri, P. and Mondovi, B. (1971) *Biochemistry (USA)*, **10**, 616
68. Fee, J. A. and Briggs, R. G. (1975) *Biochim. Biophys. Acta*, **400**, 439
69. Moss, T. H. and Fee, J. A. (1975) *Biochem. Biophys. Res. Commun.*, **66**, 799
70. Rotilio, G., Calabrese, L., Mondovi, B. and Blumberg, W. E. (1974) *J. biol. Chem.*, **249**, 3157
71. Liberman, R. A., Sands, R. H. and Fee, J. A. (1982) *J. biol. Chem.*, **257**, 336

72. Fee, J. A., Peisach, J. and Mims, W. B. (1981) *J. biol. Chem.*, **256**, 1910
73. Van Camp, H. L., Sands, R. H. and Fee, J. A. (1982) *Biochim. Biophys. Acta*, **704**, 75
74. Blumberg, W. E., Peisach, J., Eisenberger, P. and Fee, J. A. (1978) *Biochemistry (USA)*, **17**, 1842
75. Bauer, R., Demeter, I., Hasemann, V. and Johansen, J. T. (1980) *Biochem. Biophys. Res. Commun.*, **94**, 1296
76. Armitage, I. M., Uiterkamp, A. J. M. S., Chlebowski, J. F. and Coleman, J. E. (1978) *J. magn. Res.*, **29**, 375
77. Armitage, I. M. and Otvos, J. D. (1982) In *Biological Magnetic Resonance, Vol. 4* (ed. Berliner, L. J. and Reuben, J.), Plenum Press, New York, p. 79
78. Bailey, D. B., Ellis, P. D. and Fee, J. A. (1980) *Biochemistry (USA)*, **19**, 591
79. Fee, J. A., Shapiro, E. R. and Moss, T. H. (1976) *J. biol. Chem.*, **251**, 6157
80. Emptage, M. H. (1981) *Fed. Proc.*, **40**, 1798 (Abst. 1488)
81. Cass, A. E. G., Hill, H. A. O., Smith, B. E., Bannister, J. V. and Bannister, W. H. (1977) *Biochemistry (USA)*, **16**, 3061
82. Cass, A. E. G., Hill, H. A. O., Bannister, J. V., Bannister, W. H., Hasemann, V. and Johansen, J. T. (1979) *Biochem. J.*, **183**, 127
83. Dunbar, J. C., Johansen, J. T., Cass, A. E. G. and Hill, H. A. O. (1980) *Carlsberg Res. Commun.*, **45**, 349
84. Strothkamp, K. G. and Lippard, S. J. (1982) *Acc. Chem. Res.*, **15**, 318
85. Sawada, Y. and Yamazaki, I. (1973) *Biochim. Biophys. Acta*, **327**, 257
86. Forman, H. J. and Fridovich, I. (1973) *Arch. Biochem. Biophys.*, **158**, 396
87. Klug-Roth, D., Rabani, J. and Fridovich, I. (1973) *J. Am. chem. Soc.*, **95**, 2786
88. Fielden, E. M., Roberts, P. B., Bray, R. C., Lowe, D. J., Mautner, G. N., Rotilio, G. and Calabrese, L. (1974) *Biochem. J.*, **139**, 49
89. Pick, M., Rabani, J., Yost, F. and Fridovich, I. (1974) *J. Am. chem. Soc.*, **96**, 7329
90. McAdam, M. E., Fox, R. A., Lavelle, F. and Fielden, E. M. (1977) *Biochem. J.*, **165**, 71
91. McAdam, M. E., Lavelle, F., Fox, R. A. and Fielden, E. M. (1977) *Biochem. J.*, **165**, 81
92. Bugrii, G. V. and Kukhtin, V. V. (1981) *J. theor. Biol.*, **90**, 161
93. Lavelle, F., McAdam, M. E., Fielden, E. M., Roberts, P. B., Puget, K. and Michelson, A. M. (1977) *Biochem. J.*, **161**, 3
94. Fee, J. A., McClune, G. J., O'Neill, P. and Fielden, E. M. (1981) *Biochem. Biophys. Res. Commun.*, **100**, 377
95. McClune, G. J. and Fee, J. A. (1978) *Biophys. J.*, **24**, 65
96. Rigo, A., Viglino, P. and Rotilio, G. (1975) *Anal. Biochem.*, **68**, 1
97. Orsega, E. F., Argese, E., Viglino, P., Tomat, R. and Rigo, A. (1982) *J. electroanal. Chem.*, **131**, 257
98. Rigo, A., Viglino, P. and Rotilio, G. (1975) *Biochem. Biophys. Res. Commun.*, **63**, 1013
99. Hodgson, E. K. and Fridovich, I. (1973) *Biochem. Biophys. Res. Commun.*, **54**, 270
100. Hodgson, E. K. and Fridovich, I. (1975) *Biochemistry (USA)*, **14**, 5294
101. Bray, R. C., Cockle, S. A., Fielden, E. M., Roberts, P. B., Rotilio, G. and Calabrese, L. (1974) *Biochem. J.*, **139**, 43
102. Koppenol, W. H. (1981) In: Ref. 8, p. 671
103. Cudd, A. and Fridovich, I. (1982) *FEBS Lett.*, **144**, 181
104. Cudd, A. and Fridovich, I. (1982) *J. biol. Chem.*, **257**, 11443
105. Malinowski, D. P. and Fridovich, I. (1979) *Biochemistry (USA)*, **18**, 5909
106. Borders, C. L. and Johansen, J. T. (1980) *Carlsberg Res. Commun.*, **45**, 185
107. Borders, C. L. and Johansen, J. T. (1980) *Biochem. Biophys. Res. Commun.*, **96**, 1071
108. Burton, D. R. (1980) In: *ESR and NMR of Paramagnetic Species in Biological and Related Systems* (ed. Bertini, I. and Drago, R. S.), D. Reidel, Holland, p. 151
109. Villafranca, J. J., Yost, F. J. and Fridovich, I. (1974) *J. biol. Chem.*, **249**, 3532
110. Villafranca, J. J. (1976) *FEBS Lett.*, **62**, 230

111. Koenig, S. H. and Brown, R. D. (1983) In: *Coordination Chemistry of Metalloenzymes* (ed. Bertini, I., Drago, R. S. and Luchinat, C.), D. Reidel, Holland, p. 19
112. Slykhouse, T. O. and Fee, J. A. (1976) *J. biol. Chem.*, **251**, 5472
113. Fee, J. A., McClune, G. J., Lees, A. C., Zidovetzi, R. and Pecht, I. (1981) *Isr. J. Chem.*, **21**, 54
114. Gaber, B. P., Brown, R. D., Koenig, S. H. and Fee, J. A. (1972) *Biochim. Biophys. Acta*, **271**, 1
115. Boden, N., Holmes, M. C. and Knowles, P. F. (1979) *Biochem. J.*, **177**, 303
116. Terenzi, M., Rigo, A., Franconi, C., Mondovi, B., Calabrese, L. and Rotilio, G. (1974) *Biochim. Biophys. Acta*, **351**, 230
117. Bertini, I., Luchinat, C. and Messori, L. (1981) *Biochem. Biophys. Res. Commun.*, **101**, 577
118. Rigo, A., Stevanato, R., Viglino, P. and Rotilio, G. (1977) *Biochem. Biophys. Res. Commun.*, **79**, 776
119. Bertini, I., Luchinat, C. and Scozzafava, A. (1980) *J. Am. chem. Soc.*, **102**, 7349
120. Rotilio, G. (1983) In: Ref. 111, p. 147
121. Bertini, I., Luchinat, C. and Scozzafava, A. (1983) In: Ref. 111, p. 155
122. Marwedel, B. J., Kosman, D. J., Bereman, R. D. and Kurland, R. J. (1981) *J. Am. chem. Soc.*, **103**, 2842
123. Strothkamp, K. G. and Lippard, S. J. (1981) *Biochemistry (USA)*, **20**, 7488
124. Viglino, P., Rigo, A., Raineri, G. A. and Rotilio, G. (1979) *J. magn. Res.*, **34**, 265 265
125. Rigo, A., Viglino, P., Argese, E., Terenzi, M. and Rotilio, G. (1979) *J. biol. Chem.*, **254**, 1759
126. Viglino, P., Rigo, A., Argese, E., Calabrese, L., Cocco, D. and Rotilio, G. (1981) *Biochem. Biophys. Res. Commun.*, **100**, 125
127. Fee. J. A. and Ward, R. L. (1976) *Biochem. Biophys. Res. Commun.*, **71**, 427
128. Cass, A. E. G. and Hill, H. A. O. (1980) In: Ref. 31, p. 290
129. Fee, J. A. and DiCorleto, P. E. (1973) *Biochemistry (USA)*, **12**, 4893
130. Lawrence, G. D. and Sawyer, D. T. (1979) *Biochemistry (USA)*, **18**, 3045
131. Barrette, W. C., Sawyer, D. T., Fee, J. A. and Asada, K. (1983) *Biochemistry (USA)*, **22**, 624
132. Lange, R. and Debey, P. (1979) *Eur. J. Biochem.*, **94**, 485
133. Jones, P. and Dunford, H. B. (1977) *J. theor. Biol.*, **69**, 457
134. McAdam, M. E., Fielden, E. M., Lavelle, F., Calabrese, L., Cocco, D. and Rotilio, G. (1977) *Biochem. J.*, **167**, 271
135. Green, M. R., Hill, H. A. O. and Turner, D. R. (1979) *FEBS Lett.*, **103**, 176
136. Rabani, J., Klug-Roth, D. and Lillie, J. (1973) *J. Phys. Chem.*, **77**, 1169
137. McCord, J. M. and Fridovich, I. (1969) *J. biol. Chem.*, **244**, 6049
138. Beem, K. M., Rich, W. E. and Rajagopalan, K. V. (1974) *J. biol. Chem.*, **249**, 7298
139. Lippard, S. J., Burger, A. R., Ugurbil, K., Pantoliano, M. W. and Valentine, J. S. (1977) *Biochemistry (USA)*, **16**, 1136
140. Calabrese, L., Cocco, D. and Desideri, A. (1979) *FEBS Lett.*, **106**, 142
141. Beem, K. M., Richardson, D. C. and Rajagopalan, K. V. (1977) *Biochemistry (USA)*, **16**, 1930
142. Calabrese, L., Cocco, D., Morpurgo, L., Mondovi, B. and Rotilio, G. (1976) *Eur. J. Biochem.*, **64**, 465
143. Cass, A. E. G., Galdes, A., Hill, H. A. O., McClelland, C. E. and Storm, C. B. (1978) *FEBS Lett.*, **94**, 311
144. Jones, C. W. (1982) *Bacterial Respiration and Photosynthesis*, Nelson, London

ADDENDA

Since the completion of this chapter considerably more structural data on superoxide dismutases has appeared. Further refinement of the diffraction data on the cuprozinc enzyme by the Duke group has shown the occurrence of an ordered water structure in the region of the active site.[1] In the paper given in reference 2 the authors use computer graphics to superimpose electrostatic field vectors on to the molecular skeleton and clearly show that the substrate is 'guided' into the active site.

Two x-ray diffraction studies on the iron superoxide dismutases have been reported in more detail. The enzyme from *Ps. ovalis* has had its structure determined to a resolution of 2.9 Å[3] whilst that from *E. coli* was at 3.1 Å[4]. As intimated by the partial amino acid sequence data, the structures of the two proteins are essentially identical. Furthermore the structures bear no resemblance to that of the cuprozinc enzyme. Although in neither case is the resolution good enough to identify unambiguously the ligands, at least one histidine residue appears to be involved. A further interesting observation is that both proteins seem to possess an additional, organic, prosthetic group.

A second iron-containing superoxide dismutase has been shown to be reconstitutable as an active mangano-enzyme.[5]

Additional references

1. Tanier, J. A., Getzoff, E. D., Richardson, J. S. and Richardson, D. C. (1983) *Nature*, **306**, 284
2. Getzoff, E. D., Tanier, J. A., Weiner, P. K., Kollman, P. A., Richardson, J. S. and Richardson, D. C. (1983) *Nature*, **306**, 287
3. Ringe, D., Petsko, G. A., Yamakura, F., Suzuki, K. and Ohmori, D. (1983) *Proc. nat. Acad. Sci.* (USA), **80**, 3879
4. Stallings, W. C., Powers, T. B., Pattridge, K. A., Fee, J. A. and Ludwig, M. L. (1983) *Proc. nat. Acad. Sci.* (USA), **80**, 3884
5. Gregory, E. M. and Dapper, C. H. (1983) *Arch. Biochem. Biophys.*, **220**, 293

5

Structure and Chemistry of Cytochrome P-450

Ralph I. Murray, Mark T. Fisher, Peter G. Debrunner and Stephen G. Sligar

Two decades have passed since the discovery in liver microsomes of a haem-protein that forms a reduced-CO complex with the absorptive maximum of the Soret at 450 nm (Klingenberg, 1958; Garfinkel, 1958) and the identification of this protein as a new cytochrome: pigment cytochrome, P-450 (Omura and Sato, 1962, 1964a). In the intervening years, the study of cytochrome P-450 dependent monoxygenases has expanded exponentially. From the first crude attempts to solubilise a P-450 (Omura and Sato, 1963, 1964b) to the determination of the primary, secondary, and tertiary structure of cytochrome P-450$_{cam}$ by amino acid sequencing (Haniu et al., 1982a,b) and x-ray crystallography (Poulos et al., 1984) our understanding of this unique family of proteins has been advancing on all fronts. Since, perhaps, the greatest understanding of the structure and mechanism of P-450s has come from concentrated study of P-450$_{cam}$ of the *Pseudomonas putida* camphor-5-*exo*-hydroxylase, this review will concentrate on findings with P-450$_{cam}$; attention will be drawn to parallel and contrasting examples from other P-450s as appropriate.

INTRODUCTION

Cytochrome P-450s are the substrate and oxygen reactive components found at the terminus of a truncated electron transport chain in a family of monoxygenases. They are present in prokaryotes, lower eukaryotes, and probably all higher eukaryotes. P-450s have been identified in plants, insects, and fish, as well as in many mammalian tissues and cell types. The ubiquitous distribution of P-450s is equalled by the variety of molecules which are their substrates, including lipophilic substrates in intermediary metabolism such as steroids, bile acids, prostaglandins, and the D-vitamins. P-450 dependent monoxygenases are key components in the peripheral metabolic pathways of bacteria and are the primary

157

means by which lipophilic xenobiotics are detoxified and solubilised. The cyto-chrome P-450s are also agents responsible for conversion of relatively innocuous molecules into active carcinogens.

Two classes of P-450 monoxygenases are defined by the character of the electron transport chain (figure 5.1). Bacterial and mitochondrial systems have

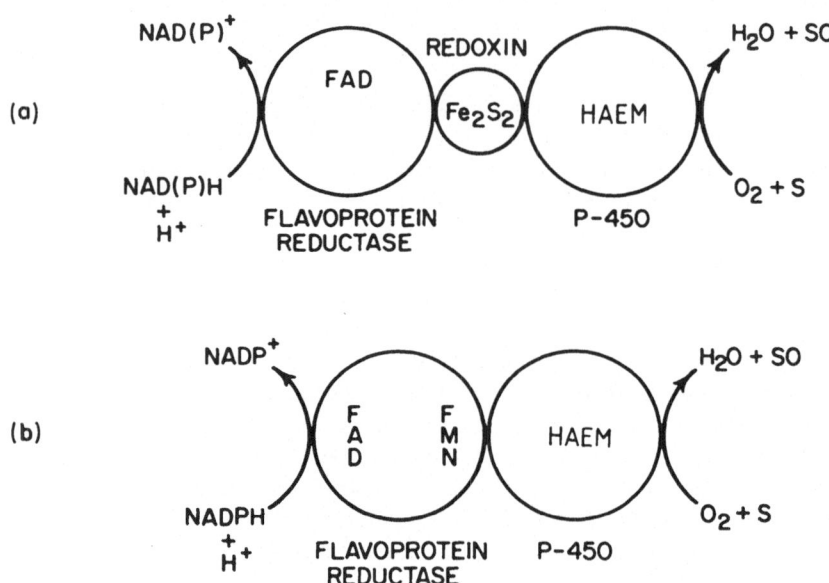

Figure 5.1   Schematic diagram of (a) bacterial/mitochondrial and (b) microsomal cyto-chrome P-450 electron transfer proteins

three protein components, utilise NAD(P)H as the electron donor, and contain an FAD-flavoprotein reductase which receives two electrons from NADPH. Single electrons are then passed via the iron sulphur centre ($Fe_2S_2Cys_4$) of a redoxin protein component to the P-450 haem active centre. In microsomal monoxygenases NADPH provides two electrons to the FAD prosthetic group of a more complicated FAD/FMN-flavoprotein reductase. The electrons are then passed singly through the FMN prosthetic group to the haem of cytochrome P-450.

The generalised reaction cycle for P-450 monoxygenases is illustrated sche-matically in figure 5.2, and has been recently reviewed in depth by White and Coon (1980). The reaction cycle begins with ferric P-450 binding substrate. A one electron reduction followed by O_2 binding generates the ternary ferrous-substrate-O_2 complex, which can auto-oxidise with loss of superoxide to regene-rate ferric P-450 (Sligar *et al.*, 1974; Lipscomb *et al.*, 1976). With the addition of a second electron, auto-oxidation to peroxide and ferric P-450 can occur (Esta-brook, 1982; Werringloer, 1982), or reaction can continue with formation of

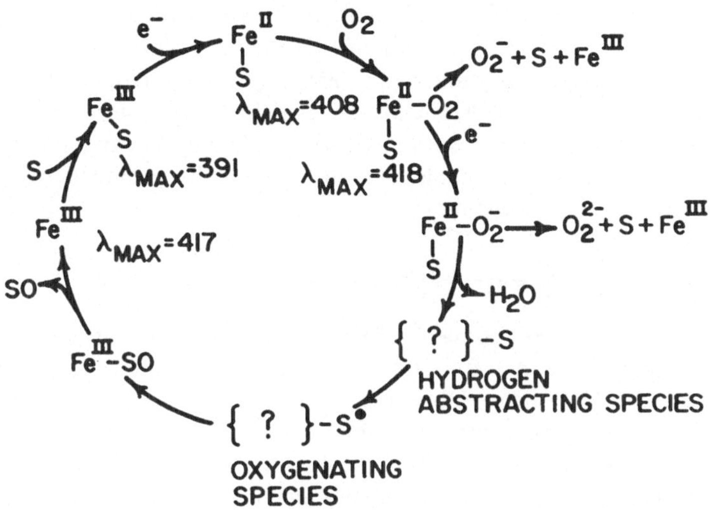

Figure 5.2 Proposed catalytic cycle for bacterial cytochrome P-450$_{cam}$. Fe^{II} and Fe^{III} refer to ferrous and ferric P-450 haem, S refers to the presence of substrate (not coordinated to the haem prosthetic group), S· to a substrate carbon radical intermediate and SO to oxygenated substrate. λ_{MAX} refers to the position of the Soret maximum

product wherein diatomic oxygen is split with one atom of oxygen being ultimately reduced to H_2O and the second atom incorporated into substrate. The nascent product subsequently dissociates, regenerating ferric P-450 and the cycle can begin again.

Although excellent progress has been made in the purification and the chemical and physical characterisation of many membrane associated P-450s, our understanding of the structure and mechanism of these fascinating proteins would have been severely restricted in the absence of data obtained from cytochrome P-450$_{cam}$ of the camphor-5-*exo*-hydroxylase of *Pseudomonas putida*. P-450$_{cam}$ was discovered during study of the plasmid-linked peripheral metabolic pathway for camphor degradation that allows *P. putida* to grow with camphor as its sole source of energy and carbon (Bradshaw *et al.*, 1959; Gunsalus *et al.*, 1965; Hedegaard and Gunsalus, 1965; Katagiri *et al.*, 1968). While most P-450s are available in relatively limited quantities and are tightly membrane bound, a strain of *P. putida* was isolated from which large quantities of P-450$_{cam}$ could be recovered from the supernatant of cell lysates. P-450$_{cam}$ is thus available in large quantities, and shows excellent solubility and stability in the absence of detergents, glycols, and reducing agents. With an unusually stable haem group, P-450$_{cam}$ was purified to homogeneity and is the only P-450 to have been crystallised (Yu and Gunsalus, 1970; Yu *et al.*, 1974). Although P-450$_{cam}$ was first crystallised as the substrate complex (Yu *et al.*, 1974; Gunsalus and Wagner, 1978), this crystalline form of the enzyme (orthorhombic I) was unsuitable for high-resolution x-ray diffraction. More recently, a crystal modification was prepared (orthorhombic II) that shows great promise for high resolution x-ray diffraction studies on several enzymatic intermediates (Gunsalus *et al.*, 1980; Poulos *et al.*, 1982). Thus, the camphor-5-*exo*-hydroxylase offers an ideal system for mechanistic studies of the P-450 dependent monoxygenases. How well it has served as a model system for the other members of the P-450 family will also become evident.

STRUCTURE OF CYTOCHROME P-450

Cytochrome P-450s were so named because the Soret band of the reduced-CO complex, typically found at 420 nm with other haemproteins, appeared at 450 nm; the spectrum also exhibits a new peak near 350 nm, and has diminished α and β bands. Recognising that these absorption differences implied a difference in the haem ligation, and noting that the 450 nm absorption could be shifted to a 420 nm maximum by treatment of the enzyme with mercurial sulphhydryl reagents, one axial ligand to the haem was tentatively identified as a cysteine thiolate (Mason *et al.*, 1965). A strenuous effort to document the presence of this thiolate ligand followed. Comparison of the spectral characteristics of model haem-thiolate complexes showed that a thiolate would produce the characteristic P-450 spectrum (e.g. Stern and Peisach, 1974; Koch *et al.*, 1975a,b; Chang and Dolphin, 1975). The discovery that chloroperoxidase isolated from *Caldaryomyces fumago* matched many spectral features of cytochrome P-450 temporarily confused the issue because chloroperoxidase could not be unambiguously demonstrated to have a free sulphhydryl group in the denatured protein (Champion *et al.*, 1975a,b; Chiang *et al.*, 1975). To date, the ligand structure of chloroperoxidase has not been completely resolved but mercaptide ligation is no longer ruled out (Cramer *et al.*, 1978). On a theoretical note, iterative extended Hückel calculations provided confirmation that a metal-ligand structure of $Fe^{II}N_4(CO)RS^-$ should have a hyperporphyrin spectrum — exhibiting a Soret absorption split with bands at 350-380 nm and 440-480 nm (Hanson *et al.*, 1976). Thiol coordination is unlikely since the calculations predict a normal Soret band for the protonated ligand, RSH. More refined calculations show that a large fraction of the negative charge on the thiolate is transferred to the porphyrin via covalent σ and π bonds with the haem iron (Jung, 1983). More recently, a synthetic ferrous protohaem with a stoichiometric thiolate ligand provided by covalent attachment and confirmed by ^{1}H- and ^{13}C-NMR was shown to have the optical properties of cytochrome P-450s (Traylor *et al.*, 1981). Epr spectroscopy of the cobalt analogue of P-450$_{cam}$ suggests that a thiolate is the most likely proximal ligand in ferrous and oxygenated ferrous P-450$_{cam}$ (Wagner *et al.*, 1981). Comparison of the uv-visible, mcd, and epr spectra of sulphur-donor ligand complexes of P-450$_{cam}$ and myoglobin has provided strong indirect evidence for a thiolate ligand (Sono *et al.*, 1982), and resonance Raman spectroscopy (Champion *et al.*, 1982), EXAFS spectroscopy

(Hahn *et al.*, 1982), and x-ray crystallography (Poulos *et al.*, 1984) have provided definitive, direct, evidence that cysteine is the proximal ligand of the haem iron in the ferric substrate complex of P-450$_{cam}$.

In addition to the unique absorption of the reduced-CO complex, cytochrome P-450s also display characteristic spectral perturbations upon binding a variety of small molecules, including substrates and inhibitors (Schenkman *et al.*, 1981). These transitions of the Soret and charge transfer bands are linked to modulation of the d-electron spin-state equilibrium and coupled porphyrin π-π^* transitions and are divided into a few empirically defined classes which correlate with the structure and chemical character of the molecule (Horie, 1978; Schenkman *et al.*, 1981). The variety of substrates and chemical transformations of P-450-catalysed reactions is enormous: hydroxylation, epoxidation, N- and O-demethylation, N-oxidation, S-oxidation, and C-C bond cleavage are ubiquitous but can be considered as occurring through similar metal–oxygen intermediates. Thus, as the number and variety of discovered P-450 reactions continued to increase it was recognised that these enzymes had many major chemical and physical characteristics in common. Could we expect to find that these common characteristics were reflections of common structural elements?

Primary structure and the concept of P-450 families

Since the successful development of procedures for the purification of viable P-450s from mitochondria and microsomes, the chemical and physical characterisation of these proteins has been vigorously pursued. Early comparisons showed that while P-450s were distributed over a range of molecular weights (e.g. P-450$_{cam}$:46 000, Dus *et al.*, 1970; P-450$_{LM2}$ and P-450$_{LM4}$:49 000 and 56 000, Haugen and Coon, 1976), they had similar amino acid compositions featuring an unusually large percentage of hydrophobic residues (>50%) and a distinct overall acid character (Dus *et al.*, 1974, 1977). These comparisons were later extended to include P-450$_{SCC}$ isolated from bovine adrenal cortex (Dus *et al.*, 1980). While immunoprecipitation and reaction inhibition assays were used to define the differences between cytochrome P-450 isozymes from phenobarbital induced rat liver, the assays also indicated the proteins shared certain antigenic identities (Thomas *et al.*, 1976). Noting that differences in substrates, electron transfer chains and degree of association with membranes would require variations in primary structure, Dus addressed structural studies to that portion of the P-450s most likely to retain the structure of a common ancestor: that of the haem binding domain (Dus, 1976, 1982a). With active-site directed photo-affinity labelling used to incorporate a radioactive marker in a stabilised haem domain, selected chemical and enzymatic digestions allowed isolation of oligo-peptides (~ 50 residues), apparently containing the haem binding site. Chemical, immunochemical, and physical characterisation followed and made possible the first structural comparison of the haem domains from different P-450s (Dus *et al.*, 1974, 1977, 1980; Swanson and Dus, 1979; Hanukoglu *et al.*, 1981).

Concurrent with analysis of the haem domain, other laboratories were engaged in peptide mapping and protein sequencing of several P-450s. Initial reports for N- and C- terminal sequences for three rat liver isozymes showed little or no homology (Botelho *et al.*, 1979), thus confirming that the P-450 isozymes arose from distinct genes rather than from post-translational modifications of a common precursor protein. Comparison of amino terminal sequences with that of a fourth P-450 isozyme from rat liver also showed a lack of homology (Botelho *et al.*, 1982). It should be noted, however, that these terminal sequences included only 15–20 amino acid residues. For proteins of 400 residues, the lack of homology in these small regions can make little statement about the existence of homologies in total protein structure. Comparison of N-terminal sequences of P-450$_{LM-2}$ and P-450$_{LM-4}$ showed them to be similar to N-terminal sequences of several 'preproteins' which are not found in the corresponding mature forms (Haugen *et al.*, 1977). The first documentation of sequence homology, utilising peptide mapping to select similar peptides, identified one pair in which the two isozymes differed in only 1 of 13 amino acid residues (Ozols *et al.*, 1981).

In the last year the total sequence has been published for P-450$_{cam}$ (Haniu *et al.*, 1982a,b) and for P-450$_b$ from phenobarbital induced rat liver (Fuji-Kuriyama *et al.*, 1982), as well as two partial sequences for P-450$_{LM-2}$ (Black *et al.*, 1982; Heinemann and Ozols, 1982). The sequence of P-450$_b$ was determined by sequencing cDNA and establishing the correct reading frame using the known amino terminal sequence of Bothelo *et al.* (1979). Comparison of this sequence to the compositions of haempeptides from P-450$_{cam}$, P-450$_{LM-2}$ and P-450$_{SCC}$ showed one region from residues 427–467 which was noted as a possible candidate for the P-450$_b$ haempeptide. Comparison of the P-450$_b$ and P-450$_{LM-2}$ sequences identified three regions with substantial homology (Black *et al.*, 1982) including one which contains a cysteine residue and is also homologous to the section of P-450$_{cam}$ containing Cys-134 which has been identified in the x-ray structure as the haem-liganding cysteine. Identification of a haempeptide from P-450$_{cam}$ as residues 103–154 narrowed the choice to Cys-134 or Cys-146 (Dus and Murray, 1982). Differential labelling with iodoacetic acid led to the choice of Cys-134 (Haniu *et al.*, 1983) which has been confirmed by the x-ray structure. A large fraction of the amino acid backbone can be traced in the unrefined, 3 Å electron density map, with an α-helical segment comprising residues 134–148 particularly visible (Poulos *et al.*, 1984). This assignment, based on the known sequence (Haniu *et al.*, 1982a,b), identifies Cys-134 as the axial ligand of the haem iron. That Cys-134 of P-450$_{cam}$ is the haem ligand would argue in support of the assignment of Cys-147 and Cys-152 as the haem ligands for P-450$_{LM-2}$ and P-450$_b$, respectively (Black *et al.*, 1982). Recently, Byron Kemper at the University of Illinois has isolated six full-length cDNA clones from +/− phenobarbitol induced total poly-A message from rabbit liver, and has documented extensive sequence homology (Leighton *et al.*, 1984).

From the sequence data to date it seems safe to conclude that while major structural differences exist among the cytochrome P-450s, there are certain functionally significant, highly conserved regions which support the concept that the diverse P-450s are in fact a closely related family of proteins. This subject has recently been reviewed in depth by Dus (1982b).

Secondary structure

Application of the method of Chou and Fasman (1978) to the completed sequence of P-450$_{cam}$ has resulted in a predicted secondary structure consisting of 46% α-helix, 16% β-pleated sheet, 21% β-turn, and 38% random coil (Haniu *et al.*, 1982b). Although the current resolution (3 Å) of the x-ray crystallographic structure of P-450$_{cam}$ does not allow confirmation of these secondary structural assignments, it does show this soluble protein to be asymmetrical with maximum dimensions along three coordinate axes of 55 Å x 55 Å x 39 Å (Poulos *et al.*, 1984). Based upon comparison of the amino acid composition of P-450$_b$ with the amino acid compositions and topologies of other membrane proteins, Heinemann and Ozols (1982) proposed that P-450$_b$ consists of a series of α-helical segments embedded in the membrane connected by small, exposed polar sections, with a carboxy-terminal region largely exposed on the membrane surface and containing the active site. This surface portion would include their projected haem-binding domain; it would also place the haem-binding domain suggested by Black *et al.* (1982) on the opposite surface of the membrane from the large, hydrophilic carboxy terminal segment.

Structure of the haem-binding site

The haem of P-450$_{cam}$ is located near the surface of the protein but is surrounded by α-helical segments, with no portion of the porphyrin directly accessible at the surface (Poulos *et al.*, 1984). Earlier spin label experiments suggesting the haem to be several angstroms below the surface of the protein are therefore corroborated (Armes *et al.*, 1981). The x-ray structure also shows the haem to have a chiral orientation like that found for haemoglobin (Perutz *et al.*, 1968) and myoglobin (Kendrew *et al.*, 1961). A preliminary fit of an idealised, planar haem group to the corresponding electron density of the unrefined 3 Å map places the haem normal within 2.5° of the crystal z-axis, an orientation that is consistent with single crystal epr and linear dichroism data. This is also apparently the orientation found for the rat liver P-450s by chemical derivatisation with a suicide substrate (Ortiz de Montellano *et al.*, 1983). Due to the formation of a strong coordinate bond between the pyridyl nitrogen and ferric haem, metyrapone is a general inhibitor of cytochrome P-450s. Using spin-labelled analogues of this and related compounds, attempts have been made to map qualitatively the active site of cytochrome P-450 (Griffin *et al.*, 1974; Ruf and Nastainczyk, 1976).

While the fifth ligand to the haem iron of P-450s has been substantially confirmed as a cysteine thiolate, the sixth ligand in the low spin complex has not yet been identified. This may in part be due to the lability of the sixth ligand coordination in the ferric low-spin P-450 and to the potential dominance of spectral characteristics by the proximal thiolate bond. On the basis of proton nmr solvent relaxation experiments Griffin and Peterson (1975) proposed an oxygen atom of H_2O coordinated to the iron at a distance of ~ 2 Å. A careful comparison of the epr spectra of haem proteins and model compounds led to the suggestion that imidazole is the ligand *trans* to the cysteine thiolate (Chevion *et al.*, 1977). In this study, however, only nitrogen donor ligands were examined. Ruf *et al.* (1979) used the pH dependence of electronic and epr spectra to conclude that a hydroxyl was a better candidate for the sixth ligand than imidazole. At the same time, a generalised treatment of solvent proton-nmr relaxation rates led Philson *et al.* (1979) to conclude that water and several amino acids with exchangeable protons (ROH:Tyr, Ser, Thr; RNH_2:Lys, Arg; $RCONH_2$:Asn, Gln) were acceptable candidates for the sixth ligand. Histidine was eliminated because the exchangeable proton was found to be 2.6–2.9 Å from the haem while the imidazole proton would have to be at a distance of 4.8 Å. Second derivative electronic spectroscopy was used to resolve the absorptions of aromatic amino acids of P-450$_{LM2}$ and showed simultaneous titration of a tyrosine hydroxyl and spectral transition of the Soret absorption as pH was increased (Ruckpaul *et al.*, 1980). Based upon electronic and epr spectral changes as a function of pH, Rein *et al.* (1980) proposed detailed variations in ligand structure with pH, oxidation state and added CO. Recently, several groups have approached the problem of identifying the sixth ligand by titrating native P-450 with exogenously added small molecule ligands analogous to the suggested native ligands. Shimizu *et al.* (1981) tested nitrogen and sulphur donors using a sensitive ratio of Soret vs mcd bands. They concluded that a sterically hindered nitrogen ligand was the best fit of the models tested, though it was not an outstanding fit on an absolute scale. Comparison of the electronic spectra from models prepared by addition of O, N, or S donor ligands to P-450$_{LM2}$ and P-450$_{LM4}$ led White and Coon (1982) to conclude that the sixth ligand is an oxygen donor. In another detailed study, Dawson *et al.* (1982) used the combined analysis of electronic, mcd, and epr spectroscopies to narrow the choices to water or alcoholic amino acids (Ser, Thr, Tyr), with a hydroxyl group polarisation enhanced by hydrogen bonding to a weakly coupled nitrogen contributed by the peptide backbone or a proximal histidine. Yoshida *et al.* (1982) also showed that the sixth ligand of rabbit liver P-450 is most likely a hydroxyl donor, using exogenous ligand complexes of rabbit liver P-448 as the model system. In their experiments they found that the intensity of the α-band is inversely proportional to the field strength of the sixth ligand. While addition of nitrogenous ligands to either P-450 or P-448 gave identical red shifted spectra, 1-propanol added to P-448 elicited a spectrum identical to that of native P-450. Thus, although the identity of the sixth ligand remains uncertain, the evidence obtained since the previous review by White and Coon (1980) continues to support the strong suggestion of water or an alcoholic oxygen donor.

Electronic structure of the haem active centre

A variety of spectroscopic methods have been applied to cytochrome P-450$_{cam}$ in an effort to characterise the electronic state of the active centre and to draw inferences, if possible, about the nature of the ligands and their spatial arrangement. Although the x-ray studies presently in progress will eventually reveal the complete three dimensional structure of the protein, an understanding of the enzymatic reaction requires in addition some knowledge of the electronic structure and the dynamics of the active site that is beyond the resolution of the crystallographer. Spectroscopic methods ranging from magnetic resonance to the Mössbauer effect can provide this additional information. Spectroscopy measures the energy emitted or absorbed in transitions between different energy levels, and from these one attempts to deduce, with the help of well known physical laws, the state of the system being studied. Depending on the type of spectroscopy and the size and complexity of the system of interest the theoretical interpretation may be straightforward or exceedingly difficult. To illustrate this point consider epr and optical spectroscopy. Epr measures the Zeeman splitting of unpaired electrons in an applied field. The magnetic moment of these electrons is specified by three g-values and, possibly, three angles of orientation. To the extent that the electrons are localised on the haem iron and more specifically occupy the Fe 3d orbitals a ligand field model with few adjustable parameters is adequate to explain the data (Palmer, 1983). A slightly extended version of the same model applies to Mössbauer spectroscopy, which uses the electric and magnetic moments of ^{57}Fe nuclei to probe the charge distribution and internal magnetic field arising from the valence electrons of the iron. Optical spectroscopy, on the other hand, which is by far the most widely used method, measures the excitations of the π-electron system on the porphyrin ring. Since the electrons are delocalised over the whole haem group and communicate, via the iron, with the axial ligands, the calculation of the energies and transition probabilities of this multi-electron multi-centre system is very difficult. Further complications arise from the coupling to the vibrational modes, which is measured explicitly in the resonance Raman spectra. While quantum chemical approximations have been developed to the point where meaningful predictions of optical spectra can be made (Loew, 1983; Jung, 1983), the calculations are more reliable if the effect of a limited change in parameters is considered. The latter approach is illustrated by the comparison of the energy levels in a ferrous haem CO complex with either a nitrogen base, a thiol or a thiolate fifth ligand (Hanson *et al.*, 1976). The calculations showed convincingly that thiolate, but not thiol, could account for the anomalous Soret band of carbonmonoxy cytochrome P-450.

The great practical value of optical spectroscopy lies in its use as a diagnostic tool. Spectral changes signal physico-chemical changes in the haem environment, and optical titrations can be interpreted without ambiguity if not more than two species are in equilibrium. Spectral similarities, on the other hand, do not imply similar haem coordination. In fact, several states of cytochrome P-450 display absorption spectra that resemble those of corresponding states in other haem proteins, whereas others do not, in particular the CO and thiol adducts, which show 'hyper' spectra with split Soret bands. Optical spectroscopy therefore cannot 'prove' that thiolate is the intrinsic iron ligand of P-450, but numerous model studies are at least consistent with the assumption of thiolate ligation in all states of cytochrome P-450. For the camphor system this implies via x-ray studies that the deprotonated form of Cys-134 is the permanent intrinsic iron ligand.

Cysteinate coordination allows rationalisation of many distinctive features of cytochrome P-450. A simple estimate shows that a cysteine-haem complex with ideal tetrahedral bond angles and distances will bring the Cβ hydrogens very close ($\leqslant 2.5$ Å) to the pyrrole nitrogens. Some distortion of the ideal geometry is therefore likely, and it remains to be seen if the porphyrin ring is not planar, if the iron–sulphur bond is not along the haem normal, if the Fe-S-Cβ angle is larger than 109.5°, or if a combination of these possibilities occurs. A preliminary fit of a planar haem and an ideal α-helix containing residues 134–148 to the unrefined 3 Å electron density map of the camphor complex of cytochrome P-450$_{cam}$ yields an angle of 12° between the haem normal and the Fe-S vector and a Fe-S-C angle of 113° (T. Poulos, private communication). In any case, a substantial π admixture to the iron–sulphur bond can be expected, which involves a suitable linear combination of the sulphur lone pair and the iron dπ orbitals. The ultimate interaction of the sulphur lone pairs with the porphyrin π electrons is responsible for the hyper spectrum of the CO complex. The same model explains other unusual features of P-450, which will be taken up later, in particular the low symmetry of the electron distribution on the iron as evidenced by epr and Mössbauer data, and the unusually large π^* density on the Fe(II) porphyrin deduced from resonance Raman spectra.

Comparative studies of haem complexes of known structure have led to a set of semi-empirical rules that correlate optical features with ligand coordination, geometry, charge and spin state of the iron (Scheidt and Gouterman, 1983). According to these rules, low-spin iron is positioned close to the haem plane and has two strong-field axial ligands. High-spin iron, on the other hand, is typically five-coordinate and is displaced from the haem plane towards the axial ligand. The low-spin to high-spin transition is accompanied by a blue shift of the Soret band for the ferric haem. These rules are satisfied by model compounds with one or two benzene thiolate ligands (Tang *et al.*, 1976; Byrn and Strouse, 1981) and are assumed to apply to cytochrome P-450 as well. Accordingly, the low-spin ferric iron of substrate free P-450$_{cam}$ should have a sixth, axial ligand in addition to cysteinate. The nature of this ligand, however, is still controversial. As described earlier, proton relaxation and endor measurements have identified an exchangeable proton that is covalently linked to one of the axial ligands and has a distance of 2.7 Å from the iron. This observation rules out histidine as the sixth ligand but is compatible with a bound water, an alcohol, or possibly an amine or amide group of the protein. Neither one of the presently favoured candidates, water or alcohol, are considered strong field ligands, though.

The intrinsic sixth iron ligand is readily displaced by nitrogen bases or thiols, which produce a red shift of the Soret band, but little change in the epr spectra. The common feature of these substituents is their π base character, and it appears that the iron effectively couples the donor π orbitals to the porphyrin π system thus affecting the optical $\pi-\pi^*$ transitions without major change in the population of the iron 3d orbitals. The electronic state of the iron in substrate-free P-450$_{cam}$ has been characterised by epr and Mössbauer measurements. The g-values observed in frozen solution, $\bar{g} = (1.91, 2.26, 2.45)$ (Lipscomb, 1980), are very close to the single crystal values listed in table 5.1. Crystallisation,

Table 5.1 Principal Values and Orientations of the g-tensors in P2$_1$2$_1$2$_1$ Single Crystals[a] of Camphor Free and Camphor Complexed Cytochrome P-450$_{cam}^{bc}$

	Camphor free			Camphor complex					
	$g_1 =$ 1.91	$g_2 =$ 2.24	$g_3 =$ 2.46	$g_1 =$ 1.97	$g_2 =$ 2.23	$g_3 =$ 2.40	$g_1 =$ 7.75	$g_2 =$ 3.93	$g_3 =$ 1.80
Angle[d]									
(x, i)	78°	166°	98°	78°	156°	110°	115°	25.5°	95°
(y, i)	12°	78°	89°	12°	78°	87°	155°	115°	90°
(z, i)	89°	98°	8°	88.5°	110°	20°	92°	85.5°	5°

[a]Poulos *et al.* (1984).
[b]From Devaney, P. W. (1980) Ph.D. Thesis, University of Illinois, Urbana, IL (unpublished).
[c]The orientation of the g-tensor quoted is the one that best matches the orientation of the haem coordinate system specified by orthogonal axes passing through the pyrrole nitrogens. Preliminary haem coordinates are based on an unrefined electron density map at 3 Å resolution (Poulos *et al.*, 1984; T. Poulos, private communication). The four molecules in the unit cell of the P2$_1$2$_1$2$_1$ crystals have g-tensors related to each other by 180° rotations about the crystal *x*, *y* and *z* axis, respectively.
[d]The angle between the crystal axes *x*, *y* and *z* and the principal axes *i*, *i* = 1, 2, 3, of the g-tensor is given.

therefore, does not modify the haem environment significantly. In fact, it renders it more homogeneous as judged by the decrease in the g-strain contribution to the epr line width. As judged by linear dichroism data the haem normal must have essentially the same orientation in these crystals as in those of the camphor complex. Thus, the angle between the haem normal and the direction of the largest g-component is estimated to be $(18 \pm 3)°$. The other two principal axes of the g-tensor are then close to the haem plane, an orientation that is found in other low-spin haem proteins and a model compound as well (Palmer, 1983; Byrn and Strouse, 1981). Compared to all other low-spin haems, cytochrome P-450 and thiolate-coordinated models are uniquely distinguished by their small deviation of the g-values from the spin-only value, $g_s = 2$. This group property is readily interpreted in terms of ligand field theory (Palmer, 1983). The theory approximates the bonding interactions of the iron with the ligands by an electrostatic potential of appropriate symmetry and strength. The Fe(III) is assumed to have a $(3d)^5$ configuration, and the energy levels of the 3d orbitals in the assumed potential are readily calculated. In the low-spin state, the five 3d electrons are distributed over the three t_{2g} orbitals, whose energies depend on the axial and rhombic components Δ and V of the potential and on the spin orbit coupling constant λ. Given Δ/λ and V/λ, the ground state wave function can be calculated as a linear combination of t_{2g} orbitals, from which the g-tensor follows. The unique feature of P-450s is the large rhombic splitting between the antibonding d_π type orbitals d_{xz} and d_{yz}. As a consequence, the unpaired electron is predominantly (87% in P-450$_{cam}$) in the highest orbital d_{yz}, the orbital angular momentum is small, and g-values are close to $g_s = 2$, the value for pure spin. Compared to other low-spin haem proteins, cytochrome P-450$_{cam}$ has an unusually small spin lattice relaxation rate T_1^{-1}, a fact that can be traced to the small orbital angular momentum which couples the spin only weakly to the lattice vibrations. The reason for the large rhombicity lies in the strong π admixture of the axial iron sulphur bond, which renders the iron $3d_\pi$ orbitals inequivalent. The lower of the two d_π orbitals, d_{xz}, is almost fully occupied by two spin paired electrons, while the highest orbital, d_{yz}, is occupied by the last, unpaired electron. The ligand field parameters deduced from the g-tensor also account for the Mössbauer spectra of P-450$_{cam}$ if the antibonding d_π orbitals are allowed to be more delocalised than the nonbonding d_{xy} orbital (Sharrock *et al.*, 1976).

It should be pointed out that the model as presented here assumes that the axes of the cubic, tetragonal and rhombic ligand field potentials coincide. By symmetry, these axes must be the principal axes of the g-tensor as well. The $(18 \pm 3)^\circ$ deviation of the g_3 axis from the haem normal suggests that the assumed coincidence of axes is an approximation, and a slightly more complex model will have to be used (Rhynard *et al.*, 1979).

Binding of camphor or other 'Type I' substrates to P-450$_{cam}$ leads to a partial conversion of the haem iron from low-spin to high-spin. The spin transition is accompanied by a blue shift of the Soret band, as found in other haem proteins. The isosbestic points are independent of the substrate, which binds near the haem but does not directly interact with it. The extent of the spin transition depends on the substrate, on temperature, cations and pH, as will be discussed elsewhere. The 3 Å electron density map is compatible with a five-coordinate haem iron but does not yet reveal its exact position relative to the haem plane. Single crystal epr measurements yield the magnitudes and orientations of the high-spin and low-spin g-tensors listed in table 5.1. The high-spin g-values differ by less than 1.3% from those observed in frozen solution, $g = (7.85, 3.97, 1.78)$ (Lipscomb, 1980). The best match of the principal axes frame to a preliminary haem-geometry puts g_3 within 4° of the haem normal and g_1, g_2 within 4° of the iron–pyrrole nitrogen directions.

The large anisotropy of the high-spin g-tensor is another unique feature of P-450 cytochromes and thiolate haem models (Tang *et al.*, 1976). It can be rationalised by an extension of the arguments invoked to explain the low-spin g-tensor, that is the π admixture in the axial iron–sulphur bond. In the high-spin case, however, the energy splitting of the d_π orbitals does not affect the $^6A_1$

ground state of the ferric ion directly, but has a profound indirect effect on it via second order spin–orbit interaction with the excited $^4T_1$ state. Specifically, the energy splitting between the single electron orbitals d_{xy}, d_{xz} and d_{yz} lifts the three-fold orbital degeneracy of the $^4T_1$ state, and the second order spin-orbit interaction splits the spin sextet into three Kramer doublets. The result is the familiar zero field splitting described by the Hamiltonian $\mathcal{H}_z = D(S_z^2 - 35/12) + E(S_x^2 - S_y^2)$, which specifies the three Kramers doublets completely. For small E/D the g-tensor of the epr active ground doublet is given by $g_1 \simeq 6(1-4\ E/D)$, $g_2 \simeq 6(1+4\ E/D)$, $g_3 \simeq 2$. In metmyoglobin and related haem proteins the rhombic term E is negligible, and the g-tensor of the ground doublet is axially symmetric, $\overleftrightarrow{g} = (6,6,2)$. In P-450, on the other hand, the g-tensor is highly rhombic. In fact, the ratio $E/D \simeq 0.09$ assumes the largest value known in any haem protein. Quantitatively, the model predicts that $g_1 = 7.85$ and 1.91 in high and low-spin P-450, respectively, should be observed in the same direction with respect to the haem if the π-admixture in the iron–sulphur bond is unchanged. The experimental results are clearly compatible with this prediction.

Unexpected structural information has come from resonance Raman experiments on isotopically labelled high-spin P-450$_{cam}$. Champion and coworkers identified an iron–sulphur stretching band at 351 cm^{-1} on the basis of the mass shifts observed on substitution of ^{54}Fe, $\Delta_{Fe} = 2.5$ cm^{-1}, and ^{34}S, $\Delta_s = -4.9$ cm^{-1}, respectively. It turns out that a two body oscillator model cannot account for both mass shifts. Thus, coupling to a third atom, the β carbon of cysteine, has to be included. The resulting three body model suggests an Fe-S-Cβ angle of 125–145°, which is considerably larger than the angle of 100° found in a P-450 model compound. Since one would expect sp^3 hybridisation with a bond angle of 109° in both cases, additional constraints have to be invoked that reduce the angle in the model and increase it in the protein. Crystal packing effects and van der Waals interactions might affect the geometry of the model compound, whereas the peptide backbone might impose constraints in the protein. An Fe-S-Cβ angle near 120° is conducive to sp^2 hybridisation and allows strong π admixture in the Fe-S bond. The larger Fe-S-Cβ angle of P-450$_{cam}$ may in fact account for the larger rhombicity, $E/D \simeq 0.09$, of its epr signal compared to $E/D = 0.05$ of model compounds.

In addition to the high-spin species just discussed, two low-spin species can be observed in the epr spectra of camphor-saturated P-450$_{cam}$ in frozen solution. One of these species has a novel g-tensor, $\overleftrightarrow{g} = (1.97, 2.24, 2.42)$, which presumably represents the low-spin form of the camphor complex. The other one has a g-tensor $\overleftrightarrow{g} = (1.91, 2.26, 2.45)$, like that of the camphor free protein (Lipscomb, 1980). The suspicion thus arises that the $\overleftrightarrow{g}$ 1.91 signal is actually due to molecules with no bound camphor rather than to a low-spin fraction of the camphor complex. Substrate binding and spin equilibria may be intertwined, and some ill defined non-equilibrium state may be frozen in at the temperatures of the epr experiments.

The measurements on single crystals of camphor complexed P-450$_{cam}$ resolve some of these questions. They show the novel g 1.97 signal assigned to the camphor complex (Lipscomb, 1980) but no trace of the g 1.91 species, which must be due to the camphor free protein. Apparently, no substrate dissociation occurs in the short time it takes to freeze the crystals. At 4.2 K, roughly 30% of the iron has low-spin, and the g-tensor, which is listed in table 5.1, is almost identical with that of the g 1.97 species in frozen solution.

It was pointed out earlier that low-spin P-450s have the smallest difference between maximum and minimum g-values of any known haem protein. The g 1.97 species is even more extreme in this respect than the g 1.91 species, and an explanation of the g-values in terms of the standard ligand field model (Palmer, 1983) leads to inconsistencies. Often an orbital reduction factor k is introduced as a third adjustable parameter in addition to the axial and rhombic ligand field terms. Applied to P-450 cytochromes and model systems this factor k assumes rather unphysical values (Chevion *et al.*, 1978), and a more consistent model is clearly needed. Attempts to include admixtures of quartet and sextet states show great promise for both high and low spin P-450, but models with fewer parameters should also be considered (Ristau, 1980; Ristau and Rein, 1983).

It is clear from table 5.1 that the g_1 principal axes of the low-spin g-tensors are practically parallel, while the g_2 and g_3 axes are $12°$ apart. The principal axes of the high-spin g-tensor, which coincide within $4°$ with the haem coordinate system, deviate by $13°$ to $20°$ from the low-spin principal axis. Lacking any evidence to the contrary it is tempting to assume that the haem orientation in the crystal is the same for high-spin and low-spin species. Linear dichroism data on camphor free and camphor complexed crystals show that this approximation holds for the haem normals within five degrees. The largest g values, g_3 =2.46 and g_3 = 2.40 then are observed at $6°$ and $18°$ from the haem normal, respectively, and the smallest values, g_1 = 1.91 and g_1 = 1.97 are observed at $1.9°$ and $2.7°$ from the haem plane, respectively and at $10°$ and $9°$ from the x-axis of the haem. In other low-spin ferric haem complexes the principal axis of g_{max} deviates up to $20°$ from the haem normal (Palmer, 1983), and in a bithiolate complex the deviation is $12°$ (Byrn and Strouse, 1981). In this preliminary interpretation the orientation of the low-spin g-tensor is, thus, not unusual; a more detailed analysis has to await further x-ray data.

One-electron reduction of the camphor complex leads to a high-spin ferrous haem state in which the iron must be five coordinate. The spin S=2 character is expected on the basis of the optical spectrum and has been unambiguously demonstrated for all temperatures by Mössbauer spectroscopy, nmr and magnetic susceptibility (Champion *et al.*, 1975a,c; Keller *et al.*, 1972). The resonance Raman spectra indicate a high degree of electron donation from the axial cysteinate via the iron to the antibonding porphyrin π^* orbitals. The added π^* density decreases the interatomic forces in the haem ring and shifts several prominent bands to lower frequencies. The oxidation state marker band ν_4 at 1344 cm^{-1}, in particular, is 10–15 cm^{-1} lower in frequency than in other ferrous porphyrins (Ozaki *et al.*, 1976; Champion *et al.*, 1978). Unusual depolarisation ratios observed in the resonance Raman spectra also give a hint that the chromophore has lower symmetry than D_{4h}, the symmetry that is appropriate for other haem proteins (Champion *et al.*, 1978).

Since $(3d)^6$ Fe(II) is epr silent, one has to turn to Mössbauer spectroscopy for a characterisation of the electronic state of the iron. Alternatively, one can substitute cobalt for iron in the haem group and study the epr of the resulting P-450 analogue. In strong magnetic fields the reduced camphor complex of P-450$_{cam}$ shows Mössbauer spectra with well resolved electric and magnetic hyperfine splittings. The spectra have been parametrised in terms of a spin Hamiltonian, and a crystal field calculation with adjustable axial and rhombic potential terms reproduced the data quite well. As far as the Mössbauer spectra are concerned, cytochrome P-450$_{cam}$ is strikingly similar to chloroperoxidase and different from all other ferrous haem proteins studied (Champion *et al.*, 1975a,b). The zero field splitting represented by $D(S_z^2 - 2) + E(S_x^2 - S_y^2)$ is large and highly rhombic, $D/k = 20K$, $E/k = 3K$, compared to deoxymyoglobin, which has $D/k \simeq 7.5K$, $E \simeq 0$. Most importantly, the electric field gradient at the iron is rhombic too, and is rotated by 70° from the z-axis defined by the zero field splitting. Although the orientation of the haem with respect to the various tensors deduced from the Mössbauer data is not known, the crystal field calculation makes it plausible that the z-axis of the zero field splitting represents the haem normal, which is an axis of the cubic potential, while the other cubic axes pass through the pyrrole nitrogens. The axial and rhombic potentials then have to be rotated away from the cubic axes so that the local symmetry at the iron is reduced to C_1. Although a ligand field of rhombic symmetry models the epr and Mössbauer data of ferric P-450$_{cam}$ reasonably well, a local C_1 symmetry at the iron is compatible with all the data. To explain the low symmetry, it is sufficient to invoke a tilt of the iron–sulphur bond away from the haem normal.

Additional information about the haem environment can be obtained from cobalt substituted P-450$_{cam}$, which is epr active in the reduced form (Wagner *et al.*, 1981). The cobalt analogue has limited, but stereospecific enzymatic activity and is therefore a useful model system. The epr spectra of reduced Co-P-450$_{cam}$ are similar to those of other five coordinate haems with strong base axial ligands, apart from a slight rhombicity of the g-tensor and the absence of nitrogen superhyperfine splitting found in complexes with axial nitrogen base. The cobalt substituted camphor complex forms a relatively unstable O_2 adduct whose epr spectrum is similar to that of the myoglobin and horse radish peroxidase cobalt oxy analogues. The ^{59}Co hyperfine structure is considerably better resolved in oxy P-450$_{cam}$, however, a fact that is explained by the lack of superhyperfine interaction with an axial nitrogen base. The hyperfine splitting of a thiol coordinated oxy cobalt haem differs sufficiently from that of the protein to rule out thiol as a ligand in reduced P-450. Thus, all epr evidence is consistent with thiolate coordination in reduced P-450$_{cam}$ and its oxy adduct (Wagner *et al.*, 1981). Model studies also support an RS$^-$ ligand in both states (Berzinis and Traylor, 1979; Budyka *et al.*, 1981).

Most spectral properties of oxy-cytochrome P-450$_{cam}$ are 'normal' in the sense that they differ little from those of oxymyoglobin. This statement holds for the optical absorption, which reflects the electronic state of the porphyrin, as well as for the ^{57}Fe Mössbauer or the cobalt epr spectra, which reflect the electronic state of the metal. Apparently little negative charge is transferred from the thiolate to the metal or to the porphyrin. Model calculations reproduce this behaviour but suggest that the O_2 acquires a more negative charge of $g = -0.7$ than the $g = -0.48$ predicted for oxyhaemoglobin (Jung, 1983). The larger negative charge should lower the O_2 stretching frequency and would tend to destabilise oxy P-450$_{cam}$ relative to oxyhaemoglobin, a trend that agrees with observation.

High field Mössbauer spectra of oxy P-450$_{cam}$ indicate a diamagnetic, low-spin ferrous state of the iron. As is observed in all oxygenated haem complexes, the quadrupole splitting is anomalously large, negative and strongly temperature dependent. The similarity in the Mössbauer parameters of all oxy haems indicates that the electronic state of the iron is dominated by the Fe–O_2 bond and suggests that the geometry of the bond is similar. Various attempts have been made to explain the anomalously large and temperature dependent quadrupole splittings of the oxy haems, but none of the models has found general acceptance.

The Mössbauer spectra of carbonmonoxy P-450$_{cam}$ resemble those of CO myoglobin; the complex is low-spin ferrous and has a small, almost temperature independent quadrupole splitting (Sharrock *et al.*, 1976; Champion, 1975a). The anomaly in the optical absorption of the P-450-CO complex thus does not noticeably affect the state of the iron, although the iron must mediate the interaction of the thiolate ligand with the porphyrin π system. Several molecular orbital calculations have been undertaken to explain the characteristic hyper spectrum of the CO adducts (Hanson *et al.*, 1976; Loew *et al.*, 1980; Jung, 1983). While the model assumptions of the various calculations differ, they all agree in the conclusion that the admixture of the sulphur $3p\pi$ electrons to the porphyrin π system via the iron $3d\pi$ orbitals is responsible for the split Soret band. The calculation also shows that the charge distribution near the iron differs little from that in a 'normal' CO adduct like myoglobin-CO and, thus, explains the fact that P-450 and myoglobin CO complexes have similar Mössbauer spectra. The results of molecular orbital calculations depend very strongly on the assumed ligand geometry; since the true structure of the haem complex is not yet known, the results of the model calculation should be considered as preliminary.

The axial iron ligand, Cys 134, is likely to remain deprotonated in the CO adduct of substrate complexed P-450$_{cam}$ as indicated by a comparison of ^{13}C nmr shifts (δ) of several ^{13}CO haem complexes. The shift in ^{13}CO-P-450$_{cam}$ (δ = 200.3 ppm) is lower by 7.4 ppm than in ^{13}CO myoglobin (δ = 207.7 ppm). A similar difference of 8.5 ppm is found between the shifts in the ^{13}CO complexes with a mercaptide (δ = 197 ppm) and an imidazole (δ = 205.5 ppm) but not between the complexes with a mercaptan (δ = 204.7 ppm) and the imidazole (Berzinis and Traylor, 1979).

Recent low temperature ftir measurements of the CO stretching frequencies revealed interesting differences between camphor free and camphor complexed P-450$_{cam}$. In the substrate free CO adduct, five different CO stretching modes are observed with temperature dependent intensities and frequencies ranging from 1910 cm^{-1} to 1966 cm^{-1}. In the camphor complex, on the other hand, only two bands are seen in the range of 1932 to 1941 cm^{-1}; at room temperature the higher frequency band at 1939 cm^{-1} has 94% of the intensity. Apparently the CO binding site is more open, flexible or disordered in the absence of substrate and therefore allows the CO complex to assume several distinct conformations. Camphor binding, on the other hand, tightens the haem pocket and restricts the number of conformations to two (Jung, 1984 (personal communication)).

Conformations and internal equilibria of cytochrome P-450

Haem spin state and redox potential

The cytochrome P-450$_{cam}$ monoxygenase is an ideal system in which to study the regulation of electron transfer between redox centres on separate protein molecules. Electrons are transferred from NADH to the haem of P-450$_{cam}$ through the action of two additional proteins (cf. figure 5.1). An FAD-flavoprotein, putidaredoxin reductase, receives electrons from NADH and passes them in single equivalents to the iron–sulphur centre of the second protein component, putidaredoxin, from which they are subsequently passed to the P-450 haem group.

The mechanism for regulating these interprotein electron transfer events has been a subject of intense interest. The binding of camphor to P-450$_{cam}$ has been shown to elicit a blue shift in the wavelength maximum of the Soret band and a shift in the epr signals of the porphyrin-chelated iron (Katagiri *et al.*, 1968; Tsai *et al.*, 1970) which has been interpreted as a change in the thermal equilibrium of the iron spin-state distribution at room temperature from a primarily low spin d-electron configuration (S=1/2) to a primarily high spin configuration (S=5/2; Sligar, 1976). It is important to realise that direct binding of substrate to the metal is not responsible for the observed optical and magnetic changes. All spectroscopic data are consistent with camphor-bound P-450$_{cam}$ containing a five-coordinate ferric ion with cysteinyl thiolate as the sole axial ligand. Camphor free P-450$_{cam}$ is predominantly a low spin, hexacoordinate iron with a cysteine thiolate and perhaps H_2O as axial ligands. Thus, substrate modulation of the haem-iron spin state must result from substrate-dependent changes in the conformation of the protein that alter the immediate ligand environment around the metal, and the term 'spin state equilibrium' refers not only to the thermodynamics of the d-shell electron distribution but must also include the changes in protein conformation and ligand structure that induce the redistribution of the d-shell electrons.

Linked to this spin state equilibrium in cytochrome P-450$_{cam}$ is a corresponding shift in the redox potential of the haem centre from -300 mV in the absence of camphor to -173 mV in the presence of substrate (Sligar, 1976). The significance of this transition is clear when one notes that the potential of the bound electron donor (reduced putidaredoxin) is -196 mV. In the absence of substrate, it is thermodynamically unfavourable for P-450$_{cam}$ to receive electrons from putidaredoxin. This prevents waste of reduction equivalents and formation of harmful reduced oxygen products. In the presence of camphor, P-450$_{cam}$ proceeds normally through the catalytic cycle with most of the reducing equivalents coupled to product formation. Similar spin state changes have been found for hepatic microsomal P-450s which receive electrons from an FAD/FMN-flavoprotein reductase without an intermediate iron–sulphur protein. Microsomal and solubilised rat liver P-450 exhibited a temperature dependent spin equilibrium (Cinti *et al.*, 1979). Spin state transitions elicited by so-called hepatic 'Type-I' substrates (Sligar *et al.*, 1979) and the presence of lipid (Gibson *et al.*, 1979) were shown to modulate the P-450 redox potential (Sligar *et al.*, 1979; Gibson *et al.*, 1980; Backes *et al.*, 1980, 1982). The effect of substrate on the spin equilibrium and reduction potential of P-450$_{LM-2}$ (Rein *et al.*, 1979), and the effect of the composition of P-450$_{SCC}$-bound membrane vesicles on the spin state and activity of the protein (Lambeth *et al.*, 1980) are in keeping with these results. On the other hand Guengerich (1983) has found some rat liver P-450 isozymes for which the spin state/redox potential correlation is not so strong. For quantitative study of the regulation of electron transfer process via the induced alteration of the spin equilibrium upon binding substrate, it is necessary to independently vary and quantitate the spin, redox, and substrate binding equilibria. This can be accomplished with cytochrome P-450$_{cam}$ through the use of various substrate analogues.

Investigation of the interaction of the substrate analogue, 5-*exo*-bromo-camphor (5BC), with P-450$_{cam}$ demonstrated that high affinity binding to the protein could occur independently of a shift in the low spin/high spin equilibrium and its concomitant shift in wavelength of the Soret maximum (Gould *et al.*, 1981). The binding affinity of 5BC (2.9 μM) is virtually identical to that of camphor under identical conditions (2.4 μM) but where camphor saturated P-450$_{cam}$ is 94% high spin and has a redox potential of -173 mV, 5BC saturated P-450$_{cam}$ has a redox potential of -246 mV and is only 46% high spin. It is of interest to note that a calculated fraction of 43% is obtained if one assumes that the -246 mV redox potential is the weighted average for the mixture of high and low spin P-450$_{cam}$ elicited by 5BC, suggesting that control of the redox potential resides primarily in the spin equilibrium. This question has been investigated in this laboratory using a variety of substrate analogues all of which bind with Type-I spectral characteristics. Two experimental quantities can be determined in binding studies with the various substrate analogues. Standard analysis of the optical titration data yields an observed dissociation constant for the binding interaction which represents the total Gibbs free energy available for conformational, redox, and other internal equilibrium changes. The maximal absorbance change (ΔA_{max}) observed in the difference spectrum when the cytochrome is saturated with each substrate analogue is an independent quantity that measures the ability of the substrate analogue to induce the conformationally linked change in the thermal spin-state equilibrium. As projected from experiments with 5BC, the dissociation constant and extent of maximal spectral change are indeed independent quantities. These two observable parameters can be related to fundamental microscopic equilibrium constants definable in this system by a four state coupling model (Sligar and Gunsalus, 1976; Sligar, 1976):

$$P450_{HS} \xleftrightarrow{\quad K_4 \quad} P450_{HS}^S$$

$$K_1 \updownarrow \qquad\qquad\qquad \updownarrow K_2$$

$$P450_{LS} \xleftrightarrow[K_3]{\quad\quad} P450_{LS}^S$$

The various $K_i (i = 1-4)$ represent the binding and spin state equilibrium constants

$$K_1 = \frac{[P450_{HS}]}{[P450_{LS}]}, K_2 = \frac{[P450_{HS}^S]}{[P450_{LS}^S]}, K_3 = \frac{[P450_{LS}][S]}{[P450_{LS}^S]}, K_4 = \frac{[P450_{HS}][S]}{[P450_{HS}^S]}$$

where the subscripts HS and LS refer to the total high ($S = 5/2$) and low ($S = 1/2$) spin forms of the ferric cytochrome and the superscript 'S' denotes the substrate bound form of the cytochrome. Free energy conservation and microreversibility lead to a multiplicative relationship of equilibrium constants; $K_1 K_3 = K_2 K_4$. In the above model it can be shown that the observed spectral dissociation constant for substrate binding, K_D^{obs}, and the total extent of spectral change observed on titration, ΔA_{max}, can be related to the various microscopic equilibrium constants:

$$K_D^{obs} = \frac{1 + K_1}{1 + K_2} \cdot K_3$$

$$\Delta A_{max} = [\text{P-450}]_{total} \, \Delta\epsilon \cdot \frac{K_1 - K_2}{(1 + K_1)(1 + K_2)}$$

where $\Delta\epsilon = (\epsilon_{HS}^{390} + \epsilon_{LS}^{390}) - (\epsilon_{HS}^{417} + \epsilon_{LS}^{417})$. Implicit in this thermodynamic linkage model is the fact that all species are in free equilibrium, i.e. substrate binds to both the low and high spin conformations of the cytochrome, and that there are two major states of the haemprotein. This latter point is consistent with the experimental observation of a clean isosbestic point at 406 nm on titration of various cytochrome P-450s with different substrates (Schenkman et al., 1981) and ions (Philson, 1977).

Table 5.2 gives the observed dissociation constants determined for various substrate analogues binding to cytochrome P-450$_{cam}$. The ΔA_{max} values for any

Table 5.2

Substrate or analogue	K_D, μM	K_2	%HS	K_3, μM	K_4, μM	E_{obs}, mV	E_7, mV	K_5, μm
TMCH	143.0	0.112	10	147.0	110.0	-282	-223	62.8
5-OH camphor	16.0	0.85	46	27.3	2.70	-218	-198	0.57
norcamphor	370.0	1.00	50	683.0	57.3	-235	-217	25.7
d-fenchone	21.6	1.33	57	46.4	2.93	-235	-221	1.50
d-3-bromocamphor (endo isomer)	48.0	1.56	61	113.0	6.10	–	–	–
thiocamphor	0.934	1.63	62	2.27	0.12	–	–	–
d-5-ketocamphor	147.0	1.91	66	395.0	17.4	–	–	–
1-camphoroquinone	7.5	2.58	72	24.8	0.81	-220	-212	0.29
d-camphor	2.9	15.00	94	42.8	0.24	-173	-171	0.0177
adamamtanone	3.5	56.00	98	184.0	0.28	–	–	–
camphorcarboxylic acid	710.0	56.00	98	37334.0	56.0	–	–	–
Camphenilone	25.0	1.27	56	52.4	3.46	-237	-222	1.88
no substrate	–	0.084	7.7	–	–	-303	-238 (E_6)	–
d-5-bromocamphor	2.9	0.85	46	4.95	0.49	-246	-226	0.31

*Abbreviations used are: TMCH, tetramethylcyclohexanone; 5-OH-camphor, 5-exo-hydroxycamphor; K_3 values were calculated from $K_D = K_3$ $(1 + K_1)/(1 + K_2)$ and a value $K_1 = 0.084$ at $20°$ C. $E_{obs} = -(RT/F)$ ℓn (K_{obs}) where $(RT/F) = 25.5$ mV at $20°$ C. K_7 is calculated from the observed oxidation-reduction equilibrium constant, K_{obs}, via $K_{obs} = (1 + K_2)(K_7/K_2)$. $E_7 = -(RT/F)$ ℓn (K_7). K_5 is calculated from $K_5 K_6 = K_4 K_7$ where K_6 is obtained from the substrate free redox potential according to the text.

two substrate analogues can be related via their respective microscopic equilibrium constants:

$$\frac{\Delta A_{max}'}{\Delta A_{max}} = \left(\frac{K_1 - K_2}{K_1 - K_2'}\right)\left(\frac{1 + K_2'}{1 + K_2}\right)$$

Thus, knowing K_1 for the substrate free enzyme and K_2 for one substrate such as camphor, the value of K_2' for a different analogue can be easily obtained without recourse to temperature dependent difference spectroscopy (Sligar, 1976). The values for K_2' are included in table 5.2 using the camphor saturated cytochrome as a reference point (Sligar, 1976). Using this value of K_2' and the observed dissociation constant, K_3 and K_4 can be calculated as described earlier. K_3 and K_4 are the microscopic binding constants characterising substrate analogue binding to the low-spin and high-spin enzyme forms, respectively, and are also included.

Table 5.2 illustrates that K_2 and K_D are indeed independent quantities and that a variety of interactions of the substrate analogue with cytochrome P-450 are possible. For example, thiocamphor binds significantly more tightly to the enzyme than does camphor, but the thermal spin equilibrium of the protein has only 62% of the high-spin configuration when this camphor analogue is bound, whereas 94% of the camphor bound protein is high spin. Thermodynamically, this indicates that thiocamphor binds less selectively than camphor to the two spin conformations of cytochrome P-450$_{cam}$. This selectivity is expressed by the ratio K_3/K_4, where K_3 and K_4 are the microscopic equilibrium constants for substrate binding to low-spin and high-spin enzyme, respectively. The larger is K_3/K_4, the larger will be the extent of conversion of the enzyme to the high spin form through substrate binding. Table 5.2 also shows that some substrate analogues induce an essentially complete change to the high spin form even though they bind less tightly to the enzyme than does camphor. For example, camphor carboxylic acid-saturated enzyme is 98% high spin and yet the dissociation constant is significantly larger than for camphor, 710 μM vs 2 μM.

As discussed previously, the redox potential for the reduction of Fe(III)-P450 to Fe(II)-P450 is at least qualitatively linked to the spin state equilibrium constant (Sligar, 1976) in many systems. Adding reduction equilibria to the substrate binding and spin state equilibria as discussed above would generate $2^3 = 8$ total states. Mössbauer spectroscopy has suggested that ferrous P-450, even in the absence of substrate and at helium temperatures, is exclusively in the high spin configuration (Sharrock *et al.*, 1976). Hence, a minimal six state model would be

In this scheme the additional equilibrium constants K_5, K_6, and K_7 are defined as:

$$K_5 = \frac{[P450_{RED}][S]}{[P450_{RED}^S]} , K_6 = \frac{[P450_{HS}]}{[P450_{RED}]} , K_7 = \frac{[P450_{HS}^S]}{[P450_{RED}^S]}$$

These definitions of reduction equilibria are analogous to those described earlier. The overall redox potential observed, E_{obs}, is related to a corresponding equilibrium constant, K_{obs}, via $E_{obs} = (RT/F) \ln (K_{obs})$, where K_{obs} is the ratio of total oxidised to total reduced species, and can be expressed in terms of the microscopic equilibrium constants via:

$$K_{obs} = \frac{[P450_{LS}] + [P450_{HS}] + [P450_{LS}^S] + [P450_{HS}^S]}{[P450_{RED}] + [P450_{RED}^S]}$$

$$K_{obs} = \frac{(1 + 1/K_2)(1 + [S]/K_4)}{(1 + [S]/K_5)} \cdot K_7$$

Again, microreversibility demands that with these definitions $K_5 K_6 = K_4 K_7$. If the behaviour of the system is considered in the presence of saturating amounts of the various substrates, four forms of cytochrome P-450 must be considered in a minimal thermodynamic model. These are the low-spin, ferric, substrate-bound P-450 (abbreviated $P450_{LS}^S$); high-spin, substrate-bound, ferric-P-450 (abbreviated $P450_{HS}^S$) and the corresponding ferrous species. We originally suggested a simplified model where the change in the observed redox potential of cytochrome P450 was due solely to an alteration of the haem iron spin state; this would correspond to K_7 being independent of the particular substrate bound to the enzyme (Sligar *et al.*, 1979). Again, the measured redox potential of the protein will always be the logarithmic ratio of total reduced and total oxidised forms (either high or low spin). If the sole effect of substrate binding is on the cytochrome oxidation–reduction potential, then E_7 in table 5.2 should be the same for all substrates. As can be seen from table 5.2, significant deviations from the simplified assumption are apparent in several cases. In particular, the spin state equilibrium constant for cytochrome P450$_{cam}$ saturated with either 5-*exo*-bromocamphor or 5-*exo*-hydroxycamphor are nearly identical (table 5.2) and yet the measured redox potentials differ by 28 mV. In addition, our original simplified model is obviously incomplete since the free energies associated with the ferrous spin state equilibrium must be included for thermodynamic completeness. Despite these irregularities, there is obviously a strong correlation between the ferric spin state equilibrium constant and the overall cytochrome redox potentials. The values of E_7 in table 5.2 are surprisingly constant for most of the substrate analogues tested.

These quantitative discrepancies in spin state and redox potential correlations imply that other equilibria are present. The original suggestion of a linear relationship for all substrates was based on the assumption that the effect of the substrate or analogue on the redox potential results only from the substrate analogue-dependent change in the spin state equilibrium constant, K_2 (Sligar and Gunsalus, 1976). Clearly, other factors must be included since the incremental change in potential for incremental spin state free energy is a factor of two larger than expected. The simplified model predicts that K_7 should be constant and insensitive to the particular substrate analogue bound to the enzyme. The results show that this assumption is not rigorously valid. Thus, the energetics of reduction of high spin, ferric cytochrome $P450_{cam}$ are determined not only by the effect of substrate binding on the ferric spin equilibrium linked conformational change, but also by a contribution of the total substrate binding energy to the alteration of the inherent oxidation/reduction free energy of high spin cytochrome. The molecular nature of this additional interaction is presently unknown, although local variations in dielectric constant and ionic factors are known drastically to alter biological oxidation/reduction equilibria (Schejter *et al.*, 1980). Ionic strength, specific ions, and pH effects have a pronounced impact on substrate binding, spin state equilibrium, and redox potential (Sligar, 1976; Hintz and Peterson, 1981; Ebel *et al.*, 1977).

It has been suggested that the oxidation–reduction potential of the haem centre and the spin state of the ferric protein are linked to internal protein equilibria. Such spin state changes in haem proteins are usually correlated with changes in the nature of the axial ligands to the iron, which could in principle involve either local structural variations or more extensive rearrangement throughout the protein. The fact that the temperature jump spin relaxation of m^o is slow by electronic standards and relative to m^{os} and several other haem proteins (Sligar, 1976; Cole and Sligar, 1981) also suggests a significant structural difference between m^o and m^{os}. Pressure studies (Hui Bon Hoa and Marden, 1982) have implicated a relatively large volume decrease in the high-to-low spin state transition of substrate-bound P-450. Recent studies in this laboratory of the dependence of pressure-induced spin state transitions of $P-450_{cam}$ on the concentration of substrate suggest that the observed transitions are due to pressure-induced displacement of substrate from the protein active site as opposed to a change in the spin state equilibrium constant of substrate bound cytochrome as was suggested by Hui Bon Hoa and Marden (1982). Our pressure results yield a calculated volume change of 75.8 $Å^3$/molecule for substrate association indicating that a substantial conformational change occurs in cytochrome $P-450_{cam}$ upon binding camphor. CD spectra of $P-450_{cam}$ also show significant differences in the presence and absence of camphor. In order directly to probe the structural equilibria linked with substrate binding and the ferric spin state, the solution structure of $P-450_{cam}$ in the presence and absence of camphor was determined by small angle x-ray scattering with the results expressed in terms of the radius of gyration (Rg) of the protein (Lewis and Sligar, 1983). Changes in the radius of gyration of a macromolecule upon binding of ligands and substrates have been observed for numerous other proteins, including haemoglobin (Conrad *et al.*, 1969), arabinose-binding protein (Newcomer *et al.*, 1981), hexokinase (McDonald *et al.*, 1979), and phosphoglycerate kinase (Pickover *et al.*, 1979).

The radius of gyration of cytochrome P-450$_{cam}$ was found to be identical in both substrate bound and free states : 23.9 ± 0.2 Å. It is clear that if P-450$_{cam}$ undergoes a conformational change upon binding camphor the change is smaller than can accurately be detected by solution-scattering methods. Thus, the structural changes inferred from CD, T-jump spin relaxation, and pressure studies most likely involve only local rearrangements near the haem. This conclusion is further supported by the observations that high and low spin forms of P-450$_{cam}$ coexist in the crystal state and that the orthorhombic II crystal, prepared from camphor free P-450$_{cam}$, remains intact when substrate is diffused in and out and on reduction and CO binding (Gunsalus *et al.*, 1980; Poulos *et al.*, 1982).

The asymmetry in structure of any molecule can be estimated by comparing its measured radius of gyration with that of a spherical particle of the same molecular weight and partial specific volume. A spherical protein with a partial specific volume of 0.74 cm^3/g and a molecular weight of 46 814 (Haniu *et al.*, 1982a,b) would have a radius of gyration of 18.5 Å. Thus, the frictional coefficient, f = Rg(observed)/Rg(sphere) is found to be about 1.3 for cytochrome P-450. If P-450$_{cam}$ is modelled as an ellipsoid of revolution, this value for f indicates either a prolate ellipsoid with an axial ratio of about 3 or an oblate ellipsoid with an axial ratio of about 0.25. These ratios correspond to actual dimensions of 33 x 33 x 98 Å(prolate) or 75 x 75 x 19 Å(oblate). Additionally, the two possible cylinders which fit the data for P-450$_{cam}$ are one of diameter 30 Å and length 75–80 Å and one of diameter 68–70 Å and height 14–15 Å. Comparison of these structures with the asymmetrical structure of 55 x 55 x 39 Å previously noted from the preliminary x-ray crystallographic structure suggests there may be significant differences between the crystalline and solution structures of P-450$_{cam}$. It must be noted, however, that the dimensions of the asymmetric shape described by the x-ray structure represent only the dimensions that are the longest found in each plane. Thus, the two sets of data may be found compatible after further analysis.

The substrate binding site of cytochrome P-450$_{cam}$

The importartce of cysteine to cytochrome P-450 has been appreciated since the general acceptance of a thiolate as fifth haem ligand. That other cysteine residues were of importance to the structure and function of P-450$_{cam}$ became evident from alkylation studies with N-ethylmaleimide (Lipscomb *et al.*, 1978). Titrations carried out in the presence and absence of camphor showed that reaction rate of one to three sulphhydryl groups decreased by an order of magnitude when camphor was present. This suggested that at least one cysteine residue was protected by camphor, either directly or through steric effects. With regard to substrate binding, Banerjee and Dec (1979) demonstrated that camphor did not bind to P-450$_{cam}$ through an enol intermediate. Subsequently, Murray and Dus (1980) proposed addition of a cysteine thiolate to the carbonyl group of camphor to yield a thiohemiketal which would allow transient covalent coupling to the protein and could account for the protection of a sulphhydryl group by the substrate. In keeping with the use of active-site directed affinity reagents to study the structure of cytochrome $P-450_{cam}$, a series of affinity reagents derived from camphor and targeted to a sulphhydryl group near the carbonyl group of bound substrate was prepared (Dus, 1982b). Recently, a model for the function of a specific cysteine residue in the $P-450_{cam}$ active site was proposed on the basis of studies with the affinity label isobornyl bromoacetate (Murray *et al.*, 1982). Concomitant with second order covalent coupling, as determined by quantitation of S-carboxymethyl cysteine in acid hydrolysates of the protein, isobornyl bromoacetate produced a 'Type-I' spectral transition which underwent first order time dependent decay. Further addition of d-camphor produced a stable 'Type-I' spectrum indicating the active site of $P-450_{cam}$ was open and intact. These results were interpreted as a conformational change linked to movement of the cysteine bound substrate. The model proposed that this reaction mimics the occurrence in native $P-450_{cam}$ of camphor binding to the sulphhydryl *via* a thiohemiketal followed by the cysteine mediated translocation of nascent product to a second, low affinity site for subsequent release. Unlike the reversible thiohemiketal, the thioether bond of the affinity reagent can not be reversed and the label is trapped in the low affinity site while the high affinity substrate binding site is empty and available for binding a molecule of camphor. Using isobornyl-$[2-^{14}C]$-bromoacetate, Dus and Murray (1982) suggested Cys-56 as the substrate binding residue. Application of similar substrate and inhibitor derivatives to $P-450_{SCC}$ and $P-450_{11\beta}$ (Dus *et al.*, 1982) and to $P-450_{LM-2}$ and $P-450_{LM-4}$ (Murray and Dus, unpublished observations) indicate that these other P-450s from diverse origins also contain a sulphhydryl group, favourably placed at the active site. The nature of the conformational change accompanying translocation, and the applicability of this model to other P-450s with carbonyl-bearing substrates remains to be determined.

The placement of camphor in the active site or $P-450_{cam}$ has been clarified by the three-dimensional x-ray structure. The substrate is centred 2.7 Å off the haem normal, with the closest carbon estimated to be 4.4 Å from the haem iron. No cysteine is seen near the camphor 2-keto group, but rather a strong histidine-ketone hydrogen bond is found. Noting that in oxygenated haem complexes the outer oxygen is 3 Å from the iron and 0.92 Å off the haem normal, the location of the substrate inferred from the electron density map appears suitable for the hydroxylation reaction.

OXYGEN ACTIVATION BY CYTOCHROME P-450

While the intermediates of the cytochrome P-450 reaction cycle have been well documented through the formation of oxygenated ferrous haem (Gunsalus and Sligar, 1978), the individual steps which include the most interesting chemistry have not yet been unambiguously defined. The various reactions occurring in this final segment of the reaction cycle have been the subject of extensive inquiry in our laboratory and that of others. These steps must necessarily include the activation of the iron–dioxygen complex, formation of an iron-oxo hydroxylating intermediate, and activation and hydroxylation of substrate. Each of these chemical steps will be reviewed in this section, again with particular emphasis on recent results obtained with the camphor monooxygenase.

Although most organic compounds are thermodynamically unstable with respect to oxidation by O_2, there is a significant kinetic barrier to these reactions arising from the triplet ground state nature of molecular dioxygen. Because the rate for the needed triplet-singlet inversion is slower than the lifetime of the molecular collision complex, the overall reaction rate of oxygen with organic compounds is quite low. To produce a more reactive oxygen species one must either generate singlet oxygen, chemically reduce dioxygen to a species which is more reactive, or otherwise alter the electronic distribution of O_2 by complexation with a transition metal catalyst. Nature has apparently focused on these latter two mechanisms (Hill, 1979). A ubiquitous system reductively activating atmospheric dioxygen are the cytochrome P-450 dependent mixed function oxidases. With a haem protein oxygen binding and activation site and pyridine nucleotide dehydrogenation by a flavoprotein reductase and subsequent electron transfer to the P-450 haem prosthetic group, atmospheric dioxygen is split producing a water molecule and a co-oxidised substrate by oxygen atom transfer.

Efforts over the past decade have resulted in the suggested reaction cycle of cytochrome P-450 discussed earlier and depicted in figure 5.2. Here, the two-electron transfer reactions are separated by dioxygen addition to ferrous haem. The iron-bound, two-electron-reduced dioxygen species is formally equivalent to hydrogen peroxide bound to ferric haem. This intermediate has never been spectrally observed in any cytochrome P-450 system, although McCandlish *et al.* (1980) acquired spectral evidence in a porphyrin model system for a structure at the redox level of peroxo-Fe^{3+} consisting of a three-member ring composed of the iron and two oxygens. The one-electron-reduced, oxygenated form of the enzyme is quasi-stable, however, and has been examined by several spectroscopic techniques. In the bacterial system this intermediate has been stabilised by Douzou and coworkers using low temperature techniques in substrate bound as well as substrate free protein preparations (Eisenstein *et al.*, 1977). Although this species is actually generated by reduction of the ferric haem followed by oxygen binding, a ferric haem-superoxide anion configuration has been considered as an alternative (Sharrock *et al.*, 1973). One decay pathway of this intermediate appears to be the release of superoxide, with the regeneration of ferric enzyme (Sligar *et al.*, 1974; Estabrook *et al.*, 1971; Estabrook and Werringloer, 1977; Ullrich and Kuthan, 1980). With at least a partial negative charge resting on the distal oxygen this species would be expected to be quite nucleophilic. This nucleophilicity is aided by the proximity effect contributions from the oxygen atom lone pairs, the contributions of the metal-porphyrin system, and possible assistance from the lone pair sulphur electrons on the back-side cysteine ligand to the haem centre. The two-electron reduced iron-oxo complex could be expected to display even stronger nucleophilic distal oxygen character. It was originally suggested by Hamilton that such a nucleophilic species, with the iron formally in a ferric or ferrous form, might be capable of attack on an electrophilic carbonyl centre, such as provided by a carboxylate or amide side chain. Subsequent loss of water from the tetrahedral intermediate so generated would form an iron-bound peracid species capable of subsequent substrate hydroxylation (Hamilton, 1974). Such a reactivity of metal coordinated dioxygen is apparently operating in the α-ketoglutarate coupled dioxygenases (Abbott and Udenfreund, 1974). The observation that peracids and peroxides are effective exogenous oxidants for P-450 catalysed substrate hydroxylations is consistent with the hypothesis of an intermediate state generated by reactivity of the distal oxygen.

The first preliminary experimental evidence for the nucleophilic reactivity of oxygenated haem derives from isotope tracer experiments using the camphor P-450 system (Sligar *et al.*, 1980). In addition to P-450, O_2, reducing equivalents and substrate, the camphor hydroxylase also requires an effector molecule to carry out substrate oxygenation at maximal velocity (Tyson *et al.*, 1970; 1972). Such an 'effector' role is also indicated in the analogous adrenal 11-β hydroxylase system. The natural effector for this camphor hydroxylase is apparently putidaredoxin, the iron–sulphur protein which also transfers electrons from the flavoprotein to the P-450 haemprotein. Certain specific organic molecules, such as dimercaptooctanoic acid (reduced lipoic acid), are also capable of functioning as effectors in this system with electrons supplied by photoreduction (Lipscomb *et al.*, 1976). If the role of this effector is to provide an electrophilic target for the oxygenated haem adduct then one would expect to see incorporation of one of the atoms of atmospheric dioxygen into the carboxyl moiety during turnover. Hence, when the reaction is conducted in an $^{18}O_2$ atmosphere, the carboxyl group of lipoic acid should contain ^{18}O. Such a result was observed under single turnover conditions (Sligar *et al.*, 1980). The acylation event implied by this result is presumably required to weaken the O–O bond and facilitate generation of the actual active oxygen hydroxylating species. It must be remembered, however, that these results apply only to the model reaction with P-450 and dihydrolipoic acid instead of putidaredoxin as an effector. A logical experiment to confirm the general applicability of this nucleophilic mechanism in the normal reconstituted P-450 systems is to identify the *in vivo* electrophilic centre, and demonstrate incorporation of ^{18}O into this centre during single turnover condition. Such an approach is rife with the technical difficulties associated with isolating and identifying heavy isotope incorporation into protein, and could be better approached by looking for labelled oxygen release to water. If P-450 catalysis involves acylation of haem bound dioxygen, then after a single turnover of the system using oxygen-18 dioxygen, water will be released from an unlabelled position in the tetrahedral intermediate. Only after multiple turnovers when oxygen label has equilibrated with the acylating electrophilic centre will label appear in water. This is in contrast with the accepted oxygen stoichiometry of the P-450 mixed function oxidase systems which dictates that for *each* O_2 molecule, one oxygen atom will appear in the oxygenated product, while the other is reduced to water. Since all oxygen-18 studies with P-450 systems to date have only examined label distribution in the product after multiple turnovers, there is as yet no experimental evidence that distinguishes between the classical oxygen stoichiometry and that predicted by a model involving an acyl-peroxide-iron intermediate generated by nucleophilic catalysis. Recently, using an isotope ratio mass-spectrometer, O'Leary and colleagues have quantitated the production of oxygen-18 water at the parts-per-million level (Shaw *et al.*, 1981). Such a technique applied to single turnovers of a P-450 monooxygenase could distinguish between these two models. We are approaching the problem in our laboratory using radioactive oxygen-15 generated *in situ* with a Van de Graff generator and oxygen-18 pertubations of carbon-13 chemical shifts. It has also been previously reported (Lipscomb *et al.*, 1976) that dimercaptopropanol could serve as an effector for the camphor hydroxylase system. Since this compound has no obvious electrophilic centre, its efficiency as an effector seemed anomalous. Even though the overall reaction of oxygenated P-450 with dimercaptopropanol to generate hydroxycamphor is not inhibited by catalase, experiments by D. Heimbrook at Yale University suggest that product formation with this effector is mediated by active site generated hydrogen peroxide. Various organic dyes, reduced by pyridine nucleotide, borohydride, or photolytic methods will also react with oxidised bacterial cytochrome P-450 in the presence of an oxygen source to generate hydroxycamphor (Sligar, unpublished results). Although, in some cases, product formation is not inhibited by catalase, hydrogen peroxide is presumably generated in an active site environment near the P-450 haem that is inaccessible to catalase. Further evidence that the effector function is related to chemical activation of a pre-existing state and not to redox transfer is the observation that apo-cytochrome b_5, with no sulphur or prosthetic group, is an efficient effector. It has been recently demonstrated, however, that organic dyes such as phenazine methosulphate can catalyse second electron transfer although at slow rates presumably independent of hydrogen peroxide generation (Eble and Dawson, 1984).

Various experiments using the camphor hydroxylase system provide indirect evidence as to the origin of the effector function. In single turnover reactions, oxidised putidaredoxin will serve as an effector without back electron transfer, hence this protein has some chemical function in the reaction in addition to redox transfer. The highly homologous iron–sulphur protein from the adrenal mitochondria P-450 system, adrenodoxin, conducts efficient electron transfer reactions with the bacterial P-450, but is unable to catalyse efficient product formation, again suggesting an additional role for the iron–sulphur protein. This incredible specificity in iron–sulphur/cytochrome coupling is also evident in the lack of cross reactivity between the components of the *Pseudomonas* linolool hydroxylase and those of the *Pseudomonas* camphor hydroxylase (Gunsalus, 1983, unpublished observations). Further evidence of the chemical involvement of putidaredoxin (Pd) in camphor hydroxylation comes from selective protein modification studies. Treatment of putidaredoxin with carboxypeptidase A removes only the C-terminus amino acid, tryptophan, and the penultimate amino acid, glutamine. The modified protein, upon reconstitution with P-450 and flavoprotein yields a significantly lower overall V_{max} of the hydroxylase catalysis (Sligar *et al.*, 1977). Detailed spectroscopic (epr, optical, cd) studies indicate that the iron–sulphur centre of putidaredoxin is unperturbed by this C-terminus cleavage. These experiments imply that either the carboxylate of the terminal tryptophan or the amide of the penultimate glutamine may be functioning in electron transfer and/or oxygenase catalysis, perhaps as the electrophilic target for the iron–oxygen species. Thus, in the overall reaction cycle of cytochrome P-450, an initial chemical reaction unique to these monooxygenases is the possible nucleophilic reactivity of oxygenated haem. The next conceptual step in oxygen activation involves scission of the oxygen–oxygen bond.

One of the primary controversies in the chemistry of monooxygenation is the nature of the hydroxylating iron–oxo species formed by cleavage of the O–O bond. This intermediate must be capable of oxygenating a variety of substrates, ranging from unactivated alkanes to olefins and aromatic systems. One of the initial suggestions for this species is formally analogous to the 'Compound I' intermediate of the peroxidases (Dolphin, 1981). Extending the iron-bound peracid model described earlier to subsequent reaction steps, this intermediate would derive from heterolytic cleavage of the oxygen–oxygen bond, generating a carboxylate anion and, formally, an iron-bound six electron oxygen atom, or 'oxene'. From this iron-oxo complex, one could envisage subsequent substrate hydroxylation occurring by a variety of alternative mechanisms (figure 5.3) (Lichtenberger *et al.*, 1976; Groves and McClusky, 1976): (a) *Direct insertion*,

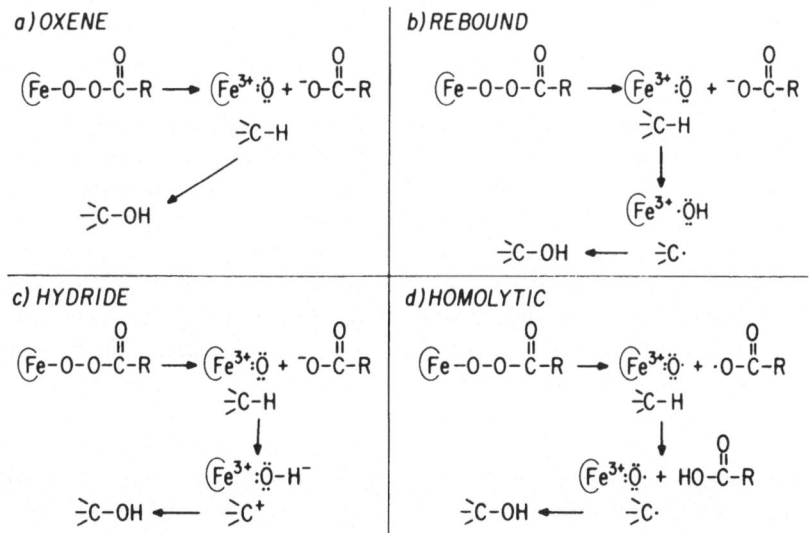

Figure 5.3 Hypothetical chemical mechanisms for cytochrome P-450 dependent oxygenations. See text for discussion

In which the 'oxene' inserts itself directly into a carbon–hydrogen bond with P-450 thus serving as an 'oxene transferase' (Ullrich, 1977), (b) *Oxygen rebound* (Groves *et al.*, 1978), in which the iron-oxo species abstracts a hydrogen radical from the substrate, followed by radical recombination to generate the alcohol (as will be discussed later, a modification of this model might incorporate an intermediate amino acid radical as the species responsible for hydrogen abstraction from the substrate), and (c) *Hydride abstraction*, in which a hydride is abstracted from the substrate, generating a substrate carbonium ion, which then condenses with the iron-bound hydroxyl to generate product. As will be discussed in various sections of this manuscript, there is substantial evidence in both the bacterial and hepatic P-450 systems that a stepwise mechanism as envisaged in the 'oxygen-rebound' process is operational, although there is evidence that distinct enzyme species may be involved in the hydrogen abstraction and oxygen transfer steps.

An alternative to mechanisms invoking a compound I-type iron-oxo intermediate is a more recently proposed scheme in which the enzyme-bound peracid or peroxide undergoes homolytic oxygen–oxygen bond scission, generating a carboxyl radical and an iron-oxo species formally equivalent to compound II of the peroxidases (figure 5.3d) (White and Coon, 1980; White *et al.*, 1980). This mechanism provides for hydroxylation via abstraction of a hydrogen atom from the substrate by the acetoxy or alkoxy radical of the exogenous oxidant followed by radical recombination between the substrate and the iron–oxygen species. Both classes of mechanisms, heterolytic and homolytic, are feasible on theoretical grounds, and a great deal of effort has been expended in recent years to resolve this question. As is so often the case, however, the results are not always consistent, and a definitive solution to this controversy remains elusive. The evidence supporting each of these two distinct pathways for oxygen–oxygen bond scission will now be discussed.

Due to the more complicated nature of the complete pyridine nucleotide and dioxygen dependent reaction, many investigations into the mechanisms of oxygen cleavage in the cytochrome P-450 have been carried out using exogenous oxidants to bring both redox equivalents and oxygen to the haem centre. Three major classes of oxidants are used in these studies: peroxides, peracids, and those compounds with a single oxygen atom, such as the various N-oxides and iodosobenzene. Peroxides and peracids both contain an oxygen–oxygen bond, and so, on a superficial level, better mimic the normal O_2 reaction. The chemistry of heteroatom bond cleavage such as must occur with the single-oxygen oxidants, provides additional information on the nature of the hydroxylating species formed at the iron. Several fundamental differences exist between the intermediate iron-oxo states generated by heterolytic versus homolytic mechanisms. The heterolytic mechanism predicts the generation of an iron-oxo species which is formally equivalent to the addition of an oxygen atom to ferric P-450 and which is the same with all exogenous oxidants used, and, by inference with that formed in the $NAD(P)H/O_2$-supported reaction. In contrast, the homolytic mechanism predicts the generation of an oxidant radical for each oxidant used, presumably leading to some oxidant-dependent variation in hydroxylation rates and/or regio- or stereo-specificity. The experimental handles of regio- and stereo-selectivity and the chemistry of non-enzymatic porphyrin model systems can be used to distinguish various modes of oxygen–oxygen bond scission and discern the individual steps involved in substrate oxygenation.

There is ample evidence in the chemical literature demonstrating both heterolytic and homolytic cleavage of O–O bonds. Alkyl peroxides, for instance, readily undergo homolytic oxygen–oxygen bond scission, and the generation of hydroxyl radicals by hydrogen peroxide with a transition metal catalyst may be the basis of Fenton chemistry (Groves and McClusky, 1976; Ingold, 1969). It is also clear, however, that these peroxides are at least formally capable of heterolytic cleavage. With a single-oxygen donor, such as iodosobenzene, radical cleavage of the iodine–oxygen bonds is difficult to envisage and a more rational expectation is electron transfer to the positive iodine nucleus producing iodobenzene and the six valence electron 'oxene' oxygen atom. In recent epr studies with chromium(III)-substituted porphyrins, it was unambiguously demonstrated that iodosobenzene breakdown proceeds in this fashion to generate a Cr(V)-oxo species (Groves and Haushalter, 1981). In similar experiments using manganese-substituted porphyrin model systems, an analogous Mn(V)-oxo formulation is most consistent with the spectral data obtained (Groves *et al.*, 1980a). Both metalloporphyrin systems are capable of performing substrate oxygenation to varying degrees, and so effectively mimic the P-450 reaction. It has been suggested that other modes of bond scission may be operating with this type of oxidant, however. Iodobenzene diacetate, for instance, which contains no obviously labile oxygen, will also generate oxygenated substrate when mixed with ferric P-450 which suggests that a radical pathway may be operating (Gustafsson *et al.*, 1979), although recent data from our laboratory suggests that diacetate may undergo rapid hydrolysis to generate iodosobenzene. We conclude from these studies that the hydroxylating capability of iodobenzene diacetate in P-450 systems is probably derived from its ability to undergo rapid hydrolysis to iodosobenzene, and its reactivity in P-450 systems thus does not provide evidence for a radical mechanism.

Several lines of experimentation allow the heterolytic mechanism for the iodosobenzene–porphyrin model systems to be extended to the case of other exogenous oxidants used with both model porphyrins and cytochrome P-450 itself. Recent investigations by Groves have included iron-porphyrin systems, and clearly suggest a similarity between peracid and iodosobenzene reaction mechanisms (Groves *et al.*, 1981). In these investigations iodosobenzene and *m*-chloroperbenzoic acid (m-CPBA) were each found to generate distinct spectral intermediates, which were interconvertible by acid/base titration. Thus, the two species are apparently at the same formal oxidation state, with protonation altering the actual electronic distribution. A major characteristic of these iron-oxo species is their ability to undergo exchange with water prior to carrying out oxygen transfer to a substrate molecule. Such exchange has been observed in the epoxidation of norbornene by a peracid iron–porphyrin system and in an iodosobenzene-supported chromium-porphyrin model system (Groves *et al.*, 1981; Groves and Kruper, 1979). An entirely analogous result in a P-450-catalysed oxygenation was observed in the iodosobenzene-supported hydroxylation of camphor by cytochrome P-450$_{cam}$ (Heimbrook and Sligar, 1981). Thus, although iodosobenzene does not undergo appreciable oxygen exchange with water at pH 7 on the experimental timescale, virtually all of the product alcohol contains oxygen derived from solvent water. Such a result clearly indicates that an exchangeable iron–oxo intermediate is generated during the catalytic cycle and implies that the P-450 iron–oxo species generated by iodosobenzene is the same as that characterised in the porphyrin model studies. It is, of course, possible that the P-450 haem is catalysing the oxygen atom exchange with water prior to formation of the active oxygen intermediate. Neither the peracid, peroxide, nor the NADH/O$_2$-supported hydroxylations exhibited oxygen exchange during camphor hydroxylation, suggesting, with the above caveat, that their reaction mechanisms may be fundamentally different from that supported by iodoso-benzene (Heimbrook and Sligar, 1981; Nordbloom *et al.*, 1976). Such a difference could reflect a distinctly different active oxygen species in the catalytic cycle. More likely, however, the differences in the rates of oxygen transfer to substrate and generation of the iron–oxo species could allow sufficient time for oxygen exchange with solvent water. Evidence for a different kinetic mechanism has been obtained, and involves the inhibition of hydroxycamphor formation at increasing concentrations of camphor (Heimbrook and Whitcombe, 1981). These data demonstrate that iodosobenzene must bind or react with the enzyme before substrate association, either in a compulsory, ordered-ternary complex or by a ping-pong mechanism (Segel, 1975). Thus, the camphor-dependent inhibition results from partitioning of the enzyme into a dead-end enzyme-camphor complex. A ping-pong mechanism provides an attractive explanation for understanding the exchange data, in which iodosobenzene binds, donates a single oxygen with the required dissociation of iodobenzene before camphor can bind. In this model, ample time would exist for oxygen to exchange before substrate hydroxylation. No such substrate inhibition is observed where cumene hydroperoxide is used as exogenous oxidant in camphor hydroxylation. Kinetic data suggest the productive formation of a ternary enzyme–oxidant–substrate complex with this oxidant. With substrate capture of the metal coordinated oxygen, exchange with water might not occur fast enough to exchange oxygen label into the product, and is, in fact, not observed (Heimbrook and Sligar, 1981). Additional support for a ping-pong mechanism in iodosobenzene-supported camphor hydroxylation is derived from equilibrium binding experiments (Heimbrook and Sligar, 1984) where it is observed that both iodosobenzene and iodobenzene are competitive inhibitors of camphor binding to P-450$_{cam}$. These results indicate that camphor cannot enter the enzyme active site while either iodosobenzene or iodobenzene are bound, and provides strong support for a ping-pong mechanism with this oxidant. Cumenol shows no evidence of competition with camphor for the P-450 active site, consistent with a ternary complex kinetic mechanism for cumenehydroperoxide-supported camphor oxygenation (Koop and Hollenberg, 1980). Thus, the unique solvent oxygen exchange observed in iodosobenzene-supported camphor hydroxylation is apparently due to this oxidant's unique kinetic mechanism, rather than the formation of a fundamentally different iron-oxo species. Such a mechanism for explaining similar oxygen exchange data in a hepatic P-450 system was proposed more recently (MacDonald *et al.*, 1982). One important ramification of a ping-pong mechanism in these reactions is that the iodobenzene skeleton cannot directly participate in substrate oxygenation, since it is not present in the active site at the time of the oxygenation reaction. This conclusion is inconsistent with the radical mechanism as illustrated in figure 5.3, and so provides tacit support for a heterolytic mechanism.

Numerous other lines of evidence support the existence of a discrete, 'Compound I-like' intermediate in the P-450 reaction cycle. It has been reported that hepatic cytochrome P-450 catalyses reversible iodosobenzene deoxygenation, thus suggesting that the iron-oxo complex generated by oxygen transfer to the metal stores both oxidizing equivalents, and argues against any homolytic scission of the iodine-oxygen bonds (Burka *et al.*, 1980). In addition, both epr and optical spectroscopy suggest the existence of common intermediates occurring with each of the three classes of exogenous oxidants (Ullrich, 1977; Lichtenberger and Ullrich, 1977). Recent studies on the deuterium isotope effects of camphor hydroxylation, discussed later in this chapter, also show similarities between the various oxidants and lend support to such a hypothesis (Gelb *et al.*, 1982a). More recently definitive evidence for the formation of a 'Compound I' type discrete transition metal-oxo complex in a cytochrome P-450 system has been obtained in our laboratory. Using a manganese-porphyrin reconstituted P-450$_{cam}$, the formation of a new spectral species on reaction with iodosobenzene was demonstrated. This species appears identical to the reported Mn(V)-oxo species of the corresponding Mn(III)-tetraphenylporphinato chloride reaction (Gelb *et al.*, 1982b). This result provides an important link between inorganic model systems and the native protein investigations. The intermediate is chemically competent in olefin epoxidation, with loss of the Mn-oxo spectral intermediate occurring concomitantly with dehydrocamphor epoxidation, an activity recently reported for ferric cytochrome P-450$_{cam}$ (Gelb *et al.*, 1982c). Interestingly, the manganese substituted protein will not hydroxylate camphor in the presence of iodoso-benzene, for reasons which are not yet clear. The redox potential of Mn-P-450 is substantially lower than that of the corresponding iron-porphyrin system which prevents pyridine nucleotide coupled electron transfer through flavoprotein reductase and putidaredoxin. With exogenous oxidant supported oxygenations, however, the two separate electron transfer steps are avoided and the two-electron oxidized metal-oxo centre generated by iodosobenzene and Mn(III)-protein might be expected to catalyse alkane hydroxylation as well as olefin epoxidation. Either different electron distributions of the active metal–oxygen centre are involved in epoxidation versus hydroxylation, or the radical recapture rate for manganese versus iron prevents alcohol production. Taken together, the correlation between the higher valence states of metallo-porphyrin systems and cytochrome P-450 itself is striking, and implies, at least in the case of iodosobenzene supported oxygenations, the generation of a compound I-type species via a heterolytic process. Clearly, direct comparison can only be made between Mn-P-450 and the Mn-model systems as the actual electronic structure of the high valent transition metal–oxo complexes of manganese and iron may be decidedly different.

The possibility of a homolytic oxygen–oxygen bond cleavage in cytochrome P-450 systems is a more recent proposal (White and Coon, 1980; White *et al.*, 1980) and provides an alternative pathway for the generation of a substrate radical intermediate in the oxygenation process. Investigations indicating the possibility of O–O bond homolysis also utilise exogenous-oxidant supported reactions. One of the more detailed studies on the subject involves the correlation of spectral intermediates and product formation rates via a systematic variation of the substituents on toluene substrates and cumene hydroperoxide oxidants (Blake and Coon, 1981), where various electron donating or withdrawing substituents on either the substrate or the peroxide had a dramatic and predictable influence on the reaction rate. Since these effects are expected only when the substituent affects a developing charge in the rate limiting step, it is suggested that the peroxide and substrate are both involved in the same rate-limiting step or that the rates for the two separate steps in which they participate are nearly equal (as might be expected in a heterolytic mechanism if oxygen donation to the iron by the peroxide and subsequent substrate oxygenation occurred at equal rates). Using mathematical models, the latter possibility was dismissed by the authors, thus implying that both the peroxide and substrate were participating in the same rate-limiting reaction. Such a direct role for the oxidant could be envisaged as hydrogen abstraction from the substrate by the cumyloxy radical generated after homolysis of the cumenehydroperoxide. As was pointed out informally by T. Bruice at the University of California-Santa Barbara, however, the range of hydroxylation rates employed in these studies was not sufficiently broad to rule out a stepwise, heterolytic mechanism. Other studies have documented oxidant-dependent regioselectivity in substrate oxygenations (Groves *et al.*, 1980b), but, since it is difficult to separate steric displacement perturbations from the differences one might expect from the generation of a variety of different radical abstracting species, these studies are not as conclusive. Further evidence suggesting that cytochrome P-450 has the ability to catalyse oxygen–oxygen bond homolysis was recently presented (White *et al.*, 1980). Homolytic scission of a peroxyacid conceptually generates a carboxyl radical, which, under certain conditions, might be expected to undergo radical induced decarboxylation. With perphenyl acetic acid, this decarboxylation generates a benzyl radical, which could then recombine with the iron–oxo species to form benzyl alcohol. In studies with both hepatic and bacterial cytochrome P-450 systems and various substrates, catalytic production of benzyl alcohol was observed in all cases. Such a finding unambiguously demonstrates that each P-450 tested is capable of inducing homolysis of this peracid. Through more recent investigations (McCarthy and White, 1983a), it has been shown that P-450 is a unique haemprotein in its ability to catalyse the decarboxylation of perphenylacetic acid. In all these investigations, however, the authors correctly point out that homolysis may not be a prerequisite for substrate hydroxylation but simply a side reaction unrelated to product formation. Evidence supporting this interpretation has been obtained very recently by White through elegant kinetic studies of hydroxylation and decarboxylation (McCarthy and White, 1983b).

In summary, great strides have been made in recent years but a definitive conclusion as to the precise mechanism of oxygen activation in cytochrome P-450 is still elusive. Extending the detailed chemical characterisations of high valence metallo-porphyrin model systems to the P-450 system is apparently valid according to numerous experiments, and provides strong evidence for a compound-I type intermediate active in substrate oxygenation. In addition, several studies suggest that all oxidants proceed via a common mechanism, thus implicating this intermediate as a general characteristic of P-450-type reactions. In contrast, the data of Blake and Coon support a homolytic mechanism in the cumenehydroperoxide-supported hydroxylations, while the decarboxylation studies initiated by White demonstrate the P-450 has the ability to catalyse the homolysis of peracid O–O bonds. Making any absolute conclusion at this time would be presumptuous, since it is impossible completely to dismiss any of these experiments out of hand. For these reasons, the first step in the hydroxylation mechanism, the activation of dioxygen, continues to receive extensive study.

SUBSTRATE ACTIVATION BY CYTOCHROME P-450

The unique aspects of P-450 catalysis lie in the conversion of the well-characterised oxygenated haem adduct $Fe^{2+}O_2$ to chemically active species which are capable of oxygen atom transfer to substrate. As discussed earlier, two distinct steps are envisaged: the activation of haem bound dioxygen through O–O bond lysis to generate an intermediate iron–oxo complex chemically competent in hydrogen abstraction and the subsequent attack of this iron–oxo species on the substrate molecule through hydrogen abstraction followed by radical recombination to generate the alcohol product. The first aspect of this sequence was discussed in the previous section of this review. Here, the focus will be on the substrate molecule. Since these events are intimately linked in the overall catalytic process, studies using various substrate analogues will also provide critical information on the oxygen activation step.

Due to its high degree of regioselectivity and the well characterised haem active centre, our laboratory has focussed our investigation primarily on the bacterial camphor monoxygenase system. Over the past few years, we have used a variety of camphor analogues in this system in an effort to gain mechanistic information on cytochrome P-450-dependent oxygenations. The scope of such studies include: definition of the pathway for substrate activation by cytochrome P-450 with regard to the mode of carbon–hydrogen bond breakage, determination of the absolute stereochemistry, regioselectivity, and heavy hydrogen atom isotope effects in camphor hydroxylation, and documentation of an epoxidation activity by this system heretofore thought only active in alkane hydroxylation. Each of these areas will be discussed separately.

Perhaps the most revealing study of the nature of substrate activation by cytochrome $P-450_{cam}$ employs the tricyclic camphor analogue pericyclocamphanone. Spectral titration studies demonstrate that this substrate binds to the enzyme with the same affinity as the normal substrate camphor. Furthermore, the enzyme catalyses the facile oxygenation of pericyclocamphanone to give both 6-*exo* and 6-*endo*-hydroxypericyclocamphanone in a reaction requiring both pyridine nucleotide (NADH) and oxygen. Since the overall structure of camphor and pericyclocamphanone are very similar, one might expect the hydroxylation reaction to occur on the 5-carbon (as is the case with camphor) and yet only the 6-alcohols are made. Interestingly, no detectable hydroxylation of the normal substrate camphor at carbon-6 by cytochrome $P-450_{cam}$ occurs. The inability of the enzyme to hydroxylate at carbon-5 of pericyclocamphanone strongly suggests that substrate activation by cytochrome P-450 involves the formation of a planar or nearly planar intermediate, since such a species would be difficult to generate at this position due to the severe ring strain in the molecule. Such a planar intermediate could be either a radical or carbonium ion, since both of these species prefer a nearly planar arrangement of atoms as opposed to carbanions which are generally believed to prefer pyramidal (sp^3) geometry. Carbonium ions are normally formed only with difficulty at a severely strained position, as is best noted by examining the rates of solvolysis of bridge-head-substituted compounds. These rates span an extremely large range (10^{14}) depending on the strain involved in planarising the system (Fort and Schleyer, 1964). Reactions involving the generation of carbon radicals are subject to similar rate effects, although the rates span a much narrower range (10^3) than those observed for carbonium ions (Humphrey *et al.*, 1968).

In order to try to force hydroxylation to occur at the tertiary carbon, we also studied the reaction of 6-ketopericyclocamphanone with cytochrome $P-450_{cam}$. With this substrate, the 6-position is blocked from oxygenation. When 6-keto-pericyclocamphanone is added to the reconstituted $P-450_{cam}$ system, rapid NADH oxidation occurs but no oxygenated products are detected, and nearly all electron equivalents from NADH reduce oxygen to form hydrogen peroxide. Thus, 6-ketopericyclocamphanone appears to serve as an excellent uncoupler of the $P-450_{cam}$ monoxygenase system by binding to the enzyme and inducing a low-to-high spin state interconversion, hence allowing efficient electron transfer to occur (Sligar, 1976), but forcing all reducing equivalents to appear in peroxide since the substrate itself is unable to be oxygenated.

Model studies on the oxygenation of another cyclopropylcarbinyl system, norcarane (bicyclo[4.1.0]heptane), have been carried out by Groves *et al.* (1980a) employing chloro(tetraphenylporphinato)manganese(III) and iodoso- benzene. Oxygenation of norcarane occurs adjacent to the cyclopropyl ring (analogous to the 6-position of pericyclocamphanone) to give the 2-*exo* and 2-*endo* norcaranols. Additional products, most notably 3-(hydroxymethyl)- cyclohexene, were also generated, as were the corresponding alkyl chlorides. The rearrangement products are a result of the formation of a carbon radical at the 2-position of norcarane with subsequent ring opening to give the cyclo- hexenyl products. The formation of alkyl chlorides results from chlorine abstrac- tion by the radical intermediates from the dichloromethane solvent. These studies demonstrate that the carbon radical intermediate exists long enough to undergo both rearrangement and abstraction from solvent. Formation of a carbonium ion at the 2-position of norcarane is unlikely, since these species are known to rearrange to cycloheptenyl products which were not observed under these reaction conditions. In the camphor hydroxylase system, one might expect to find rearrangement products resulting from the formation of either a radical or carbonium ion on the 6-carbon of pericyclocamphanone during catalysis, but no such products were observed. This lack of rearrangement suggests that the substrate intermediate is rapidly captured by the enzymatic addition of the hydroxyl group before rearrangement can occur. The kinetics of ring opening of a pericyclocamphanone radical would be expected to be similar to the well documented nortricyclo-norbornenyl rearrangement which shows a rate in both directions on the order of 10^8 s^{-1} (Wong and Griller, 1981). Thus the capture of the active intermediate in pericyclocamphanone oxygenation must be extremely fast, such as occurs in the capture of cyclo- propylcarbinyl radicals by ligand transfer with CuBr$_2$ (Jenkins and Kochi, 1972). The rapidity with which this capture reaction occurs is also suggested by model studies with chloro(tetraphenylporphinato)Fe(III) and iodosobenzene, which, in contrast to the Mn(III)TPPCl catalysed reaction, forms no alkyl chlorides (Groves *et al.*, 1979), implying that oxygen addition to activated reactants occurs before ligand transfer from the solvent. Norcarane oxygenation studies with hepatic P-450$_{LM2}$ gave only the 2-*exo* and 2-*endo* norcaranols, again demonstrating the rapid capture of an oxidised substrate intermediate. Another indication of the rapid capture of oxidised intermediates comes from model epoxidation studies on *cis*-stilbene. Epoxidation of *cis*-stilbene with Mn(III)TPPCl and iodosobenzene gave a mixture of *cis* and *trans*-stilbene oxides, demonstrating that the substrate intermediate generated by oxidative attack of the manganese-oxo species can freely rotate (Groves *et al.*, 1980a). However, the analogous reaction with Fe(III)TPPCl yields only *cis*-stilbene oxide, implying that oxygen transfer from the metal to the substrate occurs before rotation can occur (Groves *et al.*, 1979). It would be interesting to examine the stereospecificity of *cis*-stilbene epoxidation by cytochrome P-450 and to compare the results with those of the model studies outlined above. The differering capture rates for manganese and iron metal centres may explain the lack of hydroxylation activity in manganese-substituted cytochrome P-450$_{cam}$ described above.

The experimental findings on the mode of substrate activation by cytochrome P-450 cited thus far are consistent with either a radical or carbonium ion intermediate which is rapidly captured by addition of the hydroxyl group before rearrangement can occur. Evidence from the model system tends to favour a radical state. Studies on cytochrome P-450-dependent hydroxylation of norbornane (Groves *et al.*, 1978) have suggested that a carbonium ion substrate intermediate is unlikely since the authors observed the production of the *endo* isomer of 2-norbornanol and carbonium ions on the norbornyl skeleton are known to be captured exclusively in the *exo* position. One should be cautious in interpreting these enzymatic results, however, since the exclusive *exo* capture of norbornyl cations applies to solution reactions, whereas in the enzymatic reaction *endo* capture of a cation could, conceptually, be a result of geometrical effects imposed by the enzyme on the oxygen transfer step.

The observed oxidation of pericyclocamphanone to 6-*exo* and 6-*endo* alcohols is the first example of the production of more than one product by the cytochrome P-450$_{cam}$ monoxygenase system. A study was carried out to test whether these products are made in exogenous-oxidant supported turnover of the enzyme and to examine the ratio of the isomers as a function of the oxidant used. Reconstitution of the P-450$_{cam}$ hydroxylase with NADH and O$_2$ gave an exo/endo ratio of 1.03 while addition of iodosobenzene or m-chloroperbenzoic acid (m-CPBA) to ferric P-450$_{cam}$ in the presence of pericyclocamphanone yielded 6-exo and 6-endo alcohols in ratios of 1.48 (iodosobenzene) and 1.80 (m-CPBA). The data show that the product ratios depend on the oxidant used and that the ratios for the exogenous oxidant-supported reactions are different from the reconstituted P-450$_{cam}$ system utilising NADH and O$_2$. Two explanations must be considered. Either the organic portion of the oxidant *directly* participates in the hydroxylation reaction and, thus, can alter product distributions via inductive electronic effects, or the mere presence of the fragmented oxidant at the active site of the enzyme during the hydroxylation event alters regioselectivity. An example of the former would be the participation in substrate hydrogen abstraction of a carboxyl radical generated by homolysis of the O–O bond of m-CPBA (Blake and Coon, 1981). The second, steric, effect envisages the hydrogen abstraction event occurring prior to the departure of the organic portion of the oxidant, thus leading to an alteration in the regioselectivity of the reaction. Groves *et al.* (1980b) have studied the reaction of cyclohexene with P-450 in the presence of a variety of exogenous oxidants. The ratio of cyclohexenol to cyclohexene oxide is drastically altered with changes in the oxidant; similar deuterium isotope effects were observed, however, in all of the reactions. The authors concluded from this study that steric effects play an important role in determining the regioselectivity of the oxygenation reaction.

In order to further probe the stereochemistry and regioselectivity of these reactions, we examined the P-450$_{cam}$-dependent oxygenation reactions on camphor analogues carrying a substituent in either the 5-*exo* or 5-*endo* position. In all cases studied, with an *exo*- bromo or iodo group, aketone is formed at the 5-position of camphor with oxygen isotope studies demonstrating that the 5-keto oxygen originates from atmospheric oxygen. These findings were initially unexpected since the normally hydroxylated (5-*exo*) position of camphor is blocked in some of these substrates suggesting that in these cases P-450$_{cam}$ can abstract the 5-*endo* hydrogen to generate a planar or nearly planar intermediate. The simplest interpretation of these reactions involves the formation of a *gem*-halohydrin intermediate, in the case of the bromo and iodo analogues, and the hydrate of 5-ketocamphor (*gem*-diol), in the case of the alcohols. Oxygen addition, presumably to the *exo* face of the substrate intermediate gives the *gem*-substituted species followed by breakdown to the final observed reaction products. All of these reactions occur both with the reconstituted bacterial monooxygenase system (NADH/O$_2$) and when the exogenous oxidants iodosobenzene, hydrogen peroxide, or m-CPBA react with the ferric haemoprotein. To carry out a more detailed study on the stereochemistry of camphor hydroxylation in this system, both 5-*exo* and 5-*endo*-deuterated camphors were synthesised (Gelb *et al.*, 1982a). The presence of deuterium opens the possibility for observing isotope effects in the reaction. As will be discussed below, the stereochemistry and isotope effects are fundamentally related, with a complete understanding of both phenomena required in order to gain mechanistically useful information. The overall stereochemistry of various P-450 monoxygenations has been examined in several systems (Hayano *et al.*, 1958). The stereochemical course of steroid hydroxylations by both bacterial and mammalian monoxygenase systems proceeds with almost complete retention of configuration (90–98%) at the oxidised carbon. Thus the hydrogen atom that is replaced by a hydroxyl group is preferentially removed by the enzyme. as opposed to a general Walden type inversion. Groves *et al.* (1978), in studying the hydroxylation of norbornane by purified rabbit liver P-450$_{LM2}$, reported the first example of a P-450 catalysed reaction that occurs with a partial loss of configuration (up to 25%) at the oxidised carbon. The authors suggested that a partial epimerisation of an enzyme-generated carbon radical was responsible for the observed loss of stereochemistry. In our study with deuterated camphors and the bacterial P-450$_{cam}$ system, there is a fundamental difference from the norbornyl hydroxylations by hepatic P-450. In the P-450$_{cam}$ hydroxylation of camphor, only the 5-*exo*-isomer of hydroxycamphor is made, whereas both *exo* and *endo* norbornanols are made in the hydroxylation of norbornane by the liver microsomal monooxygenase system. Thus, it is possible to study separately the overall stereochemistry of the hydrogen abstraction step which occurs prior to the stereospecific oxygen addition step. It should be pointed out that many stereochemical investigations of P-450 hydroxylations that purport to prove that oxygen transfer occurs with complete retention of configuration make severe assumptions which are often not clearly stated. Consider the hydroxylation of a secondary carbon as exemplified by the camphor 5-*exo*-hydroxylase and the formation of phenylethanol from ethylbenzene by hepatic P-450 preparations. Two isomeric products, the R-alcohol and S-alcohol, can be generated, which, due to the stepwise nature of the putative hydrogen abstraction and radical recombination events, could contain either the pro-R or pro-S hydrogens of the secondary alcohol. Since H-abstraction and O-transfer are two separate events, there are two places at which the enzyme can control stereochemistry.

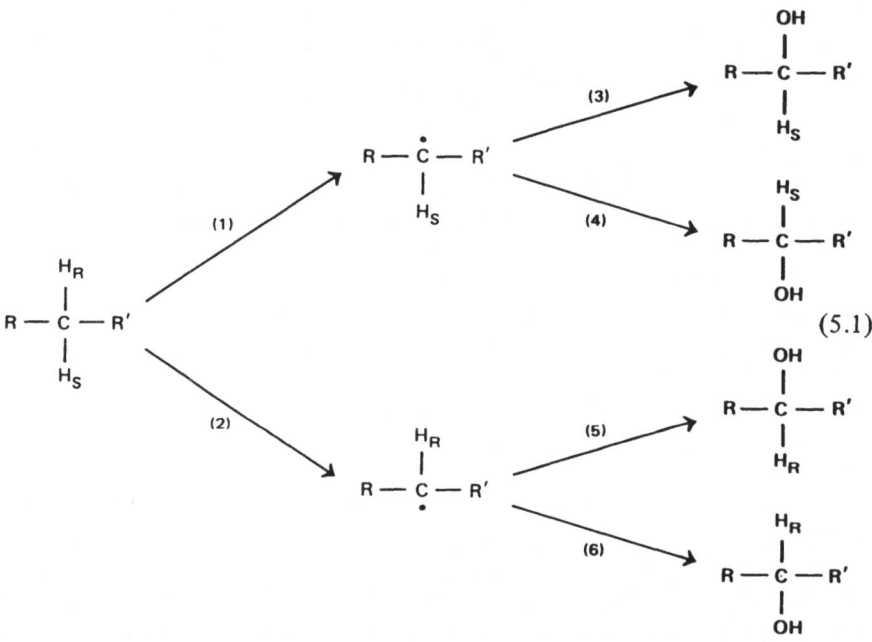

$$(5.1)$$

In equation (5.1), (1) and (2) refer to the stereospecific hydrogen abstraction of either the pro-R or pro-S hydrogens of the substrate to generate the carbon radical intermediate. The second point of stereochemical control occurs in the capture of the radical intermediate to yield either the R- or S- alcohol product. Processes (3) and (6) occur with retention of configuration in the oxygen transfer step while (4) and (5) represent an inversion. There are many examples of enzymatic stereochemical control of a prochiral centre and hence the documentation of the overall stereochemistry of hydroxylation must address both the hydrogen abstraction and oxygen radical transfer steps. In order to deconvolute the above pathways to R- and S-product, the pro-R and pro-S hydrogens must be distinguished by deuterium or tritium isotopic labelling. While thus distinguishing the four isomeric products shown, additional parameters are introduced by way of the primary and secondary isotope effects operating. In the camphor hydroxy-lase where only one isomeric alcohol is produced, a combination of deuterium and tritium labelling can resolve the stereochemistry of each step if one ignores the small secondary isotope effects. Such problems are also present in attempts to determine the stereochemical course of primary carbon hydroxylation using hydrogen–deuterium–tritium chiral methyl group analysis. It is *impossible* to ascertain the absolute stereochemistry of hydrogen abstraction and radical recombination using a single chiral methyl analysis without assuming the exist-ence of large primary intrinsic isotope effects. In many enzymic processes rigorously examined there appears to be efficient masking of the isotope effects of abstraction and recombination. A final caution to those reviewing the pub-lished oxygenase isotope effect literature concerns the inexcusable belief of some that all methyl/methylene monoxygenases are cytochrome P-450s. In particular the bacterial omega-hydroxylase has been definitely shown *not* to contain a haem prosthetic group. With these comments we will briefly review the stereochemical investigations of camphor hydroxylation by cytochrome P-450$_{cam}$.

Maximal velocities were measured in turnover studies with the deuterated camphor substrate and the reconstituted camphor hydroxylase system (NADH/O$_2$), in addition to the velocities of 5-ketocamphor formation from deuterated and non-deuterated 5-*exo*-bromocamphor. The data show that in all cases a small intermolecular isotope effect ($V_{max}^{H}/V_{max}^{D} = 1.12$–$1.23$) is measured. The hydroxylation of camphor by cytochrome P-450$_{cam}$ is a multi-step process, including electron transfer from NADH to the cytochrome via two intermediate redox proteins and all of the chemical reactions involved in the actual scission of the O–O bond and hydroxylation of substrate. Therefore, intermolecular isotope effects are expected to show considerable steady state suppression (Northrop, 1977) as is observed in these studies. This is particularly true since the Gunsalus laboratory has presented strong evidence that the major contributing factor to the rate limiting step in camphor oxygenation is transfer of the second electron from putidaredoxin to oxygenated cytochrome or a subsequent intermediate (Pederson *et al.*, 1977).

Table 5.3 presents the product deuterium content of 5-*exo*-hydroxycamphor formed with the reconstituted hydroxylase system (NADH/O$_2$) and the exo-genous oxidants iodosobenzene, m-CPBA and hydrogen peroxide. Oxidation of the product alcohol to 5-ketocamphor results in the complete loss of deuterium, confirming that the product deuterium is exclusively in the 5-*endo* position. Since, in all cases, both deuterated and non-deuterated alcohols are formed, there is unambiguous demonstration that the hydrogen abstraction step in cytochrome P-450$_{cam}$ hydroxylation of camphor proceeds with a partial loss of stereochemistry, as originally suggested by the formation of 5-ketocamphor from the 5-*exo* substituted camphor analogues. It can also be seen from table 5.3 that the product deuterium content is quite insensitive to the choice of oxidant used to support hydroxylation. This observation presents the first evidence that the

Table 5.3   Mass spectrometry of the product alcohol deuterium content

Substrate	Oxidant	Alcohol-d_1/ Alcohol-d_0 [*]	Average
5-*exo*- deuterocamphor	NADH/O_2	1.15, 1.20	1.18
	iodosobenzene	1.47, 1.50	1.48
	m-CPBA	1.58, 1.69	1.64
	H_2O_2	1.19, 1.21	1.20
5,6-*endo, endo*- dideuterocamphor	NADH/O_2	4.40, 4.62	4.51
	iodosobenzene	3.77, 3.75	3.76
	m-CPBA	4.03, 3.88	3.96
	H_2O_2	3.85, 3.70	3.78
5-*endo*- deuterocamphor	NADH/O_2	4.25, 4.52	4.39
	iodosobenzene	4.72, 4.65	4.69
	m-CPBA	4.22, 3.99	4.11
	H_2O_2	3.97, 3.90	3.94

[*]The double entries are the result of independent experimental trials (taken from Gelb *et al.*, (1982a)).

hydrogen abstracting species is the same in all of the exogenous oxidant supported systems as well as in the physiologically relevant reaction with dioxygen. The treatment of stereochemistry and intramolecular deuterium isotope effects in camphor hydroxylation has been described in detail (Gelb *et al.*, 1982a) and we give here only a brief discussion of the results. In interpreting the data in table 5.3, one must consider the intrinsic selectivity for *exo* versus *endo* hydrogen abstraction by the enzyme given hydrogen at both positions (designated by G) and the primary intramolecular deuterium isotope effects for both *exo* and *endo* abstraction (designated I_{exo} and I_{endo}). G can only be measured indirectly by distinguishing the 5-*exo* and 5-*endo* positions by deuterium substitution and then deconvoluting the data so as to correct for the primary and secondary isotope effects that are operational in the system. Also, since the geometry of the transition states for *exo* and *endo* hydrogen abstraction may be different, I_{exo} and I_{endo} will not, in general, have identical values (Melander and Saunders, 1980). These three parameters are related to the observed product deuterium ratios via: $I_{exo} = AG$, $I_{endo} = B/G$ where A and B are the ratios of deuterated to non-deuterated product alcohols when 5-*exo*-deuterated and 5-*endo*-deuterated substrates are hydroxylated (Gelb *et al.*, 1982a). As there are three unknown quantities (G, I_{exo}, I_{endo}) and two observables (A, B) we plot the allowed isotope effects as a function of G, finding that as the ratio of *exo* versus *endo* hydrogen abstraction increases (increasing G), I_{exo} becomes large while I_{endo} decreases (Gelb *et al.*, 1982a). Given only the data with the deuterated substrates, the value of G, and hence I_{exo} and I_{endo} cannot be uniquely derived; however, G can be evaluated if one performs an analogous study with tritiated camphor. Although tritium isotope effects must now be considered, the tritium and deuterium isotope effects are not independent if one *assumes* that the Swain relationship is valid for the reaction under study (Northrop, 1977). The precise free energy difference between tritium, deuterium, and hydrogen abstraction are somewhat system dependent, however, and assumptions must be made and clearly stated. The similar product deuterium ratios obtained with the exogenous oxidant and NADH/O_2 supported reactions must be discussed in light of the partial loss of stereochemistry in camphor hydroxylation. A mechanism involving the direct insertion of an oxygen atom into the C-H bond of camphor is obviously incompatible with these stereochemical results. A single, in-line trajectory, oxygen rebound mechanism for P-450-dependent catalysis (Groves and McClusky, 1976), whereby the iron-oxo species performs the hydrogen abstraction and oxygen addition events would be consistent with the results if G was much greater than unity. In this case I_{exo} would be much larger than I_{endo}. Norbornyl-centred free radicals show only a modest selectivity for *exo* capture as opposed to the predominant *exo* capture of norbornyl-centred carbonium ions (Bartlett *et al.*, 1970). If G had a value near unity, however, a second mechanism could involve generation of a radical centre, $(X\cdot)$, at a site on the protein removed from the iron-oxo porphyrin prosthetic group. In this case, the hydrogen-abstracting and oxygen addition species would be distinct. Such a higher oxidation state of cytochrome P-450 might be similar to the structure 'ES' in cytochrome c peroxidase (Yonetani *et al.*, 1966) and could be generated by two conceptual pathways. As the designation of the higher oxidation states of iron–porphyrin systems can become confusing, we suggest the following nomenclature. The porphyrin entity carries a formal charge of 4- (two propionic acid anions and two pyrrole nitrogen anions) and is designated P. Thus, the one electron oxidised porphyrin (π cation radical) is writtten as P^+. Iron is designated with its proper oxidation number and iron-bound oxygen with its formal charge. Thus, compound I in horseradish peroxidase is written as $[(P^+)Fe^{4+}O^{2-}]^{3+}$. The charge outside the brackets is written in this case as 3+ and represents the sum of the charges in the complex. Cytochrome c peroxidase, with one oxidation equivalent off the porphyrin ring, is written as $[(P)Fe^{4+}O^{2-}(X^+)]^{3+}$ where X^+ is the one electron oxidised species. Production of a $[(P)Fe^{5+}O^{2-}]^{3+}$ state by heterolysis of the O–O bond of O_2, ROOH, or RCO_3H liganded to the haem iron could generate $X\cdot$ by hydrogen abstraction or electron transfer. Alternatively, homolysis of the peroxide bond could yield $X\cdot$ by electron transfer or a concerted process. The substrate radical intermediate is then envisaged as generated via hydrogen transfer from either the *exo* or *endo* position of the substrate to $X\cdot$ with stereospecific radical recombination with the iron-bound oxygen giving the observed 5-*exo* alcohol product. The stereochemical data given above only suggest the possibility for a hydrogen abstracting species removed from the prosthetic group and direct observation of $X\cdot$ by spectroscopic means will be required to demonstrate its existence unambiguously. Movement of the substrate radical before oxygenation occurs will also explain the observed loss of stereochemistry during camphor hydroxylation. Thus, if there is a large reactivity of the iron–oxo species toward hydrogen abstraction, the selectivity of *exo* versus *endo* abstraction may be small. The newly generated hydroxyl group on the haem-iron might sterically interact with the substrate radical such that only *exo* capture is permitted. Since the hydrogen abstraction species has not yet been identified in cytochrome P-450 catalysed oxygenations, a definite conclusion from the stereochemical data cannot yet be made.

The final study involving substrate analogues in the cytochrome P-450$_{cam}$ monoxygenase system reported herein utilises 5,6-dehydrocamphor in order to ascertain whether this bacterial enzyme can efficiently catalyse an epoxidation reaction. It has been well established that hepatic microsomal P-450 systems will catalyse the epoxidation of a variety of olefinic substrates (Hayano *et al.*, 1958). The chemical nature of epoxidation and its similarity to hydroxylation is not yet fully understood. Cytochrome P-450$_{cam}$ was found to catalyse the epoxidation of dehydrocamphor with nearly perfect coupling of the reducing equivalents from NADH into the production of a single epoxide product (Gelb *et al.*, 1982c). Oxygen-18 studies demonstrate that the epoxide oxygen originates from atmospheric oxygen. High resolution proton nmr of the isolated product epoxide indicates that only the *exo* isomer of 5,6-epoxycamphor was formed. Thus the oxygen atom is added to the *exo* face of camphor in both hydroxylation and epoxidation reactions. Epoxide was also formed when iodosobenzene or hydrogen peroxide was added to ferric P-450 while m-CPBA gave epoxide in control reactions in the absence of enzyme.

Detailed study of the kinetics of NADH driven hydroxylation and epoxidation shows that, at least at this level of analysis, the two reactions are very similar. Many olefinic substrates have recently been shown to inactivate microsomal P-450 in a suicidal fashion (Ortiz de Montelanno and Kunze, 1981). In this regard, P-450$_{cam}$ epoxidation of dehydrocamphor is not a suicidal process as competent enzyme remains after several thousand reaction cycles per P-450 molecule. Nor can loss of cytochrome P-450, as detected by optical spectral analysis of the CO adduct, be seen after several hundred reaction cycles. This new reactivity of cytochrome P-450$_{cam}$ is under continued investigation in the hope of dissecting the active intermediates involved in substrate epoxidation reactions.

In summary, there appear to be two unique chemical reactivities that are displayed by cytochrome P-450 over that observed for many other haem proteins. The first involves the reactivity of an oxygenated haem adduct to generate a transition state with highly strained oxygen–oxygen bond suitable for scission. The second involves the generation of a species at the active site which is able to abstract hydrogen efficiently from an unactivated alkane carbon. We would like to propose a use for these two chemical reactivities in understanding yet another unique reactivity of cytochrome P-450, that of carbon–carbon bond scission in the C-20/C-22 and C-17/C-20 lyase activities observed in steroid biosynthesis. Hypothetical chemical mechanisms for these two reactions utilise nucleophilic and hydrogen abstracting activities analogous to those described above. Predictions from this model are first, that the C-20/C-22 lyase activity should not be operative if 22-keto cholesterol is used as substrate whereas C-17/C-20 lyase activity should be abolished if the 20-carbon bears a hydroxyl group instead of the normal ketone of pregnenelone. Such mutually exclusive reactivities have been observed. Second, iodosobenzene, which is known to generate a higher valence iron-oxo species isoelectronic in charge stoichiometry with $(Fe^V = O)$ directly without an intermittent nucleophilic oxygen adduct, should not support C-17/C-20 lyase activity but should catalyse the $P-450_{scc}$ dependent cleavage of dihydroxycholesterol. In collaboration with Dr Peter Hall at the Worcester Foundation, we are examining the chemical nature of carbon–carbon scission in steroidal, bacterial, and hepatic P-450 preparations.

Thus a wealth of information is available on the P-450 dependent monoxygenase systems and as such they represent one of the best studied classes of haemoproteins. Seriously lacking, however, is an understanding, at the precise molecular level, of the *Physics* of electron transfer between pyridine nucleotide, flavin, iron–sulphur and haem prosthetic groups and the *Chemistry* of metal–oxygen interaction involved in O–O bond scission of atmospheric dioxygen and substrate carbon chain functionalisation. Hence further study of these ubiquitous metalloenzymes is not only justified, but promises to yield exciting new information on a broad relevance to biochemists, biophysicists, and inorganic chemists.

REFERENCES

Abbott, M. T. and Udenfreund, S. (1974) In: *Molecular Mechanisms of Oxygen Activation* (ed. Hayaishi, O.), Academic Press, New York, p. 167

Armes, L. G., Yasunobu, K. T., Gunsalus, I. C. and Piette, L. H. (1981) *Fed. Proc., Fed. Am. Soc. Exp. Biol.*, **40**, 1662

Backes, W. L., Sligar, S. G. and Schenkman, J. B. (1980) *Biochem. Biophys. Res. Commun.*, **97**, 860

Backes, W. L., Sligar, S. G. and Schenkman, J. B. (1982) *Biochemistry*, **21**, 1324

Banerjee, S. and Dec, G. (1979) *Biochem. Biophys. Res. Commun.*, **88**, 833

Bartlett, P. D., Fickes, G. N., Haupt, F. C. and Helgeson, R. (1980) *Acc. Chem. Res.*, **3**, 177

Berzinis, A. P. and Traylor, T. G. (1979) *Biochem. Biophys. Res. Commun.*, **87**, 229

Black, S. D., Tarr, G. E. and Coon, M. J. (1982) *J. biol. Chem.*, **257**, 14616

Blake, R. C. and Coon, M. J. (1981) *J. biol. Chem.*, **256**, 12127

Botelho, L. H., Ryan, D. E. and Levin, W. (1979) *J. biol. Chem.*, **254**, 5635

Botelho, L. H., Ryan, D. E., Yuan, P.-M., Kutny, R., Shively, J. E. and Levin, W. (1982) *Biochemistry*, **21**, 1152

Bradshaw, W. H., Conrad, H. E., Corey, E. J., Gunsalus, I. C. and Lednicer, D. (1959) *J. Am. chem. Soc.*, **81**, 5507

Budyka, M. F., Khenkin, A. M. and Shteinman, A. A. (1981) *Biochem. Biophys. Res. Commun.*, **87**, 229

Burka, L. T., Thorsen, A. and Guengerich, F. P. (1980) *J. Am. chem. Soc.*, **102**, 7615

Byrn, M. P. and Strouse, C. E. (1981) *J. Am. chem. Soc.*, **103**, 2633

Champion, P. M., Lipscomb, J. D., Munck, E., Debrunner, P. and Gunsalus, I. C. (1975a) *Biochemistry*, **14**, 4151

Champion, P. M., Chiang, R., Munck, E., Debrunner, P. and Hager, L. P. (1975b) *Biochemistry*, **14**, 4159

Champion, P. M., Munck, E., Debrunner, P. G., Moss, T., Lipscomb, J. and Gunsalus, I. C. (1975c) *Biochim. Biophys. Acta*, **376**, 579

Champion, P. M., Gunsalus, I. C. and Wagner, G. C. (1978) *J. Am. chem. Soc.*, **100**, 3743

Champion, P. M., Stallard, B. R., Wagner, G. C. and Gunsalus, I. C. (1982) *J. Am. chem. Soc.*, **104**, 5469

Chang, C. K. and Dolphin, D. (1975) *J. Am. chem. Soc.*, **97**, 5948

Chevion, M., Peisach, J. and Blumberg, W. E. (1977) *J. biol. Chem.*, **252**, 3637

Chevion, M., Stern, A., Peisach, J., Blumberg, W. E. and Simon, S. (1978) *Biochemistry*, **17**, 1745

Chiang, R., Makino, R., Spomer, W. E. and Hager, L. P. (1975). *Biochemistry*, **14**, 4166

Chou, P. Y. and Fasman, G. D. (1978) *Ann. Rev. Biochem.*, **47**, 251

Cinti, D. L., Sligar, S. G., Gibson, G. G. and Schenkman, J. B. (1979) *Biochemistry*, **18**, 36

Cole, P. and Sligar, S. G. (1981) *FEBS Lett.*, **133**, 252

Conrad, H., Thomas, H. P., Vogel, H. and Mayer, A. (1969) *J. molec. Biol.*, **41**, 225

Cramer, S. P., Dawson, J. H., Hager, L. P. and Hodgsen, K. O. (1978) *J. Am. chem. Soc.*, **100**, 7282

Dawson, J. H., Andersson, L. A. and Sono, M. (1982) *J. biol. Chem.*, **257**, 3606

Dolphin, D. (1981) *Israel J. Chem.*, **21**, 67

Dus, K. M. (1976) In: *The Enzymes of Biological Membranes*, Vol. 4 (ed. Martonosi, A.), Plenum Press, New York, p. 199

Dus, K. M. (1982a) In: *From Cyclotrons to Cytochromes* (ed. Kaplan, N. O.), Academic Press, New York, p. 231

Dus, K. M. (1982b) *Xenobiotica*, **12**, 745

Dus, K. M. and Murray, R. I. (1982) *Devel. Biochem.*, **23**, 377

Dus, K., Katagiri, M., Yu, C.-A., Erbes, D. L. and Gunsalus, I. C. (1970) *Biochem. Biophys. Res. Commun.*, **40**, 1423

Dus, K., Litchfield, W. J., Miguel, A. G., van der Hoeven, T. A., Haugen, D. A., Dean, W. L. and Coon, M. J. (1974) *Biochem. Biophys. Res. Commun.*, **60**, 15

Dus, K. M., Gowert, R. and Swanson, R. A. (1977) In: *Microsomes and Drug Oxidations* (ed. Ullrich, V. *et al.*), Pergamon Press, New York, p. 95

Dus, K. M., Litchfield, W. J., Hippenmeyer, P. J., Bumpus, J. A., Obidoa, O., Spitsberg, V. and Jefcoate, C. R. (1980) *Eur. J. Biochem.*, **111**, 307

Dus, K. M., Bumpus, J. A. and Murray, R. I. (1982) In: *Microsomes, Drug Oxidations, and Drug Toxicity* (ed. Sato, R. and Kato, R.), Japan Scientific Societies Press, Tokyo and Wiley-Interscience, New York, p. 45

Ebel, R. E., O'Keeffe, D. H. and Peterson, J. A. (1977) *Arch. Biochem. Biophys.*, **183**, 317

Eble, K. S. and Dawson, J. H. (1984) *Biochemistry*, in press

Eisenstein, L., Debey, P. and Douzou, P. (1977) *Biochem. Biophys. Res. Commun.*, **77**, 1377

Estabrook, R. W. (1982). In: *Microsomes, Drug Oxidations, and Drug Toxicity* (ed. Sato, R. and Kato, R.), Japan Scientific Societies Press, Tokyo, and Wiley-Interscience, New York, p. 133

Estabrook, R. W., Hildebrandt, A. G., Baron, T., Netter, K. and Leibmon, K. (1971) *Biochem. Biophys. Res. Commun.*, **47**, 132

Estabrook, R. W. and Werringloer, J. (1977) In: *Microsomes and Drug Oxidations* (ed. Ullrich, V. *et al.*), Pergamon Press, New York, p. 748

Fort, R. C. and Schleyer, P. von R. (1964) *Chem. Rev.*, **64**, 277

Fuji-Kuriyama, Y., Mizukami, Y., Kawajiri, K., Sogawa, K. and Muramatsu, M. (1982) *Proc. nat. Acad. Sci., USA*, **79**, 2793

Garfinkel, D. (1958) *Arch. Biochem. Biophys.*, **77**, 493

Gelb, M. H., Heimbrook, D. C., Malkonen, P. and Sligar, S. G. (1982a) *Biochemistry*, **21**, 370

Gelb, M. H., Toscano, W. A., Jr. and Sligar, S. G. (1982b) *Proc. nat. Acad. Sci., USA*, **79**, 5758

Gelb, M. H., Malkonen, P. J. and Sligar, S. G. (1982c) *Biochem. Biophys. Res. Commun.*, **104**, 853

Gibson, G. G., Cinti, D. L., Sligar, S. G. and Schenkman, J. B. (1979) *Biochem. Soc. Trans.*, **7**, 1289

Gibson, G. G., Sligar, S. G., Cinti, D. L. and Schenkman, J. B. (1980) *Biochem. Soc. Trans.*, **8**, 101

Gould, P. V., Gelb, M. H. and Sligar, S. G. (1981) *J. biol. Chem.*, **256**, 6686

Griffin, B. W., Smith, S. M. and Peterson, J. A. (1974) *Arch. Biochem. Biophys.*, **160**, 323

Griffin, B. W. and Peterson, J. A. (1975) *J. biol. Chem.*, **250**, 6445

Groves, J. T. and McClusky, G. A. (1976) *J. Am. chem. Soc.*, **98**, 859

Groves, J. T. and Kruper, W. J., Jr. (1979) *J. Am. chem. Soc.*, **101**, 7613

Groves, J. T. and Haushalter, R. C. (1981) *J. chem. Soc., Chem. Commun.*, 1165

Groves, J. T., McClusky, G. A., White, R. E. and Coon, M. J. (1978) *Biochem. Biophys. Res. Commun.*, **81**, 154

Groves, J. T., Nemo, T. E. and Myers, R. S. (1979) *J. Am. chem. Soc.*, **101**, 1032

Groves, J. T., Kruper, W. J., Jr. and Haushalter, R. C. (1980a) *J. Am. chem. Soc.*, **102**, 6375

Groves, J. T., Krishnan, S., Avaria, G. E. and Nemo, T. E. (1980b) *Adv. Chem. Ser.*, **191**, 277

Groves, J. T., Haushalter, R. C., Nakamura, M., Nemo, T. E. and Evans, B. J. (1981) *J. Am. chem. Soc.*, **103**, 2884

Guengerich, F. P. (1983) *Biochemistry*, **22**, 2811

Gunsalus, I. C., Chapman, P. J. and Kuo, J.-F. (1965) *Biochem. Biophys. Res. Commun.*, **18**, 924

Gunsalus, I. C. and Sligar, S. G. (1978) *Adv. Enzymol.*, **47**, 1

Gunsalus, I. C. and Wagner, G. C. (1978) In: *Methods in Enzymology*, Vol. 52 (ed. Fleischer, S. and Packer, L.), Academic Press, New York, p. 166

Gunsalus, I. C., Wagner, G. C. and Debrunner, P. G. (1980) In: *Microsomes, Drug Oxidation and Chemical Carcinogenesis* (ed. Coon, M. J. *et al.*), Academic Press, New York, pp. 233–242

Gustafsson, J. A., Rondahl, L. and Bergman, J. (1979) *Biochemistry*, **18**, 865

Hahn, J. E., Hodgson, K. O., Andersson, L. A. and Dawson, J. H. (1982) *J. biol. Chem.*, **257**, 10934

Hamilton, G. (1974) In: *Molecular Mechanisms of Oxygen Activation* (ed. Hayaishi, O.), Academic Press, New York, p. 405

Haniu, M., Tanaka, M., Yasunobu, K. T. and Gunsalus, I. C. (1982a) *J. biol. Chem.*, **257**, 12657

Haniu, M., Armes, L. G., Yasunobu, K. T., Shastry, B. A. and Gunsalus, I. C. (1982b) *J. biol. Chem.*, **257**, 12664

Haniu, M., Yasunobu, K. T. and Gunsalus, I. C. (1983) *Biochem. Biophys. Res. Commun.*, **116**, 30

Hanson, L. K., Eaton, W. A., Sligar, S. G., Gunsalus, I. C., Gouterman, M. and Connel, C. R. (1976) *J. Am. chem. Soc.*, **98**, 2672

Hanukoglu, I., Spitsberg, V., Bumpus, J. A., Dus, K. M. and Jefcoate, C. R. (1981) *J. biol. Chem.*, **256**, 4321

Haugen, D. A. and Coon, M. J. (1976) *J. biol. Chem.*, **251**, 7929

Haugen, D. A., Armes, L. G., Yasunobu, K. T. and Coon, M. J. (1977) *Biochem. Biophys. Res. Commun.*, **77**, 967

Hayano, M., Gut, M., Dorfman, R. I., Sebek, O. K. and Peterson, D. H. (1958) *J. Am. chem. Soc.*, **80**, 2336

Hedegaard, J. and Gunsalus, I. C. (1965) *J. biol. Chem.*, **240**, 4038

Heimbrook, D. C. and Sligar, S. G. (1981) *Biochem. Biophys. Res. Commun.*, **99**, 530

Heimbrook, D. C. and Whitcombe, T. (1981) *Fed. Proc., Fed. Am. Soc. Exp. Biol.*, **40**, 1963

Heimbrook, D. C. and Sligar, S. G. (1984) In preparation

Heinemann, F. S. and Ozols, J. (1982) *J. biol. Chem.*, **257**, 14988

Hill, H. A. O. (1979) In: *Oxygen Free Radicals and Tissue Damage*, Ciba Foundation, Excerpta Medica, New York, p. 51

Hintz, M. J. and Peterson, J. A. (1981) *J. biol. Chem.*, **256**, 6721

Horie, S. (1978) In: *Cytochrome P-450* (ed. Sato, R. and Omura, T.) Kodansha Ltd., Tokyo, and Academic Press, New York, p. 73

Hui Bon Hoa, G. and Marden, M. R. (1982) *Eur. J. Biochem.*, **124**, 311

Humphrey, L. B., Hodgson, B. and Pincock, R. E. (1968) *Can. J. Chem.*, **46**, 3099

Ingold, K. U. (1969) *Acc. Chem. Res.*, **2**, 1

Jenkins, C. L. and Kochi, J. K. (1972) *J. Am. chem. Soc.*, **94**, 856

Jung, C. (1983) *Studia Biophysica*, **93**, 225

Katagiri, M., Ganguli, B. N. and Gunsalus, I. C. (1968) *J. biol. Chem.*, **243**, 3543

Keller, R. M., Wuthrich, K. and Debrunner, P. G. (1972) *Proc. nat. Acad. Sci., USA*, **69**, 2073

Kendrew, J. C., Watson, H. C., Strandberg, B. E., Dickerson, R. E., Phillips, D. C. and Shore, V. C. (1961) *Nature, Lond.*, **190**, 666

Klingenberg, M. (1958) *Arch. Biochem. Biophys.*, **75**, 376

Koch, S., Tang, S. C., Holm, R. H. and Frankel, R. B. (1975a) *J. Am. chem. Soc.*, **97**, 914

Koch, S., Tang, S. C., Holm, R. H., Frankel, R. B. and Ibers, J. A. (1975b) *J. Am. chem. Soc.*, **97**, 916

Koop, D. R. and Hollenberg, P. F. (1980) *J. biol. Chem.*, **255**, 9685

Lambeth, J. D., Kamin, H. and Seybert, D. W. (1980) *J. biol. Chem.*, **255**, 8282

Leighton, J. K., Debrunner-Vossbrink, B. A. and Kemper, B. (1984) *Biochemistry*, **23**, 204

Lewis, B. A. and Sligar, S. G. (1983) *J. biol. Chem.*, **258**, 3599

Lichtenberger, F. and Ullrich, V. (1977) In: *Microsomes and Drug Oxidations* (ed. Ullrich, V. *et al.*), Pergamon Press, New York, p. 192

Lichtenberger, F., Nastainczyk, W. and Ullrich, V. (1976) *Biochem. Biophys. Res. Commun.*, **70**, 939

Lipscomb, J. D. (1980) *Biochemistry*, **19**, 3590

Lipscomb, J. D., Sligar, S. G., Namtvedt, M. J. and Gunsalus, I. C. (1976) *J. biol. Chem.*, **251**, 1116

Lipscomb, J. D., Harrison, J. E., Dus, K. M. and Gunsalus, I. C. (1978) *Biochem. Biophys. Res. Commun.*, **83**, 771

Loew, G. H., Herman, Z. S., Rohmer, M. M., Goldblum, A. and Pudzianowski, A. T. (1980) *Ann. N.Y. Acad. Sci.*, **367**, 192

Loew, G. H. (1983) In: *Iron Porphyrins* (ed. Lever, A. B. P. and Gray, H. B.), Addison-Wesley, London, Vol. 1, p. 1

McCandlish, E., Miksztal, A. R., Nappa, M., Sprenger, A. Q. and Valentine, J. S. (1980) *J. Am. chem. Soc.*, **102**, 4268

McCarthy, M. and White, R. E. (1983a) *J. biol. Chem.*, **258**, 9153

McCarthy, M. and White, R. E. (1983b) *J. biol. Chem.*, **258**, 11610

McDonald, R. C., Steitz, T. A. and Engelman, D. M. (1979) *Biochemistry*, **18**, 338

MacDonald, T. L., Burka, L. T., Wright, S. T. and Guengerich, F. P. (1982) *Biochem. Biophys. Res. Comm.*, **104**, 620

Mason, H. S., North, J. C. and Vanneste, M. (1965) *Fed. Proc., Fed. Am. Soc. Exp. Biol.*, **24**, 1172

Melander, L. and Saunders, W. H. (1980) In: *Reaction Rates of Isotopic Material*, Wiley, New York, p. 129

Murray, R. I. and Dus, K. M. (1980) In: *Microscomes, Drug Oxidations, and Chemical Carcinogenesis*, Vol. 1 (ed. Coon, M. J. *et al.*), Academic Press, New York, p. 367

Murray, R. I., Gunsalus, I. C. and Dus, K. M. (1982) *J. biol. Chem.*, **257**, 12517

Newcomer, M. E., Lewis, B. A. and Quiocho, F. A. (1981) *J. biol. Chem.*, **256**, 13218

Nordbloom, G. D., White, R. E. and Coon, M. J. (1976) *Arch. Biochem. Biophys.*, **175**, 524

Northrop, D. B. (1977) In: *Isotope Effects on Enzyme-Catalyzed Reactions* (ed. Cleland, W. W. *et al.*), University Park Press, Baltimore, p. 122

Omura, T. and Sato, R. (1962) *J. biol. Chem.*, **237**, PC1375

Omura, T. and Sato, R. (1963) *Biochim. Biophys. Acta*, **71**, 224

Omura, T. and Sato, R. (1964a) *J. biol. Chem.*, **239**, 2370

Omura, T. and Sato, R. (1964b) *J. biol. Chem.*, **239**, 2379
Ortiz de Montellano, P. R. and Kunze, E. L. (1981) *Biochemistry*, **20**, 7266
Ortiz de Montellano, P. R., Kunze, K. L. and Beilan, H. S. (1983) *J. biol. Chem.*, **258**, 45
Ozaki, Y., Kitagawa, T., Kyogoku, Y., Shimada, H., Iizuka, T. and Ishimura, Y. (1976) *J. Biochem.*, **80**, 1447
Ozols, J., Heinemann, F. S. and Johnson, E. F. (1981) *J. biol. Chem.*, **256**, 11405.
Palmer, G. (1983) In: *Iron Porphyrins* (ed. Lever, A. B. P. and Gray, H. B.), Addison-Wesley, London, Vol. 2, p. 43
Pederson, T., Austin, R. and Gunsalus, I. C. (1977) In: *Microsomes and Drug Oxidations* (ed. Ullrich, V. *et al.*), Pergamon Press, New York, p. 275
Perutz, M. F., Muirhead, H., Cox, J. M. and Guaman, L. G. (1968) *Nature, Lond.*, **219**, 131
Philson, S. B. (1977) PhD Dissertation, University of Illinois at Urbana-Champagn, Urbana, Il
Philson, S. B., Debrunner, P. G., Schmidt, P. G. and Gunsalus, I. C. (1979) *J. biol. Chem.*, **254**, 10173
Pickover, C. A., McKay, D. B., Engelman, D. M. and Steitz, T. (1979) *J. biol. Chem.*, **254**, 11323
Poulos, T. L., Alden, R. A. and Wagner, G. C. (1984) *J. mol. Biol.*, to be published
Poulos, T. L., Perez, M. and Wagner, G. C. (1982) *J. biol. Chem.*, **257**, 10427
Rein, H., Ristau, O., Misselwitz, R., Buder, E. and Ruckpaul, K. (1979) *Acta. Biol. Med. Germ.*, **38**, 187
Rein, H., Pirrwitz, J., Friedrich, J., Ristau, O., Jänig, G.-R. and Ruckpaul, K. (1980) *Devel. Biochem.*, **13**, 577
Rhynard, D., Lang, G., Spartalian, K. and Yonetani, T. (1979) *J. Chem. Phys.*, **71**, 3715
Ristau, O. (1980) *Acta Biol. Med. Germ.*, **39**, 71
Ristau, O. and Rein, H. (1983) *Biomed. Biochim. Acta*, **42**, 673
Ruckpaul, K., Rein, H., Ballou, D. P. and Coon, M. J. (1980) *Biochim. Biophys. Acta*, **626**, 41
Ruf, H. H. and Nastainczyk, W. (1976) *Eur. J. Biochem.*, **66**, 139
Ruf, H. H., Wende, P. and Ullrich, V. (1979) *J. inorg. Biochem.*, **11**, 189
Scheidt, W. R. and Gouterman, M. (1983) In: *Iron Porphyrins* (ed. Lever, A. B. P. and Gray, H. B.), Addison-Wesley, London, Vol. 1, p. 89
Schejter, A., Aviram, I. and Goldkorn, T. (1980) In: *Electron Transport and Oxygen Utilization* (ed. Ho, C.), Elsevier, New York, p. 95
Schenkman, J. B., Sligar, S. G. and Cinti, D. L. (1981) *Pharmac. Ther.*, **12**, 43
Segel, I. H. (1975) In: *Enzyme Kinetics*, Wiley, New York, p. 506
Sharrock, M., Munck, E., Debrunner, P., Marshall, V., Lipscomb, J. D. and Gunsalus, I. C. (1973) *Biochemistry*, **12**, 258
Sharrock, M., Debrunner, P. G., Schulz, C., Lipscomb, J. D., Marshall, V. and Gunsalus, I. C. (1976) *Biochim. Biophys. Acta*, **420**, 8
Shaw, R. W., Rife, J. E., O'Leary, M. H. and Beinhart, H. (1981) *J. biol. Chem.*, **256**, 1105
Shimuzu, T., Iizuka, T., Shimada, H., Ishimura, Y., Nozawa, T. and Hatano, M. (1981) *Biochim. Biophys. Acta*, **670**, 341
Sligar, S. G. (1976) *Biochemistry*, **15**, 5399
Sligar, S. G. and Gunsalus, I. C. (1976) *Proc. nat. Acad. Sci., USA*, **73**, 1078
Sligar, S. G., Debrunner, P. G., Lipscomb, J. D., Namtvedt, M. J. and Gunsalus, I. C. (1974) *Proc. nat. Acad. Sci., USA*, **71**, 3906
Sligar, S. G., Lipscomb, J. D., Debrunner, P. G. and Gunsalus, I. C. (1974) *Biochem. Biophys. Res. Comm.*, **61**, 290
Sligar, S. G., Shastry, B. S. and Gunsalus, I. C. (1977) In: *Microsomes and Drug Oxidations* (ed. Ullrich, V. *et al.*), Pergamon Press, New York, p. 405
Sligar, S. G., Cinti, D. L., Gibson, G. G. and Schenkman, J. B. (1979) *Biochem. Biophys. Res. Commun.*, **90**, 925
Sligar, S. G., Kennedy, K. A. and Pearson, D. C. (1980) *Proc. nat. Acad. Sci., USA*, **77**, 1240
Sono, M., Andersson, L. A. and Dawson, J. H. (1982) *J. biol. Chem.*, **257**, 8308
Stern, J. O. and Peisach, J. (1974) *J. biol. Chem.*, **249**, 7495
Swanson, R. A. and Dus, K. M. (1979) *J. biol. Chem.*, **254**, 7238

Tang, S. C., Koch, S., Papaefthymiou, G. C., Foner, S., Frankel, R. B., Ibers, J. A. and Holm, R. H. (1976) *J. Am. Chem. Soc.*, **98**, 2414

Thomas, P. E., Lu, A. Y. H., Ryan, D., West, S. B., Kawalek, J. and Levin, W. (1976) *J. biol. Chem.*, **251**, 1385

Traylor, T. G., Mincey, T. C. and Berzinis, A. P. (1981) *J. Am. chem. Soc.*, **103**, 7084

Tsai, R., Yu, C. A., Gunsalus, I. C., Peisach, J., Blumberg, W., Orme-Johnson, W. H. and Beinert, H. (1970) *Proc. nat. Acad. Sci., USA*, **66**, 1157

Tyson, C. A., Tsai, R. L. and Gunsalus, I. C. (1970) *J. Am. Oil chem. Soc.*, **47**, 343A

Tyson, C. A., Lipscomb, J. D. and Gunsalus, I. C. (1972) *J. biol. Chem.*, **247**, 5777

Ullrich, V. (1977) In: *Microsomes and Drug Oxidations* (ed. Ullrich, V. *et al.*), Pergamon Press, New York, p. 218

Ullrich, V. and Kuthan, H. (1980) *Dev. Biochem.*, **13**, 267

Wagner, G. C., Gunsalus, I. C., Wang, M.-Y. and Hoffman, B. M. (1981) *J. biol. Chem.*, **256**, 6266

Werringloer, J. (1982) In: *Microsomes, Drug Oxidations, and Drug Toxicity* (ed. Sato, R. and Kato, R.), Japan Scientific Societies Press, Tokyo, and Wiley-Interscience, New York, p. 171

White, R. E. and Coon, M. J. (1980) *Ann. Rev. Biochem.*, **49**, 315

White, R. E. and Coon, M. J. (1982) *J. biol. Chem.*, **257**, 3073

White, R. E., Sligar, S. G. and Coon, M. J. (1980) *J. biol. Chem.*, **255**, 1108

Wong, P. C. and Griller, D. (1981) *J. org. Chem.*, **43**, 2327

Yonetani, T., Schleyer, H. and Ehrenberg, A. (1966) *J. biol. Chem.*, **241**, 3240

Yoshida, Y., Imai, Y. and Hashimoto-Yutsudo, C. (1982) *J. Biochem.*, **91**, 1651

Yu, C.-A. and Gunsalus, I. C. (1970) *Biochem. Biophys. Res. Commun.*, **40**, 1431

Yu, C. A., Gunsalus, I. C., Katagiri, M., Suhara, K. and Takemori, S. (1974) *J. biol. Chem.*, **249**, 94

6

Nitrogenase

D. J. Lowe, R. N. F. Thorneley and B. E. Smith

INTRODUCTION

Nitrogenase is the enzyme which converts dinitrogen to ammonia and is responsible for the 122 million metric tonnes of nitrogen which is estimated to be fixed annually by micro-organisms (Burris, 1980). The ability to fix dinitrogen is currently restricted to prokaryotes but genetic manipulation may eventually enable organelles such as chloroplasts or mitochondria to fix nitrogen in higher organisms such as plants. Nitrogenase is present in both obligate (e.g. *Clostridium pasteurianum*, Cp) and facultative (e.g. *Klebsiella pneumoniae*, Kp) anaerobes, obligate aerobes (e.g. *Azotobacter vinelandii*, Av), photosynthetic bacteria (e.g. *Rhodospirillum rubrum*, Rr), cyanobacteria (e.g. *Anabaena cylindrica*, Ab) and the bacteria which form both casual (e.g. *Azospirillum brasilense*, Azb) and symbiotic (e.g. *Rhizobium japonicum*, Rj) associations with plant root systems. This list is not exhaustive and the reader is referred to Postgate (1981) and Starr *et al.* (1981) for authoritative discussions of the microbiology of nitrogen fixation. The list above does serve to introduce the shorthand notation that will be used in this chapter for the two proteins that comprise nitrogenase. The addition of the number 1 or 2 to the code designates either the molybdenum iron (MoFe) protein (e.g. Kp1) or the iron (Fe) protein (e.g. Kp2) of for example, the nitrogenase from *K. pneumoniae*.

The genetics and molecular biology of nitrogen fixation in *K. pneumoniae* are currently understood in terms of seventeen genes which are distributed amongst eight contiguous transcriptional units. The gene products, their molecular weights and a summary of their functions are given in table 6.1. The nitrogenase protein sub-units are coded for by the *nif* genes *H*, *D* and *K*. The synthesis and processing of the FeMo-cofactor (see section on the MoFe proteins) is controlled by genes, *B*, *V*, *N* and *E*. The *F* and *J* gene products are involved in the electron transport chain to nitrogenase. The major control genes are *A* and *L* which respectively activate and inhibit the expression of other *nif* genes. Reviews by

Table 6.1 *Klebsiella Pneumoniae: nif* gene products and functions

Gene	Gene product M_r	Native protein M_r	Function of product
Q (6)	Unknown		Unknown
B (1, 2, 5, 6)	51 500 (28)		Essential for active FeMo-co-factor centre in MoFe protein (1, 4, 8)
A (2, 6)	60 000 (14) 57 000 (16) 66 000 (27)		Activates expression of other *nif* genes (2, 8, 9, 11, 21)
L (3, 6)	50 000 (14) 45 000 (16) 52 000 (27)		Inhibits expression of other *nif* genes (15, 18, 22, 23)
F (1, 2, 5, 6)	17 000 (8) 10 000 (11) (14) 20–22 000 (13)	22 000 (13)	Electron transfer to nitrogenase (1, 8, 10, 13)
M (6, 8)	28 000 (14) 27 000 (27)		Activation of Fe protein (8)
V (6)	42 000 (14) 38 000 (27)		Modifies FeMo-cofactor and substrate specificity of MoFe protein (17, 26)
S (6, 7)	45 000 (14) 38 000 (27)		Possible activation of Fe protein (8)
U (12)	22 000 (11) 25/32 000 (14) 28 000 (27)		Unknown
X (14)	18 000		Unknown
N (6, 7)	50 000 (8, 14) 40 000 (6)		Essential for active FeMo-cofactor centre in MoFe protein (8)
E (2, 5, 6)	46 000 (8) 40 000 (6)		Essential for active FeMo-cofactor centre in MoFe protein (8)
Y (14)	21 000 (14) 19 000 (29)		Unknown
K (2, 6)	60 000 (8, 14)		MoFe protein β sub-unit (8)
D (1, 2, 6)	56 000 (8) 60 000 (14)	220 000 $\alpha_2\beta_2$ (24)	MoFe protein α sub-unit (8)
H (1, 2, 6)	31 753 (19, 20)	68 000 α_2 (24)	Fe protein sub-unit (2, 8)
J (3, 5, 6, 7)	120 000 (7, 8, 14)	245 000 α_2 (25)	Electron transfer to nitrogenase (10, 13, 25)

REFERENCES

1. St. John *et al.* (1975)
2. Dixon *et al.* (1977)
3. Kennedy (1977)
4. Shah & Brill (1977)
5. Elmerich *et al.* (1978)
6. MacNeil *et al.* (1978)
7. Merrick *et al.* (1978)
8. Roberts *et al.* (1978)
9. Dixon *et al.* (1980)
10. Hill & Kavanagh (1980)
11. Houmard *et al.* (1980)
12. Merrick *et al.* (1980)
13. Nieva-Gomez *et al.* (1980)
14. Pühler & Klipp (1981)
15. Hill *et al.* (1981)
16. Ausubel & Cannon (1981)
17. McLean & Dixon (1981)
18. Kennedy *et al.* (1981)
19. Sundaresan & Ausubel (1981)
20. Scott *et al.* (1981)
21. Buchanan-Wollaston *et al.* (1981a)
22. Merrick *et al.* (1982)
23. Buchanan-Wollaston *et al.* (1981b)
24. Eady & Smith (1979)
25. Bogusz *et al.* (1981)
26. Hawkes *et al.* (1984)
27. Roberts & Brill (1980)
28. Sibold *et al.* (1983)
29. Cannon *et al.* (1979)

Postgate and Cannon (1981), Roberts and Brill (1981), Kennedy *et al.* (1981), Robson *et al.* (1983) and Eady (1984) provide detailed accounts of this rapidly expanding area.

This chapter has been necessarily selective in the topics covered and is probably biased towards the current views and interests of the Sussex group. Reviews by Nelson *et al.* (1982), Mortenson and Thorneley (1979) and Ljones (1979) and the books edited by Coughlan (1980), Newton and Orme-Johnson (1980) and Gibson and Newton (1981) serve to redress the balance. We discuss the structures and properties of the Fe protein and the MoFe protein which constitute nitrogenase and present our current working hypothesis for the mechanism of nitrogenase.

PROTECTION OF NITROGENASE FROM O_2 INACTIVATION

Both nitrogenase proteins are very unstable in the presence of O_2 (half-lives of the order of 1 min in air; and therefore great care must be taken to exclude O_2 from purified preparations. Organisms that are actively fixing N_2 have various strategems for protecting nitrogenase from O_2 (Robson and Postgate, 1980). Many that can tolerate O_2 when growing on fixed nitrogen are microaerophilic when fixing.

Most cyanobacteria develop heterocysts; these contain nitrogenase, do not possess photosystem 2 (which produces O_2), and have thicker walls than the normal vegetative cells. Other cyanobacteria utilise different forms of spatial or temporal compartmentalisation of nitrogenase and photosystem 2 (Stewart *et al.*, 1980).

The azotobacteraceae which are obligate aerobes, dramatically increase their respiration rate in order to reduce their internal O_2 concentration and have a branched respiratory chain that is weakly coupled to ATP production at high external O_2 concentrations (Yates and Jones, 1974). These organisms have a second O_2 protection mechanism that switches off their nitrogenase when they are exposed to sudden increases in O_2 concentration. This mechanism probably involves the interaction of the nitrogenase proteins with a 2Fe-2S protective protein of molecular weight either 24 000 (Haaker and Veeger, 1977; Veeger *et al.* 1980) or 14 000 (Robson, 1979). Crude preparations of Azotobacter nitrogenase are stable in air whereas the purified proteins are O_2 sensitive; adding back the protective protein stabilises them.

THE Fe PROTEIN OF NITROGENASE

Polypeptide structure

The Fe proteins of nitrogenase are α_2 dimers with molecular weights of between 58 000 and 72 000. Comprehensive descriptions can be found in the reviews by Mortenson and Thorneley (1979), Lowe *et al.* (1980) and Eady and Smith (1979) and the references therein.

All Fe proteins that have been studied have about 20% acidic and 10% basic residues. Their amino acid residue sequences are very similar (figure 6.1). There is 60-70% homology between Cp2, Kp2, Av2 and Ab2, which are respectively from a strict anaerobe, a microaerophile, an aerobe and a cyanobacterium. This degree of homology is remarkable especially since the DNA base sequences show considerable variation (only about 30% homology between Kp2 and Ab2). Thus although the DNA sequences have evolved considerably the protein sequences have not. We conclude that the interactions with the MoFe protein for electron transfer, ATP hydrolysis and substrate reduction place severe constraints on the variations of the primary structure of the Fe proteins.

Prosthetic groups and spectroscopic properties

Metal and acid-labile sulphur analyses suggest that the Fe proteins contain 4Fe and $4S^{2-}$ g.atom mol^{-1} which have been extruded from Cp2 as a 4Fe-4S cluster (Averill *et al.*, 1978). Dithionite-reduced Kp2 has a Mössbauer spectrum similar to a $[4Fe-4S]^+$ in the bacterial ferredoxins (Smith and Lang, 1974). The uv-visible spectrum is a broad featureless absorption, with intensity increasing as wavelength decreases towards 300 nm. Its intensity in the visible region increases on oxidation; this optical change has been utilised in many studies observing electron transfer to and from the Fe proteins (see below). Their cd spectra show strong negative peaks in the protein backbone region at 220 nm and there are much weaker bands in the metal cluster region at wavelengths >300 nm. Both the dithionite-reduced and dye-oxidised proteins have metal cd spectra similar to those of the 4Fe-4S ferredoxins (Stephens *et al.*, 1981). Epr spectra of reduced Fe proteins are somewhat different from those of the ferredoxins in that although the g-values ($g_{av} = 1.95$) are similar, the g_x and g_z-lines are very broad with long tails (see figure 6.2). The spectra usually have integrated intensities in the range

0.2-0.5 electrons mol^{-1}. It has been suggested (Lowe, 1978) that the unusual line-shapes and the low intensities are due to an interaction between two magnetic centres. A study of the effect of a large linear electric field on the epr *g*-values by Peisach *et al.* (1977) and Orme-Johnson *et al.* (1977) showed that the ligand field surrounding the epr active centre(s) of Cp2 and the 4Fe-4S bacterial ferredoxins are non-centrosymmetric. Thus it is probable that the Fe protein contains a single 4Fe-4S cluster similar to that in the 4Fe-4S bacterial ferredoxins although an additional paramagnetic centre may be present.

The 4Fe-4S cluster is presumably shared between the two sub-units of the Fe protein. Cluster binding cannot involve the highly conserved Cys-x-x-Cys-x-x-Cys sequence of the 4Fe-4S ferredoxins since it does not appear in the Fe protein.

Activity of the Fe protein

Quantitative interpretation of data on Fe proteins usually requires a knowledge of the activity of the fully functional proteins and, as with most proteins, this is a very difficult parameter to measure. The proteins are relatively easy to purify using the criterion of no observable contamination on electrophoresis in sodium dodecyl sulphate (SDS) polyacrylamide gels. This does not guarantee that all molecules have a full complement of metal ions or other functional groups. A lower limit for full activity is the highest activity ever obtained (*c.* 3000 nmol C_2H_4 produced min^{-1} mg $protein^{-1}$). Ljones and Burris (1978a) used the reaction between the FeII chelating reagent bathophenanthroline disulphate (BPS) and the 4Fe-4S cluster in an attempt to calculate the maximum specific activity of Cp2. This reaction is greatly accelerated by the presence of MgATP. The data were interpreted in terms of a model in which both of two MgATP binding sites on Cp2 must be occupied before the protein undergoes a conformational change allowing the Fe atoms to react rapidly with the chelator. When activity was lost, due to exposure to O_2 or incubation at $0°C$ (Cp2 protein is cold-labile), a parallel decrease in the ability to react with BPS occurred. These measurements enabled the calculation of a maximum specific activity of 2980 nmol C_2H_4 produced min^{-1} mg $protein^{-1}$ and a difference in extinction coefficient at 430 nm of $6600 M^{-1}$ cm^{-1} between oxidised and reduced Cp2. These calculations clearly showed that one principal source of damage to purified proteins (O_2-damage) is correlated with the BPS assay but ignores other possible forms of inactive protein.

One example of inactivation for *in vivo* regulation of nitrogenase activity may be observed in the non-purple photosynthetic bacteria. They are apparently unusual in that their nitrogenase can be inactivated rapidly *in vivo* in the presence of NH_4^+. This phenomenon has been studied in *Rhodospirillum rubrum* by Ludden and Burris (1979), Ludden *et al.* (1982) and Preston and Ludden (1982). At least one Rr2 sub-unit becomes modified by addition of a pentose-phosphate-adenine-like group during this inactivation. This modified sub-unit migrates differently on SDS polyacrylamide gel electrophoresis. An enzyme has been purified which *in vitro* reactivates the Rr2. The phosphate group is only liable

Multiple sequence alignment of five nitrogenase Fe‑protein sequences (Rm, Kp, An, Cp, Av). The alignment is printed vertically; it is reproduced here as three aligned blocks. Position markers 20, 40, 60, 100 and 120 are shown in the original figure.

Block I (residues 1–20)

	1	2	3	4	5	6	7	8	9	10	11	12	13	14	15	16	17	18	19	20
Rm:	H₂N‑Met	Ala	Leu	Arg	Gln	Ile	Ala	Ala	Phe	Tyr	Gly	Lys	Gly	Ile	Gly	Ser	Thr	Thr	Gln	Asn
Kp:	H₂N‑Thr	Met	Arg	Gln	Cys	Ala	Ile	Phe	Tyr	Gly	Lys	Gly	Ile	Gly	Lys	Ser	Thr	Thr	Gln	Asn
An:	H₂N‑Met Thr Asp	Glu	Asn	Ile	Arg	Gln	Ile	Ala	Phe	Tyr	Gly	Lys	Gly	Ile	Gly	Lys	Ser	Thr	Thr	Gln	Asn
Cp:	H₂N‑Met	Arg	Gln	Val	Ala	Ile	Tyr	Gly	Lys	Gly	Gly	Ile	Gly	Lys	Ser	Thr	Thr	Gln	Asn
Av:	H₂N‑Ala	Met	Arg	Gln	Cys	Ala	Ile	Tyr	Gly	Lys	Gly	Gly	Ile	Gly	Val	Ser	Thr	Thr	Gln	Asn

Block II (residues ~21–60, markers 40 and 60)

Rm:	Thr	Leu	Ala	Ala	Leu	Val	Asp	Leu	Gly	Gln	Lys	Ile	Leu	Ile	Val	Gly	Cys	Asp	Pro	Lys
Kp:	Leu	Val	Ala	Leu	Ala	Glu	Ala	Gly	Lys	Lys	Val	Met	Ile	Val	Gly	Cys	Asp	Pro	Lys
An:	Thr	Leu	Ala	Met	Ala	Glu	Ala	Gly	Gln	Arg	Ile	Met	Ile	Val	Gly	Cys	Asp	Pro	Lys
Cp:	Leu	Thr	Ser	Gly	Leu	His	Ala	Met	Leu	Gly	Thr	Lys	Ile	Leu	Gln	Ile	Gly	Cys	Asp	Pro	Lys
Av:	Leu	Val				Met	Ile	Val	Gly	Cys	Asp	Pro	Lys

Rm:	Ala	Asp	Ser	Thr	Arg	Leu	Ile	Leu	Asn	Ala	Lys	Ala	Gln	Asp	Thr	Val	Leu	His	Leu	Ala	Ala	Thr	Gly	Ser	Val	Glu	Asp	Leu	Glu
Kp:	Ala	Asp	Ser	Thr	Arg	Leu	Ile	Leu	His	Ala	Lys	Ala	Gln	Asn	Thr	Ile	Met	Glu	Met	Ala	Ala	Glu	Val	Gly	Ser	Val	Glu	Asp	Leu	Glu
An:	Ala	Asp	Ser	Thr	Arg	Leu	Met	Leu	His	Ala	Lys	Ala	Gln	Asn	Thr	Ile	Met	Glu	Met	Ala	Ala	Glu	Ala	Gly	Ala	Val	Glu	Asp	Leu	Glu
Cp:	Ala	Asp	Ser	Thr	Arg	Leu	Leu	Leu	Gly	Gly	Leu	Ala	Gln	Lys	Ser	Val	Leu	Asp	Thr	Leu	Arg	Glu	Glu	Gly	Glu	Asp	Val
Av:	Lys	Ala	Asp	Ser	Thr	Arg	Leu	Leu	Leu	Gly	Gly	Leu	Ala	Gln	Lys	Ser	Tyr	Asp	Leu	Asp	Phe	Val	Leu	Arg	Glu	Gly	Leu	Gly	Asp	Val

Block III (residues ~81–120, markers 100 and 120)

Rm:	Leu	Glu	Asp	Val	Leu	Lys	Val	Gly	Tyr	Arg	Gly	Ile	Lys	Cys	Val	Glu	Ser	Gly	Gly	Pro
Kp:	Leu	Glu	Asp	Val	Leu	Gln	Ile	Gly	Tyr	Gly	Asp	Ile	Arg	Cys	Ala	Glu	Ser	Gly	Gly	Pro
An:	Leu	His	Glu	Val	Met	Leu	Thr	Gly	Phe	Arg	Gly	Val	Lys	Cys	Val	Glu	Ser	Gly	Gly	Pro
Cp:	Leu	Asp	Val	Leu	Lys	Ile	Gly	Tyr	Gly	Gly	Ile	Lys	Cys	Val	Glu	Ser	Gly	Gly	Pro
Av:	Leu	Asp	Ser	Ile	Glu	Asn	Leu	Glu	Gln	Leu	Gly	Tyr	Cys	Val	Glu	Ser	Gly	Gly	Pro

Rm:	Glu	Pro	Gly	Val	Gly	Cys	Ala	Gly	Arg	Gly	Val	Ile	Thr	Ser	Ile	Asn	Phe	Leu	Glu	Glu	Asn	Gly	Ala	Tyr	Asn	Asp	Val	Tyr
Kp:	Glu	Pro	Gly	Val	Gly	Cys	Ala	Gly	Arg	Gly	Val	Ile	Thr	Ala	Ile	Asn	Phe	Leu	Glu	Glu	Asn	Gly	Ala	Tyr	Glu	Asp	Leu	Asp	Phe
An:	Glu	Pro	Gly	Val	Gly	Cys	Ala	Gly	Arg	Gly	Ile	Ile	Thr	Ala	Ile	Asn	Phe	Leu	Glu	Glu	Asn	Gly	Ala	Tyr	Glu	Asp	Leu	Asp	Phe
Cp:	Glu	Pro	Gly	Val	Gly	Cys	Ala	Gly	Arg	Gly	Ile	Ile	Thr	Ala	Ile	Asn	Phe	Leu	Glu	Glu	Asn	Gly	Ala	Tyr	Gln	Asp	Leu	Asp	Phe
Av:	Glu	Pro	Gly	Val	Gly	Cys	Ala	Gly	Arg	Gly	Val	Ile	Thr	Ser	Ile	Asn	Met	Leu	Glu	Gln	Leu	Gly	Ala	Tyr	Thr	Asp	Asp	Leu	Tyr

Rm: Val Ser Tyr Asp Val Leu Gly Asp Val Val Cys Gly Gly Phe Ala Met Pro Ile Arg Glu Asn Lys Ala Gln Glu
Kp: Val Phe Tyr Asp Val Leu Gly Asp Val Val Cys Gly Gly Phe Ala Met Pro Ile Arg Glu Asn Lys Ala Gln Glu
An: Val Ser Tyr Asp Val Leu Gly Asp Val Val Cys Gly Gly Phe Ala Met Pro Ile Arg Glu Gly Lys Ala Gln Glu
Cp: Val Ser Tyr Asp Val Leu Gly Asp Val Val Cys Gly Gly Phe Ala Met Pro Ile Arg Glu Gly Lys Ala Gln Glu
Av: Val Phe Tyr Asp Val Leu Gly Asp Val Val Cys Gly Gly Phe Ala Met Pro Ile Arg
       Asp Val Val Cys

140

Rm: Ile Tyr Ile Val Met Ser Gly Glu Met Met Ala Leu Tyr Ala Ala Asn Ile Ala Lys Gly Ile Leu Lys Tyr
Kp: Ile Tyr Ile Val Cys Ser Gly Glu Met Met Ala Met Tyr Ala Ala Asn Asn Ile Ser Lys Gly Ile Val Lys Tyr
An: Ile Tyr Ile Val Thr Ser Gly Glu Met Met Ala Met Tyr Ala Ala Asn Asn Ile Ala Arg Gly Ile Leu Lys Tyr
Cp: Ile Tyr Ile Val Ala Ser Gly Glu Met Met Ala Leu Tyr Ala Ala Asn Asn Ile Ser Lys Gly Ile Lys Tyr
Av: Ile Tyr Ile Val Cys Ser Gly Glu Gly Glu Ser Gln Lys Tyr

160

Rm: Ala His Ala Gly Gly Val Arg Leu Gly Gly Leu Ile Cys Asn Glu Arg His Thr Asp Arg Glu Leu Asp Leu Ala
Kp: Ala Lys Ser Gly Lys Val Arg Leu Gly Gly Leu Ile Cys Asn Glu Arg Gln Thr Asp Arg Glu Asp Glu Leu Ile
An: Ala His Ser Gly Val Arg Leu Gly Gly Leu Ile Cys Asn Ser Arg Val Asp Arg Asp Asp Glu Leu Ile
Cp: Ala Lys Ser Gly Gly Val Arg Leu Gly Gly Ile Ile Cys Asn Ser Arg Lys Val Asn Glu Tyr Glu Leu Leu
Av: Leu Gly Gly Leu Asn Ser Lys Ile Ile Cys Asn Ser Arg

180

Rm: Glu Ala Leu Ala Ala Arg Leu Asn Ser Lys Leu Ile His Phe Val Pro Arg Asp Asn Ile Val Gln His Ala Glu
Kp: Ile Ala Leu Ala Glu Glu Lys Leu Gly Thr Gln Met Ile His Phe Val Pro Arg Asp Asn Ile Val Gln Arg Ala Glu
An: Met Asn Leu Ala Glu Ala Leu Asn Thr Gln Met Ile His Phe Val Pro Arg Asp Asn Ile Val Gln His Ala Glu
Cp: Asp Ala Phe Ala Lys Glu Leu Gly Leu Ser Met Pro Arg Ser Pro Met Val Thr Lys Ala Glu

200   220

Figure 6.1 *(Caption and figure continued overleaf)*

240

```
Rm: Leu Arg Lys Met Thr Val Ile Gln Tyr Ala Pro Asn Ser Lys Gln Ala Gly Glu Tyr Arg Ala Leu Ala Glu Lys
Kp: Ile Arg Arg Met Thr Val Ile Glu Tyr Asp Pro Ala Cys Lys Gln Ala Asn Glu Tyr Arg Thr Leu Ala Gln Lys
An: Leu Arg Arg Met Thr Val Asn Glu Tyr Ala Pro Asp Ser Asn Gln Gly Gln Glu Tyr Arg Ala Leu Ala Lys Lys
Cp: Ile Asn Lys Gln Thr Val Ile Glu Tyr Asp Pro Thr Cys Glu Gln Ala Glu Glu Tyr Arg Glu Leu Ala Arg Lys
Av:                                 Tyr Asp Pro Lys Ala Lys Gln Ala Asp Glu
```

260

```
Rm: Ile His Ala Asn Ser Gly Arg Gly Thr Val Pro Thr Met Glu Thr Met Glu Glu Leu Glu Asp Met Leu Leu Asp
Kp: Ile Val Asn Asn Thr Met Lys     Val Val Pro Thr Pro Cys Thr Met Asp Glu Leu Glu Ser Leu Leu Met Glu
An: Ile Asn     Asn Asp Lys Leu     Thr Ile Pro Thr Met Glu Met Asp Glu Leu Glu Ala Leu Lys Ile Glu
Cp: Val Asp Ala Asn Glu Leu Phe     Val Ile Pro Lys Pro Met Gln Leu Glu Glu Leu Glu Glu Met Gln
Av:                                     Arg Gln Glu Leu Glu Glu Leu Leu Met Glu
                                                   Glu Glu Leu Glu Glu Leu Met Glu
```

280

```
Rm: Phe Gly Ile Met Lys Ser Asp Glu Gln Met Leu Ala Glu Leu His Ala Lys Glu Ala Lys Val Ile Ala Pro
Kp: Phe Gly Ile Met Glu Glu Glu Asp Thr Ser     Ile Ile Gly Lys Thr Ala Ala Glu Gly Asn Ala Ala Ala COOH
An: Tyr Gly Leu Leu Asp Asp Asp Thr Lys His Ser Glu Ile Ile Gly Lys Pro Ala Glu Ala Thr Arg Ser Cys
Cp: Tyr Gly Leu Met Asp Leu Glu COOH
Av: Phe Gly Ile Met Glu Val Glu Asp Glu Ser Ile Val Gly Lys Thr Ala Glu Glu Val Val COOH
```

```
Rm: His COOH
An: Arg Asn COOH
```

Figure 6.1 Comparison of amino acid sequences of the Fe protein from *K. pneumoniae* (Kp), *C. pasteurianum* (Cp), *Anabaena* 7120 (Ab), *A. vinelandii* (Av), and *Rhizobium melilote* (Rm). Identical amino acid sequences are boxed

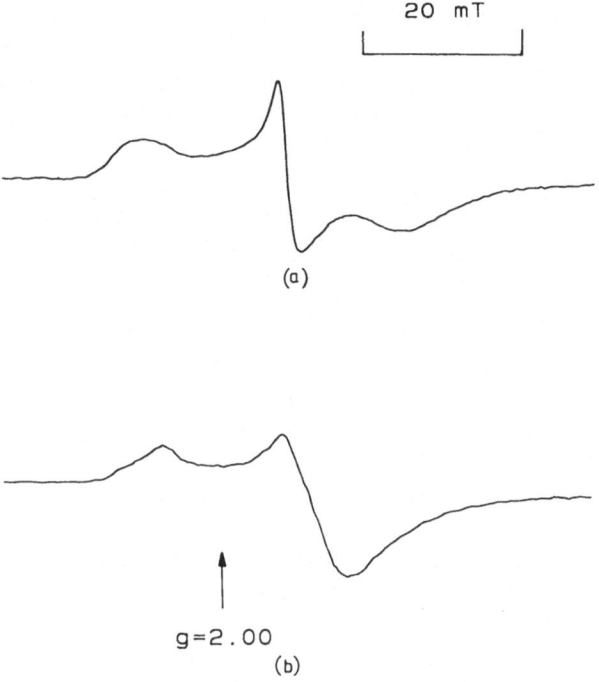

20 mT

(a)

g=2.00

(b)

Figure 6.2   Epr spectra of (a) dithionite reduced Kp2, (b) dithionite reduced Kp2 + MgATP. Spectra were recorded at *ca*. 10 K with a microwave frequency of 9.1 GHz and microwave power of 20 mW using 1.0 mT field modulation at 100 kHz

to removal by *E. coli* alkaline phosphatase after treatment with activating enzyme which converts the sub-unit back into a species co-migrating with active Rr2 on SDS polyacrylamide gel electrophoresis.

Redox properties

The consensus view has been that the Fe proteins are capable of undergoing a one-electron redox process. The redox potential (vs standard H_2 electrode) of this change has been monitored by optical and epr spectroscopy with values ranging from --240 mV for Cp2 (Walker and Mortenson, 1973) to −393 mV for Av2 (Braaksma *et al.*, 1982). A negative shift of about 100 mV for Cp2 (Zumft *et al.*, 1974) or about 40 mV for Av2 (Braaksma *et al.*, 1982) occurs on addition of MgATP. A similar effect is observed with MgADP.

The number of electrons involved in this redox process is at present the subject of some discussion in the literature (Ljones, 1979; Braaksma *et al.*, 1982; Hageman *et al.*, 1980). Thorneley *et al.* (1976) using optical observation of dithionite oxidation, reported an initial fast phase in the reduction of dye-oxidised Ac2 (activity 1400 nmol C_2H_4 min^{-1} mg protein^{-1}) corresponding to one electron per Ac2 molecule, followed by several slow phases which summed to a second electron per Ac2 molecule. The initial fast phase corresponded to full development of the Ac2 epr signal and was considered to be due to active protein. The specific activity suggests that this protein was only about half-active. Therefore one might conclude that fully active protein would show a two-electron fast phase, with no slow phases. Walker and Mortenson (1973) titrated Cp2 (specific activity 2300 nmol C_2H_4 min^{-1} mg protein^{-1}) with dyes and ferricyanide and observed spectrophotometrically 1.4–2.0 electrons per molecule Cp2. No distinction was made between slow and fast phases. The above studies have been criticised by Ljones and Burris (1978b) because the proteins were not oxidised physiologically, and therefore electrons from inactive protein were included and oxidation states other than those found in catalytically functional Fe protein may have been introduced. They used physiologically (MoFe protein and MgATP) oxidised Cp2 and obtained a value of one electron per molecule. This value was strictly per active 4Fe-4S centre using an extinction coefficient based upon the BPS assay (see above); the protein's specific activity was not given.

Another approach to electron counting is based on the shape of the redox titration curves using Nernst equation fits. Zumft et al. (1974) using epr monitoring found $n = 1.02$ (±0.02) for oxidative titration of Cp2 (±MgATP) with ferricyanide in the presence of mediator dyes. In contrast Braaksma et al. (1982), also using epr monitoring, found $n = 2$ for titration of Av2 and Av2 + MgATP and $n = 1$ for Av2 + MgADP but suggested no explanation for this variation of n. They used varying ratios of dithionite to sulphite to poise the redox potential. Zumft et al. (1974) could be criticised because ferricyanide is a relatively strong oxidising agent, although their titrations are reversible up to at least 80%. The effects of varying high concentrations of Na_2SO_3 in the experiments of Braaksma et al. (1982) throw some doubt on their analysis.

Additional complications include the possible presence of inactive protein with a different redox potential and/or an additional paramagnetic centre. How do we reconcile these apparently contradictory results? Perhaps fully active proteins contain two redox centres, for example, a 4Fe-4S and a single Fe atom, both measured by the BPS assay. This would reconcile the electron counting experiments. Two independent redox centres would give $n = 1$ type redox titration curves and any coupling would shift n towards 2. Whether the putative two centres act in concert or not in functioning nitrogenase remains to be resolved.

THE STRUCTURE OF THE MoFe PROTEIN

Polypeptide structure

MoFe proteins have been isolated and characterised from a wide range of bacteria and all are reported to have an $\alpha_2\beta_2$ sub-unit structure with molecular weights in the range 200 000–240 000 (Eady and Smith, 1979; Lowe *et al.*, 1980). Kp1 sub-units are coded for by the *nifK* and *D* genes which are on the same transcriptional unit as the *nifH* gene which codes for Kp2 (see table 6.1). The molecular weights of the sub-units have been estimated as approximately 60 000 and 50 000 from their migration on SDS polyacrylamide. These estimates are unreliable since resolution of the sub-units into two bands and their migration on the gel depends on the source of the SDS (Kennedy *et al.*, 1976).

Three distinct bands can be detected when Av1 is run on 9.5 M urea isoelectric focussing gels. However partial tryptic digestion of the protein in these bands indicated that two of them had the same polypeptide structure. It was suggested that they may have differed in metal content (Harker and Wullstein, 1981).

Single crystals of Cp1, Av1 and Kp1 suitable for x-ray diffraction studies have been prepared. The Cp1 crystals contain only one molecule per asymmetric unit and high quality diffraction data have been collected to 2.4 Å resolution (Weininger and Mortenson, 1982; Yamane *et al.*, 1982). Preliminary analysis of the data has identified a molecular dimer axis between $\alpha\beta$ pairs and indicates low resolution structural homology between the α and β chains.

All common amino acids are present in the MoFe proteins with about 20% acidic and about 10% basic amino acids (Eady and Smith, 1979; Lowe *et al.*, 1980).

Hybridisation experiments (Ruvkun and Ausubel, 1980) with DNA comprising the *K. pneumoniae nifK, D* and *H* genes indicated considerable homology with the *nif* DNA of 19 widely divergent species. More detailed examination showed that the *nifH* and *D* but not *K* genes were highly conserved.

No complete amino acid sequence for both sub-units from a single organism has yet been published but some data are available (table 6.2). The sequence data for the *nifD* polypeptides shows a region of homology near the amino terminus for all but Cp1 and considerable amino acid homology around four of the cysteine residues in Kp1, Cp1 and Av1. Furthermore a cysteine containing penta-peptide from Av1 is homologous with cys 46 to Asn 50 in Kp1. Although there is no corresponding cysteine in Cp1 there is one at position 57.

The entire sequence of the *nifK* gene from Anabaena 7120 has been reported (Mazur and Chui, 1982). The relevant Anabaena DNA was identified by its homology with the Kp1 *nifK* gene although it was 11 kbp separated from the *nifD* and *H* genes. Despite this separation of the *nifK* and *D* genes they could still be coordinately regulated since their putative promoter sequences and ribosome binding sites were apparently homologous. The only other sequence data available for the *nifK* gene product are the sequences of the tryptic peptides containing the cysteine residues of the sub-unit from Av1 (Lundell and Howard, 1981). Four of the six cysteine residues are found in identical positions in the two organisms with considerable homology of nearby amino acids. However three of the Av1 *nifK* polypeptide cysteine residues have no counterpart in the Anabaena sequence.

Table 6.2 Sequence data for MoFe proteins

Organisms	*nif* gene	Method	Comments	Reference
K. pneumoniae	D	DNA sequence	207 residues from N-terminus	Scott *et al.* (1981)
C. pasteurianum	D	Amino acid sequence	179 residues from N-terminus	Hase *et al.* (1981)
A. vinelandii	D	Amino acid sequence	19 residues from N-terminus	Lundell and Howard (1978)
A. vinelandii	D	Amino acid sequence	Amino acid sequences of tryptic peptides containing the cysteine residues	Lundell and Howard (1981)
R. Meliloti	D	DNA sequence	20 residues from N-terminus	Török and Kondorosi (1981)
Anabaena 7120	D	DNA sequence	22 residues from N-terminus	Mevarech *et al.* (1980)
Anabaena 7120	K	DNA sequence	Complete sub-unit	Mazur and Chui (1982)
A. vinelandii	K	Amino acid sequence	A.A sequences of tryptic peptides containing the cysteine residues	Lundell and Howard (1981)

The most striking feature of the above sequence data is the conservation of at least four cysteine residues in each sub-unit. 4Fe-4S clusters in simple iron-sulphur proteins are bound to the polypeptide chain through cysteine ligands to the iron atoms. Four 4Fe-4S clusters, albeit with unusual properties, are considered to be present in the MoFe protein molecule. There are enough conserved cysteine residues to bind all the postulated 4Fe-4S clusters in the MoFe proteins although the unusual properties of these clusters (see below) may be due to the influence of alternative ligands.

The metal and acid-labile sulphur atoms

The most active preparations of the MoFe proteins contain close to two Mo atoms and between 24 and 36 Fe atoms per molecule with acid-labile sulphur contents generally slightly lower than the Fe content (see Smith, 1983 and other reviews).

Evidence for an alternative nitrogen fixation system which apparently does not require Mo has been reported for Nif^- (Nif = nitrogen fixation phenotype) mutants and Nif^+ revertant strains of *A. vinelandii* (Bishop *et al.*, 1980, 1982). These strains grew on N_2 and incorporated $^{15}N_2$ but only in the absence of Mo. The normal MoFe polypeptides were apparently absent from these strains when fixing nitrogen, instead either two or four, new, NH_4^+-repressible, polypeptides were detected depending on the growth conditions. Biochemical characterisation of these polypeptides and the nitrogen fixing components of these strains should prove very interesting.

The metal and acid-labile sulphur atoms can be extracted from purified MoFe proteins. The Mo atoms and some of the Fe and acid-labile sulphur atoms can be extracted from the protein with N-methylformamide after precipitation with DMSO (Smith, 1980) or treatment with acid followed by neutralisation to the protein isoelectric point (Shah and Brill, 1977). The extracted metals chromatograph together and act as if they are part of a metal cluster, the iron–molybdenum cofactor (FeMoco), with the stoicheiometry $MoFe_{6-8} S_{4.9}$ (Shah and Brill, 1977; Smith, 1980; Burgess *et al.*, 1980a). FeMoco can activate the inactive MoFe protein found in extracts of *K. pneumoniae nifB, N* or *E* mutants (table 6.1) and the *A. vinelandii* mutant, UW45.

When the dithionite-reduced MoFe protein was denatured with an organic solvent in the presence of o-xylyl-α,α'-dithiol approximately 50% of the total Fe was extruded as 4Fe-4S clusters (Kurtz *et al.*, 1979). Earlier attempts (Wong *et al.*, 1979) using other thiols had yielded mixtures of 4Fe-4S and 2Fe-2S clusters with incomplete extrusion. However the data of Kurtz *et al.* (1979) indicated that the 4Fe-4S clusters constituted 90-103% of the non-FeMoco iron in Cp1 and Av1. The above data together with Mössbauer and redox titration data (Zimmermann *et al.*, 1978; Huynh *et al.*, 1980) can be interpreted in terms of a model (figure 6.3) in which all of the metal atoms of the MoFe protein are

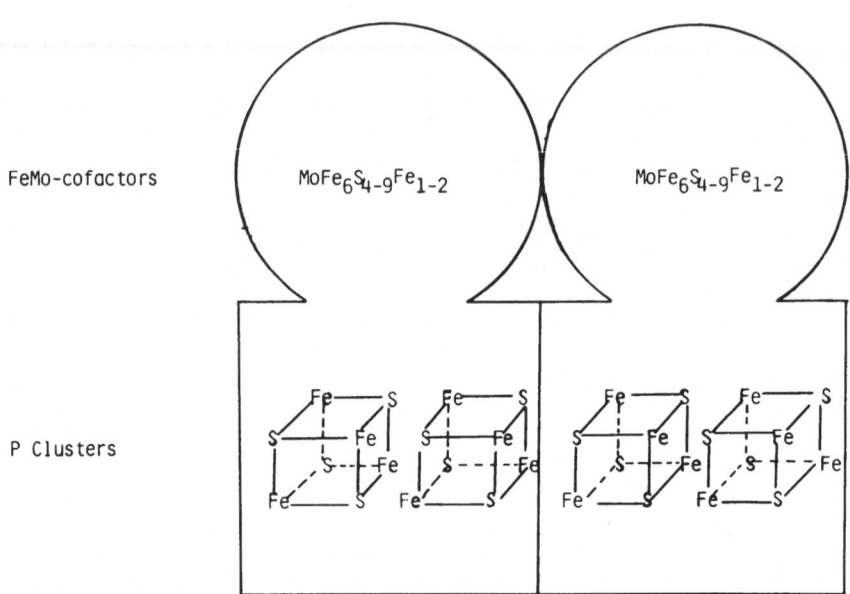

FeMo-cofactors

P Clusters

Figure 6.3 Proposed MoFe protein structure. The structure is drawn for an $\alpha_2\beta_2$ polypeptide structure containing 2 Mo and 30–32 Fe atoms per molecule. The location of the FeMo-cofactor and P clusters within the sub-units is not known

arranged as two FeMo-cofactors plus four 4Fe-4S P clusters per molecule of MoFe protein.

It should be emphasised that this model (figure 6.3) is hypothetical (see Smith, 1983 for a detailed discussion) and is open to some criticism particularly on the basis of redox titrations. The isolated MoFe proteins exhibit two regions of redox activity below 0 mV. The P clusters are oxidised at a lower potential than the FeMo-cofactors (Zimmermann *et al.*, 1978; Smith *et al.*, 1980). The latter redox step has been shown to be pH and species dependent (O'Donnell and Smith, 1978). Experiments designed to quantify the number of electrons associated with each redox step have been performed with the expectation that these numbers would reflect the numbers of each type of metal cluster present in the protein. Unfortunately such electron-counting experiments have not given uniform results. During oxidative titrations on Cp1 (Orme-Johnson *et al.*, 1981) and Kp1 (Smith *et al.*, 1980) the epr signal intensity from the FeMo-cofactors disappeared linearly as six electrons were removed from each molecule indicating that up to six individual clusters were being oxidised indistinguishably. With Av1 the oxidation took place in two steps. The first changing the Mössbauer spectrum, the optical absorbance at 700 nm, and the cd at 450 nm, and the second removing the epr signal due to the FeMo-cofactors. The results of Zimmermann *et al.* (1978), Watt *et al.* (1981) and Stephens *et al.* (1981) show that 3 or 4 electrons are required to oxidise or reduce P clusters with 1–3 electrons being required for the FeMo-cofactors. Amperometric reduction of methylene blue oxidised Av1 showed two 3-electron reduction waves (Watt *et al.*, 1980). Rationalisation of these data is required. One problem is that the protein preparations of different workers did not all contain the same complement of metal atoms (see later discussion of the maximum activity). Another problem may be that some of the redox dyes do not react reversibly with the protein. Despite the variability in the above results figure 6.3 remains the best available model and will form the basis of the ensuing discussion.

The FeMo-cofactors in the protein

Most of our information on the Mo and its environment in the MoFe protein has come from x-ray spectroscopy. The edge-spectra showed that terminal Mo=O bonds occurred only in oxygen-damaged protein. Analysis of the EXAFS data (Cramer *et al.*, 1978; S. Conradson, personal communication) from the dithionite-reduced proteins indicated that in the immediate environment of the Mo atoms there were 4–5 S atoms at 2.37 Å, 3–4 Fe atoms at 2.67 Å and perhaps two low Z atoms (C, N or O) at about 2.1 Å. Similar EXAFS spectra have been obtained from complexes containing [$MoFe_3S_4$] clusters (Newton *et al.*, 1981; Holm, 1981). However such compounds are deficient in iron relative to FeMo-cofactor and cannot substitute for it in the activation of the inactive MoFe protein from mutants.

The FeMo-cofactors in the dithionite-reduced MoFe protein exhibit characteristic rhombic epr signals with *g*-values near 4.3, 3.7 and 2.01 (figure 6.4a). Isotope substitution ([57]Fe, [95]Mo) experiments (Smith *et al.*, 1973) have shown that this epr signal is associated with iron in the protein but no interaction of the unpaired electrons with the molybdenum nuclei was observed. Nevertheless pulsed epr studies on Cp1 (Orme-Johnson *et al.*, 1981) detected weak interactions between the unpaired electrons and the nuclei of [95]Mo, ^{14}N and [1]H. In addition the epr-active centre may be extruded as FeMoCO which has a similar epr signal although with larger magnetic anisotropy (Rawlings *et al.*, 1978) (figure 6.4b). The epr signal is consistent with that expected from the $M_z = \pm\frac{1}{2}$ Kramer's doublet of an $S = \frac{3}{2}$ system (Palmer *et al.*, 1972) and integrates to 1 ± 0.1 electrons per Mo atom with each electron being associated with 6 Fe atoms, indicating a ratio of 6Fe:1Mo in FeMoco (Rawlings *et al.*, 1978). A similar conclusion was reached from [57]Fe ENDOR experiments on the MoFe protein (Hoffman *et al.*, 1982a). Other ENDOR experiments indicated that at least 5 non-exchangeable protons are coupled to the unpaired electron spins in Av1 (Hoffman *et al.*, 1982b).

The 77 K Mössbauer spectra of dithionite-reduced Kp1, (figure 6.5a) Av1 and Cp1 have been interpreted in terms of three overlapping quadrupole doublets M4, M5 and M6 together with a minor species S (Smith and Lang, 1974; Münck *et al.*, 1975; Huynh *et al.*, 1980). Species M6 becomes a broad shallow absorption, indicating magnetic hyperfine interaction at 4.2 K (figure 6.5b) and can be

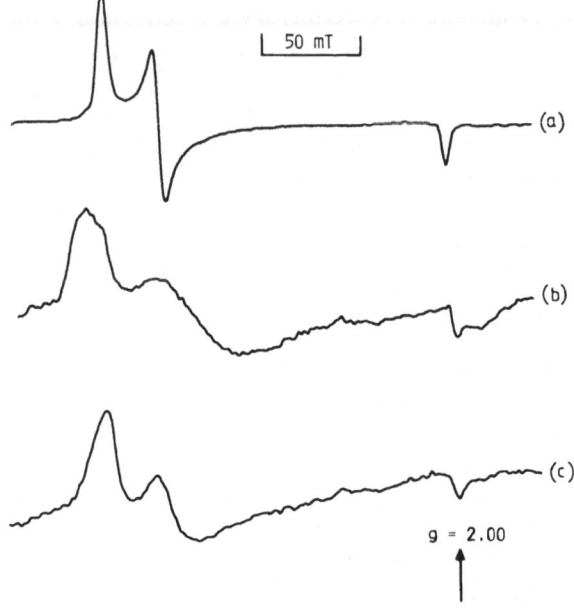

Figure 6.4 Epr spectra of (a) the FeMo-cofactors in Kp1 (50 mM BES buffer pH 5.8); (b) FeMoco as isolated in N-methylformamide and (c) FeMoco as (b) but with added thiophenol. The spectra were recorded at *c.* 10 K with a microwave frequency of 9 GHz and microwave power of 150 mW using 1.6 mT field modulation at 100 kHz

assigned to the FeMo-cofactors (Smith and Lang, 1974). This spectral species has been studied in great detail by Huynh *et al.* (1979) who defined six sub-components, three with positive and three with negative hyperfine coupling constants. This analysis also indicated that there are six spin-coupled Fe atoms associated with the FeMo-cofactors. Dye oxidation of the MoFe protein oxidises the FeMo-cofactor iron atoms to a diamagnetic Mössbauer species (Smith *et al.*, 1980; Huynh *et al.*, 1980).

Isolated FeMoco*

It must be emphasised that contrary to earlier reports (Nason *et al.*, 1971) FeMoco isolated from nitrogenase is distinct from the Mo-cofactor that is common to many other Mo-containing enzymes. FeMoco will not reconstitute activity in apo-nitrate reductase and the Mo-cofactor from xanthine oxidase cannot replace FeMoco in its activation of inactive MoFe protein (Pienkos *et al.*, 1977). Further information in Mo-cofactor is available in Coughlan (1980).

*FeMoco is used to distinguish the isolated cluster from the cofactor in the protein which is called FeMo-cofactor.

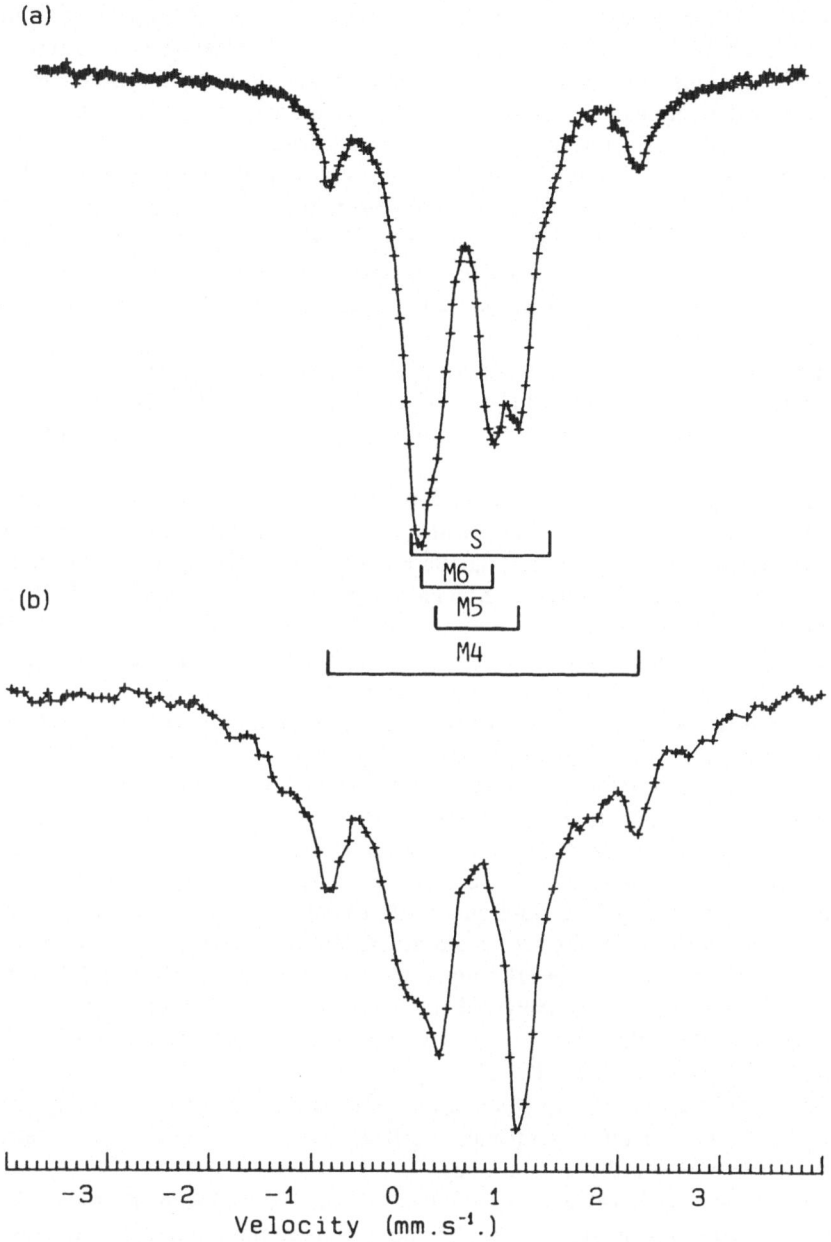

Figure 6.5 Mössbauer spectra of dithionite-reduced Kp1 with the species M4, M5, M6 and S indicated. (a) measured at 77 K in a 0.035 T magnetic field, (b) measured at 4.2 K in a 0.029 T magnetic field. In both (a) and (b) the magnetic fields were applied perpendicular to the γ-ray beam. Note that at 4.2 K species M6 becomes a broad shallow absorption indicating magnetic hyperfine interaction

Analyses of preparations of FeMoco from a number of organisms (Shah and Brill, 1977; Smith, 1980; Burgess et $al.$, 1980a) indicate a stoicheiometry of $MoFe_{7-8}$. This contrasts with the spectroscopic data on the intact MoFe proteins which indicate 6 $paramagnetic$ Fe atoms per Mo atom (see above). Some clarification of this anomaly has been afforded by the isolation of a different Mo and Fe containing cluster from Av1 (Shah and Brill, 1981). This cluster was extracted from the acid-treated protein into methylethylketone and had the stoicheiometry $MoFe_6$. It had an $S = \frac{1}{2}$ epr spectrum with g values near 2. When exchanged into N-methylformamide it exhibited the typical FeMoco epr signal (figure 6.4b) but was incapable of activating MoFe protein from the $A.$ $vinelandii$ mutant UW45.

These data support the spectroscopic interpretation that there are only six $paramagnetic$ Fe atoms per Mo in FeMoco and indicate that the one or two additional Fe atoms found in FeMoco preparations are essential for activation of inactive MoFe protein.

The Mössbauer spectra show only four Fe species in the MoFe protein M4, M5, M6 and S (figure 6.5). Species M4 and M5 constitute the P clusters (see below) and species M6, the paramagnetic Fe atoms in FeMo-cofactors. Hence the species S atoms are probably the non-paramagnetic Fe components of FeMoco. Admittedly, the Mössbauer spectrum of isolated FeMoco (Rawlings et $al.$, 1978) did not show species S, but an additional component was observed which might correspond to species S in an altered environment.

The Mo EXAFS of isolated FeMoco is very similar to that of the MoFe protein except that the analyses indicate that one S atom may have been replaced by O, N or C (S. Conradson, personal communication). The Fe EXAFS has been analysed as indicating that the nearest neighbours of the average Fe atom in isolated FeMoco are 3.4 ± 1.6 S (Cl) atoms at 2.25 Å; 2.3 ± 0.9 Fe atoms at 2.66 Å; 0.4 ± 0.1 Mo atoms at 2.76 Å and 1.2 ± 1.0 O (N or C) atoms at 1.81 Å (Antonio et $al.$, 1982). The fractional value for Mo indicates that the cluster is asymmetric and that the Fe atoms are not all equivalent. This interpretation is supported by the Mössbauer (Huynh et $al.$, 1979) and ENDOR data (Hoffman et $al.$, 1982a) which respectively indicate two and six distinct environments for the $paramagnetic$ Fe atoms. In these circumstances the interpretation of the Fe EXAFS data is clearly difficult.

The sulphur content of FeMoco is not well established, estimates range from 4 to 9 S atoms per Mo atom (Shah and Brill, 1977; Yang et $al.$, 1982; Nelson et $al.$, 1983; B. E. Smith, unpublished).

FeMoco is anionic (Newton et $al.$, 1980). Smith (1980) studied its behaviour in gel exclusion chromatography, and found that it eluted similarly to anionic metal complexes with molecular weights of 700 or 1500 depending on the method of preparation. Citrate, a reagent in one of these methods, affected the elution properties of FeMoco but not its ability to activate.

Other possible ligands used in the isolation procedures are tris buffer, Cl^-, HPO_4^{2-}, $S_2O_4^{2-}$, dimethylformamide and N-methylformamide. None of these are essential in the activation reaction (Yang *et al.*, 1982). FeMoco preparations contained no significant quantities of amino acids (Smith, 1980) or common sugars (Yang *et al.*, 1982) or, despite an earlier report to the contrary (Levchenko *et al.*, 1980), co-enzyme A or lipoic acid (Yang *et al.*, 1982).

FeMoco is extremely sensitive to oxygen. However it is more stable than [4Fe-4S] and [2Fe-2S] cluster compounds to several reagents: thiophenol (1 mole per g atom Mo) reacts and sharpens the epr spectrum (figure 6.4c) but does not affect activation ability (Rawlings *et al.*, 1978; Burgess *et al.*, 1980b); bipyridyl apparently does not react but mercurials destroy FeMoco (Rawlings *et al.*, 1978); excess EDTA and 1-phenanthroline both abolish the epr signal of FeMoco without affecting its activation ability, these reactions can be reversed by addition of Zn^{2+} or Fe^{2+} respectively (Yang *et al.*, 1982).

In summary, the anionic nature and metal content of FeMoco imply the presence of a considerable number of anionic ligands, e.g. S^{2-}, O^{2-} or OH^-. No organic ligands have yet been detected. The solvent N-methylformamide may be a ligand in most preparations but can be replaced by formamide. The complex seems able to undergo facile ligand exchange reactions but these do not affect its activation capability.

Isolated FeMoco, with borohydride as reductant, has been reported to reduce C_2H_2 and cyclopropene (Shah *et al.*, 1978; McKenna *et al.*, 1979). The activity was relatively insensitive to oxygen, was ATP-independent and in the case of cyclopropene did not yield the same products as nitrogenase. It is unlikely that intact FeMoco is the active catalyst in these reactions and hence these data have little relevance to the role of FeMo-cofactor in nitrogenase.

The P clusters

Four [4Fe-4S] clusters may be extruded from each molecule of the MoFe protein (see above). These clusters are each thought to contain one M4 and three M5 spin coupled Fe atoms (Zimmermann *et al.*, 1978; Huynh *et al.*, 1980). In the Mössbauer spectrum of the dithionite-reduced protein (figure 6.5) species M4 are characterised by parameters typical of high spin ferrous atoms in 4Fe-4S or 2Fe-2S ferredoxins although there is no indication of spin coupling to atoms with half-integer spin; species M5 have similar isomer shifts, typical of ferrous atoms but a much smaller quadrupole splitting (Smith and Lang, 1974; Münck *et al.*, 1975). Thus in the dithionite-reduced protein the P clusters are thought to be at the $[4Fe-4S]^0$ oxidation level which is extremely difficult to attain in model compounds (Holm, 1977). This assignment is supported by the observation that a transient epr signal with $g_{av} = 1.93$, characteristic of $[4Fe-4S]^{1+}$ clusters, is exhibited by the P clusters during *oxidation* of the MoFe protein (Smith *et al.*, 1983). Subsequently, at equilibrium, the oxidised P clusters are epr inactive but exhibit an extremely complex magnetic Mössbauer spectrum at 4.2 K (figure 6.6b) (Smith and Lang, 1974; Zimmermann *et al.*, 1978). The Mössbauer spectra at higher temperatures (figure 6.6a) indicate that species M4

D. J. Lowe et al.

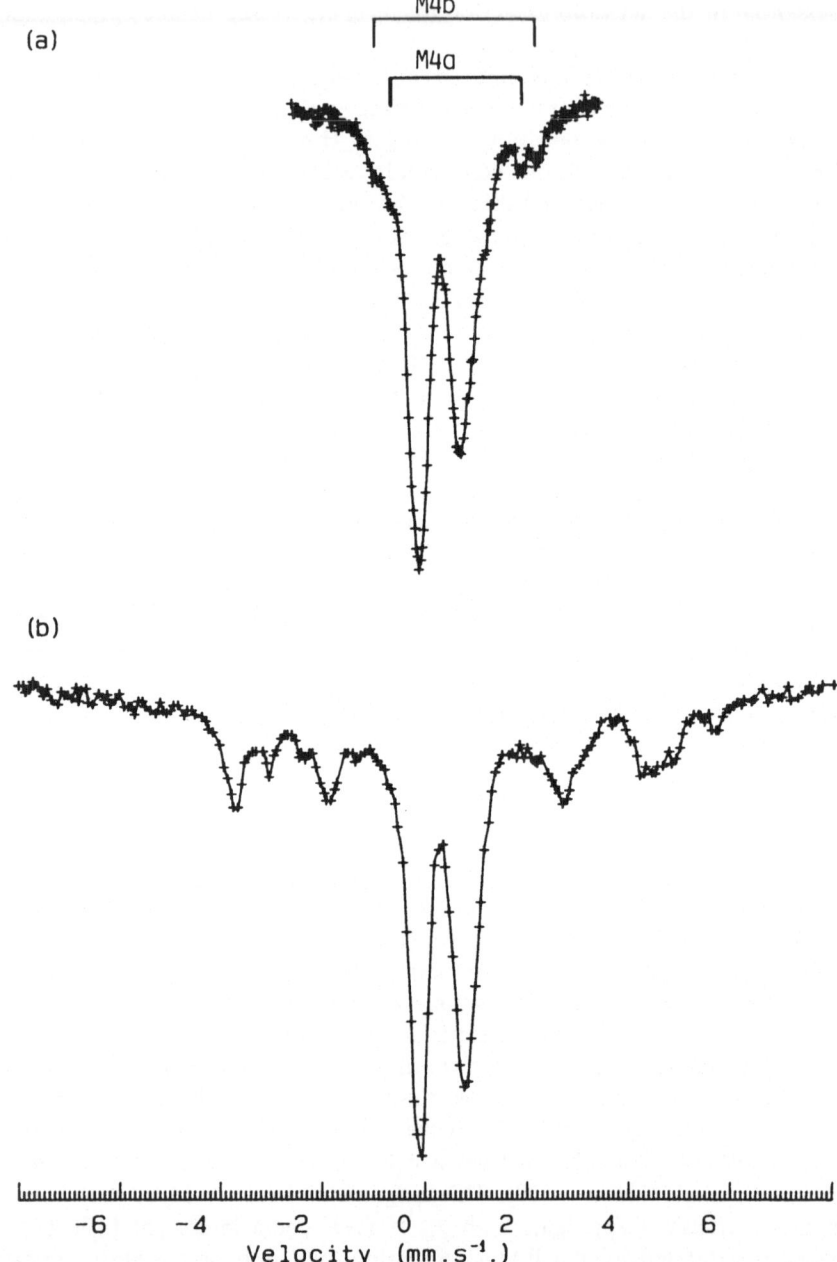

Figure 6.6 Mössbauer spectra of Kp1 oxidised to −36 mV vs SHE. (a) measured at 195 K in zero magnetic field, (b) measured at 4.2 K with a 0.029 T magnetic field perpendicular to the γ-ray beam. Note that species M4a and M4b, visible at 195 K, become part of the broad, complex, magnetic spectrum at 4.2 K indicating magnetic interaction with the oxidised species M5

was not oxidised but split into two closely related species (M4a and M4b) by interaction with oxidised M5 (Smith *et al.*, 1980). Detailed analysis of the magnetic Mössbauer spectrum for Av1 and Cp1 allowed assignment of eight sub-components, two of which were associated with M4a and M4b (Zimmermann *et al.*, 1978; Huynh *et al.*, 1980).

The above analysis of the Mössbauer spectra defined the spin state of the oxidised P clusters as half-integer and between $\frac{3}{2}$ and $\frac{9}{2}$. The magnetic circular dichroism spectrum of the oxidised P clusters showed temperature and magnetic field dependence which was interpreted in terms of a paramagnet of spin state $S = \frac{5}{2}$ or $\frac{7}{2}$ (Johnson *et al.*, 1981). Virtually identical magnetic circular dichroism spectral characteristics were observed (Hawkes, 1981) with MoFe protein isolated from *nifB* mutants of *K. pneumoniae* (Hawkes and Smith, 1983). This protein lacks FeMoco. These data demonstrate that the structure of the P clusters is essentially independent of the presence of FeMoco. Magnetic suscepti-bility experiments on Av1 have been interpreted in terms of a $S = \frac{5}{2}$ spin state for oxidised P clusters (Smith, J. *et al.*, 1982). However in the absence of detail-ed temperature dependence such data are extremely difficult to interpret.

In summary the P clusters appear to be unusual 4Fe-4S clusters. Their redox behaviour and spectral characteristics may indicate that, in contrast to the ferredoxins, cysteine residues are not the only ligands binding them to the protein.

Maximum specific activity

Most studies on the MoFe protein have used preparations which were pure in terms of the constituent polypeptides but were not fully active. Different preparations have been reported to contain 1-2 Mo and 18-33 Fe atoms/mole protein with specific activities varying between 1000 and 3000 nmole C_2H_4 produced $min^{-1} mg^{-1}$ at 30°.

The maximum reported activity is in reasonable accord with that calculated from an analysis of the relationship between the Mo content and activity. Ten preparations of Kp1, with specific activities between 1090 and 1770 nmol C_2H_4 produced $min^{-1} mg^{-1}$, gave a constant value of 275 ± 27 nmole C_2H_4 produced (ng atom Mo)$^{-1}$ (Hawkes *et al.*, 1984). Preparations of FeMoco also give activi-ties of 250-300 nmole C_2H_4 produced (ng atom Mo)$^{-1}$ after reconstitution. These values extrapolate to a specific activity of 2520 ± 250 nmole C_2H_4 produced $min^{-1} mg^{-1}$ for Kp1 of molecular weight 218 000 and containing 2 Mo atoms mole^{-1}.

The above analysis implies that the activity of a MoFe protein's preparation depends solely on its FeMoco content. However the ^{57}Fe Mössbauer spectra for Kp1 (Smith and Lang, 1974), Av1 (Münck *et al.*, 1975) and Cp1 (Huynh *et al.*, 1980) indicate that the ratio of Fe in P clusters to that in the FeMo-cofactors is constant. This has led to the 'all-or-none' hypothesis that each MoFe protein molecule contains none, exactly half, or all of its maximum metal content.

Table 6.3 Some reactions catalysed by nitrogenase

	Reference[a]
1. $N_2 + 6H^+ \xrightarrow{6e^-} 2NH_3$	Wherland *et al.* (1981)
2. $HCN + 6H^+ \xrightarrow{6e^-} CH_4 + NH_3$	Li *et al.* (1982)
3. $CH_3NC + 6H^+ \xrightarrow{6e^-} CH_4 + CH_3NH_2$	Hwang & Burris (1972)
4. $N_3^- + 7H^+ \xrightarrow{6e^-} N_2H_4 + NH_3$	Dilworth & Thorneley (1981)
5. $HCN + 4H^+ \xrightarrow{4e^-} CH_3NH_2$	Li *et al.* (1982)
6. $N_3^- + 3H^+ \xrightarrow{2e^-} N_2 + NH_3$	Dilworth & Thorneley (1981)
7. $N_2O + 2H^+ \xrightarrow{2e^-} N_2 + H_2O$	Hwang *et al.* (1973)
8. $C_2H_2 + 2H^+ \xrightarrow{2e^-} C_2H_4$	Hageman & Burris (1980)
9. $HC{=}CH\overset{CH_2}{\diagdown} + 2H^+ \xrightarrow{2e^-} \frac{1}{3} CH_2{-}CH_2\overset{CH_2}{\diagdown} + \frac{2}{3} CH_3{-}CH{=}CH_2$	McKenna & Hwang (1979)
10. $2H^+ \xrightarrow{2e^-} H_2$	Burgess *et al.* (1981)
11. $D_2 + 2H^+ \xrightarrow{2e^-} 2HD$	Burgess *et al.* (1981)

[a]The most recent relevant reference is quoted. It is not intended to indicate the original discovery of the reaction but to be a good starting point for a literature review of each substrate.

The Mo and Fe containing clusters may not be the only determinants of activity. Kimber *et al.* (1982), using an nmr technique, found that the specific activity of preparations of Kp1 was linearly related to the number of divalent metal binding sites on the protein. The bound metal ions interacted with ATP. The linear relationship extrapolated to 4 binding sites per molecule of Kp1 at a specific activity of 2900 nmol C_2H_4 produced $min^{-1} mg^{-1}$. Miller *et al.* (1980) using Kp1 of this specific activity and a gel equilibration technique found 4 MgATP binding sites per molecule.

A maximum specific activity of 2900 nmol C_2H_4 produced $min^{-1} mg Kp1^{-1}$ is the same as that calculated from pre-steady state measurements of the rate limiting step (see below).

Interactions between the prosthetic groups in the isolated protein

The FeMo-cofactors and the P clusters act independently in redox titrations (Smith *et al.*, 1980). Dithionite (in the absence of mediator dyes) reduces the FeMo-cofactors but not the P clusters of dye oxidised Av1 (Orme-Johnson *et al.*, 1977).

However a transient epr signal with $g_{av} \simeq 1.93$ was observed from rapidly frozen samples of Kp1 during oxidation by some redox dyes or ferricyanide (Smith *et al.*, 1983). The kinetics of this reaction were interpreted in terms of intramolecular electron transfer from the FeMo-cofactors to P clusters which had lost two electrons, i.e. P^{2+} clusters. The rate constant (4.1 ± 0.8 s^{-1} at $23°$) for this process was the same within experimental error as the rate of dissociation of the Kp2$_{ox}$ Kp1 protein complex which is the rate determining step in turnover (Thorneley and Lowe, 1982, 1983). A conformational change of the MoFe protein which caused protein–protein dissociation and was either triggered by, or itself triggered electron transfer within the MoFe protein could explain this coincidence (Smith *et al.*, 1983). The function of the P clusters would then be to feed electrons to the FeMoco active sites as required. Some of the epr signals observed during turnover which we discuss later and are given in table 6.4 might then come from oxidised P clusters at the $[4\text{Fe-4S}]^{1+}$ oxidation level.

Nucelotide binding

MgATP and MgADP bind to both the Fe protein and the MoFe protein and modulate the kinetics of their electron transfer reactions (see mechanism section). The determination of the number of binding sites and the binding constants is difficult because of the uncertain effects of inactive protein. Additionally some of the earlier studies did not take into account the equilibria shown in equation (6.1) (Thorneley and Willison, 1974).

$$2\text{Mg}^{2+} + \text{ATP}^{4-} \rightleftharpoons \text{Mg}^{2+} + \text{MgATP}^{2-} \rightleftharpoons \text{Mg}_2\text{ATP} \tag{6.1}$$

Another problem with measurements on the Fe proteins is that they can catalyse the disappearance of dithionite, especially in the presence of MgATP. Subsequently the proteins become oxidised (Stephens *et al.*, 1979).

Stephens *et al.* (1983) used the large cd change in the visible region which is observed on binding of either nucleotide, to show that oxidised Av2 and Cp2 bind two molecules of either nucleotide per molecule of protein ($K_D < 5$ μM).

Table 6.4 Epr signals, associated with transient forms of the MoFe protein of *K. pneumoniae*, which are only observed during turnover

Description	g values	g_{av}	Comments
III	2.073, 1.969, 1.927	1.990	CO bound tightly
IV	2.17, 2.06, 2.06	2.10	CO bound weakly
V	2.139, 2.001, 1.977	2.039	May be associated with H_2 evolution
VI	2.125, 2.000, 2.000	2.042	Ethylene bound form, decreased by acetylene
VII	4.7, 5.4		
VIII	2.092, 1.974, 1.933	2.000	

There is no observable effect of MgATP or MgADP on the cd of reduced Fe-proteins (Stephens *et al.*, 1982). However, both nucleotides affect the epr spectrum. MgATP changes the form of the epr spectrum from rhombic to axial (Zumft *et al.*, 1972; Orme-Johnson *et al.*, 1972; Smith *et al.*, 1973). Further evidence for the binding of these nucleotides is shown by the increased sensitivity to O_2 damage (Smith *et al.*, 1976a) and an altered reactivity of protein SH groups (Thorneley and Eady, 1973; Walker and Mortenson, 1973). The Fe in the protein becomes more accessible to chelation by bipyridyl (Walker and Mortenson, 1973) and BPS (Ljones and Burris, 1978a) in the presence of MgATP. Hageman *et al.* (1980) used the latter technique to obtain dissociation constants of 430 μM and 220 μM for the binding of MgATP to nominally reduced Av2.

The characteristics of MgADP binding to reduced Fe proteins are not yet clear. However MgADP certainly does bind but affects the above properties less than MgATP (e.g. Zumft *et al.*, 1973; Thorneley and Eady, 1973; Walker and Mortenson, 1973).

Kimber *et al.* (1982) demonstrated by nmr that Kp1 has four equivalent binding sites for Mg^{2+} (K_D = 2.2 ± 0.3 mM). Miller *et al.* (1980) have shown that MgATP binds to Kp1 at four equivalent sites (K_D = 0.6 ± 0.1 mM). These data, together with those discussed above for MgATP binding to the Fe-protein, suggest that MgATP may bind to both Kp1 and Kp2 in a bridging mode. The two MgATP sites on Kp2 would complement the two sites on half of the Kp1 tetramer. This model could also explain why the complexed proteins catalyse the reductant independent hydrolysis of ATP whereas the isolated proteins do not (Imam and Eady, 1980).

The mechanism of nitrogenase

The electron transfer sequence

The electron transfer sequence (equation 6.2) was demonstrated by electron paramagnetic resonance (Orme-Johnson *et al.*, 1972; Smith *et al.*, 1972, 1973; Zumft *et al.*, 1974) and Mössbauer (Smith and Lang, 1974) studies.

$$
\left.
\begin{array}{l}
\text{Ferredoxin} \\
in\ vivo \quad \text{or} \\
\text{Flavodoxin} \\
in\ vitro\ \ \text{Na}_2\text{S}_2\text{O}_4
\end{array}
\right\} \rightarrow \text{Fe protein} \rightarrow \text{MoFe protein} \rightarrow \text{substrate} \\
[\text{H}^+, \text{N}_2, \text{N}_3^-, \text{C}_2\text{H}_2]
$$

$$(6.2)$$

The reader is referred to the review by Yates (1980) for further details of the structures of flavodoxins and ferredoxins and their role in nitrogen fixation. Sodium dithionite is the most commonly used electron donor to nitrogenase *in vitro*. The predissociation $S_2O_4^{2-} \rightleftharpoons 2SO_2^{\cdot-}$ generates the active reductant $SO_2^{\cdot-}$ and accounts for the half-order dependence on dithionite ion concentration often observed in kinetic studies (Thorneley *et al.*, 1976).

The sites of substrate and inhibitor binding

Table 6.3 lists some of the small, multiple bonded substrates which together with protons are reduced by nitrogenase. All these reductions require both nitrogenase proteins and are coupled to the hydrolysis of MgATP to MgADP + P_i. Steady state kinetic studies (Hwang *et al*., 1973; Rivera-Ortiz and Burris, 1975) were used to determine K_m's and K_i's for various substrates and inhibitors. These results were interpreted in terms of five sites or modified sites for substrate and inhibitor binding. These sites need not be on separate metal centres. The ligand environment and oxidation level of a metal centre can profoundly affect the reactivity of that centre towards a potential substrate or inhibitor. For instance, the same coordination position on molybdenum may be involved in both azide and dinitrogen binding. However azide may only bind to molybdenum at a high oxidation level and dinitrogen at a lower level (Dilworth and Thorneley, 1981). The electrochemistry of transition metal complexes provides many useful examples of these effects and in particular has contributed to the understanding of the requirements for N_2, CO and H_2 binding to metal centres (Pickett, 1983).

Evidence for FeMoco being the site of N_2 binding has been provided by a study of *nifV* mutants of *K. pneumoniae* (McLean and Dixon, 1981; Hawkes *et al*., 1984). *NifV* mutant cells cannot fix N_2 but can reduce C_2H_2 to C_2H_4. This phenotype is associated with the MoFe protein, NifV⁻ Kp1. An unusual property of the nitrogenase from these mutants is that H_2 evolution is partially inhibited by CO whereas in the wild-type enzyme it is not (McLean and Dixon, 1981; McLean *et al*., 1983). The FeMoco from NifV⁻Kp1 was combined with NifB⁻Kp1 (lacking FeMoco) to form active Kp1 with the properties of NifV⁻ Kp1. The FeMoco from wild-type Kp1 conferred normal activity on NifB⁻Kp1 (Hawkes *et al*., 1984). These experiments demonstrated that the NifV⁻ phenotype was associated with the FeMoco of the NifV⁻Kp1 and that the function of the *nifV* gene product is to modify the FeMoco to generate an effective nitrogenase.

Smith (1977) has reviewed the circumstantial evidence for reducible substrate binding to the Mo of the MoFe protein. More recently this assumption has been supported by the correlation between acetylene reduction activity and Mo content (see above). Further evidence involving N_2 reduction is the steady state concentration of dinitrogen hydride intermediate (Thorneley *et al*., 1978). This showed that Kp1 (1.4 Mo per mol) bound and reduced 1.4 mol of N_2 (figure 6.7). Thus fully active Kp1 with 2 Mo atoms per mol would bind and reduce two molecules of N_2 presumably at independently functioning sites. This conclusion excludes the involvement of binuclear Mo-N_2-Mo intermediates in the enzymic reduction of N_2. Since the Mo is part of FeMo-cofactor we cannot exclude a role for Fe in binding substrates. However the reduction of N_2 bound to Mo in model complexes (Richards, 1980) supports the view that the Mo in nitrogenase binds N_2.

In the steady-state, the FeMo-cofactors are further reduced, to an integer

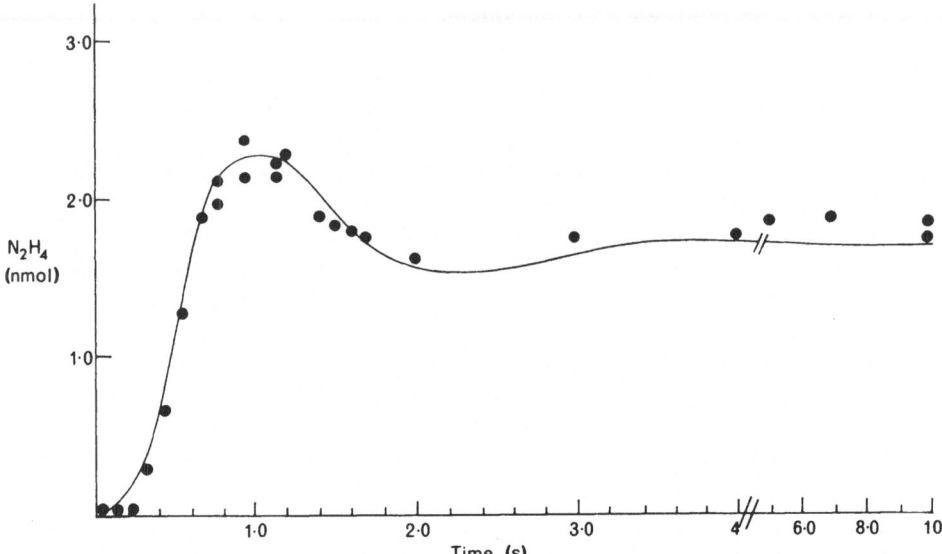

Figure 6.7 Pre-steady state kinetics of hydrazine formation from a dinitrogen hydride intermediate in dinitrogen reduction at 23°, pH 7.4. The data were obtained with the rapid quench technique. Each shot (0.38 ml) contained 10 nmol Kp1 (14 nmol Mo), 4.4 nmol Kp1, 3.4 μmol MgATP, 3.8 μmol $Na_2S_2O_4$. The curve is that computed for four slow steps prior to hydrazine formation (see schemes 2 and 4)

spin, epr-inactive oxidation level, but the P clusters appear to be unaffected. However when the temperature was lowered to 10° and the supply of reduced Kp2 limited a number of epr signals (table 6.4) from intermediates associated with the MoFe protein were detected (Lowe *et al.*, 1978). The intensities of these epr signals were sensitive to the concentration of substrates, their reduction products (Lowe *et al.*, 1978) and the inhibitor CO (Davis *et al.*, 1979). At least one of the epr signals, signal VI, which only appeared in the presence of C_2H_4, has been shown to arise from FeMo-cofactor in the $S = \frac{1}{2}$ spin state (Hawkes *et al.*, 1983). By analogy it was suggested that signals IV and V, which also have $g_{av} > 2.000$, might arise from FeMoco in the $S = \frac{1}{2}$ spin state with a modified environment. $[^{13}C]$-ethylene, $[^{13}C]$-acetylene (Lowe *et al.*, 1978) and $[^{13}C]$-carbon monoxide (Davis *et al.*, 1979) do not broaden any of the epr signals. Thus there is no evidence for the direct binding of C_2H_4, C_2H_2 or CO to the metal centres.

The Fe protein catalytic cycle

The reduction of oxidised Fe protein by $SO_2\cdot^-$ and its subsequent MgATP-dependent oxidation by the MoFe protein constitute a catalytic cycle (Scheme 1). The cycle comprises four partial reactions which have been studied for *K. pneumoniae* nitrogenase by stopped-flow spectrophotometry and rapid-quenching followed by product analysis (Thorneley and Lowe, 1982). The order of assembly of the Kp2 $(MgATP)_2$ Kp1 complex (k_1, k_{-1}) has not been established since all these reactions are too fast to be measured by the stopped-flow technique (Thorneley, 1975).

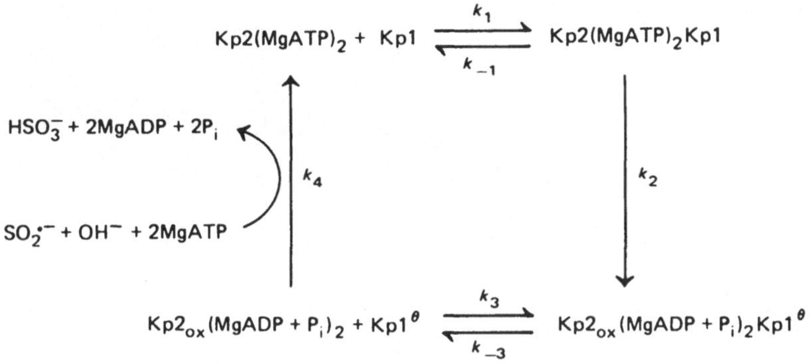

$$Kp2(MgATP)_2 + Kp1 \underset{k_{-1}}{\overset{k_1}{\rightleftharpoons}} Kp2(MgATP)_2Kp1$$

HSO$_3^-$ + 2MgADP + 2P$_i$ ↖ k_4 ↑ k_2 ↓

SO$_2^{--}$ + OH$^-$ + 2MgATP ↙

$$Kp2_{ox}(MgADP + P_i)_2 + Kp1^{\theta} \underset{k_{-3}}{\overset{k_3}{\rightleftharpoons}} Kp2_{ox}(MgADP + P_i)_2Kp1^{\theta}$$

Scheme 1

According to Scheme 1, after one cycle one electron has been transferred from Kp2 to Kp1. All the MoFe proteins investigated have an $\alpha_2\beta_2$ sub-unit structure. Therefore if Kp2 is a one electron donor, then Kp1 in scheme 1 represents an $\alpha\beta$ moiety. If the Fe proteins are two electron donors (see above) then Kp1 in scheme 1 would have to represent a $\alpha_2\beta_2$ tetramer with one electron going to each half. This is necessary for scheme 1 to be consistent with the substrate reduction cycle discussed below. The reasons for this uncertainty have been discussed above.

The electron transfer reaction (k_2) has been studied by monitoring the absorption increase at 420 nm in the Kp2 spectrum which occurs on oxidation by Kp1. Since $k_1 > 10^7 M^{-1} s^{-1}$, Kp1 and Kp2 come together to form a protein complex at a rate close to the diffusion controlled limit (Thorneley, 1975). Paradoxically, it is this diffusion controlled limit which we consider causes nitrogenase to be such a slow enzyme (see below).

The protein–protein electron transfer $(k_2 = 2 \times 10^2 s^{-1}$ at 23°) is not rate-limiting in turnover and is coupled to the hydrolysis of MgATP to MgADP + P$_i$ (Eady *et al*., 1978a; Hageman *et al*., 1980). The apparatus used to demonstrate this coupling was essentially that described by Gutteridge *et al*. (1978) for rapid-freezing of epr samples. Instead of freezing the samples of reacting enzyme as they emerged from a needle jet, a chemical quench was used to stop the reaction in small vials fitted with rubber closures. The gas phase was subsequently analysed for H_2 or C_2H_4, and the liquid phase for N_2H_4, NH_4^+ or P$_i$. Data which show the coupling of electron transfer from Kp2 to Kp1 to the hydrolysis of MgATP is shown in figure 6.8. Hageman *et al*. (1980), using a similar technique, calculated a stoicheiometry of one electron transferred per 2 MgATP hydrolysed. Thorneley and Lowe (1983) have shown that the rate limiting step for nitro-

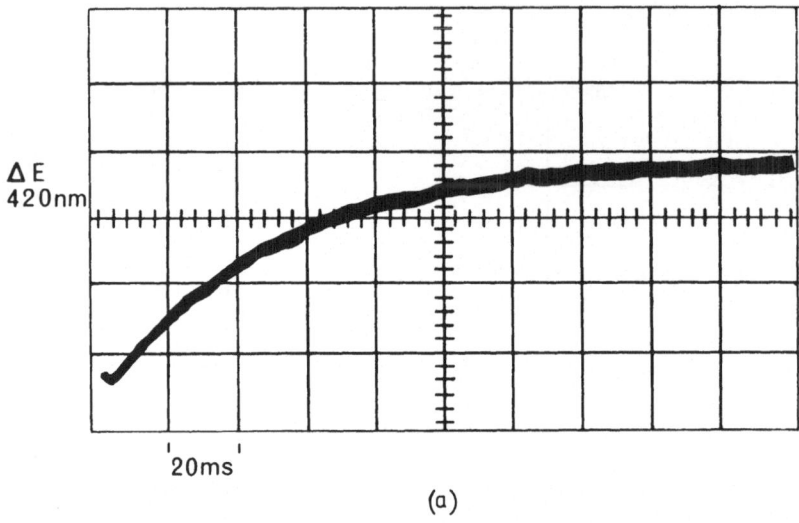

(a)

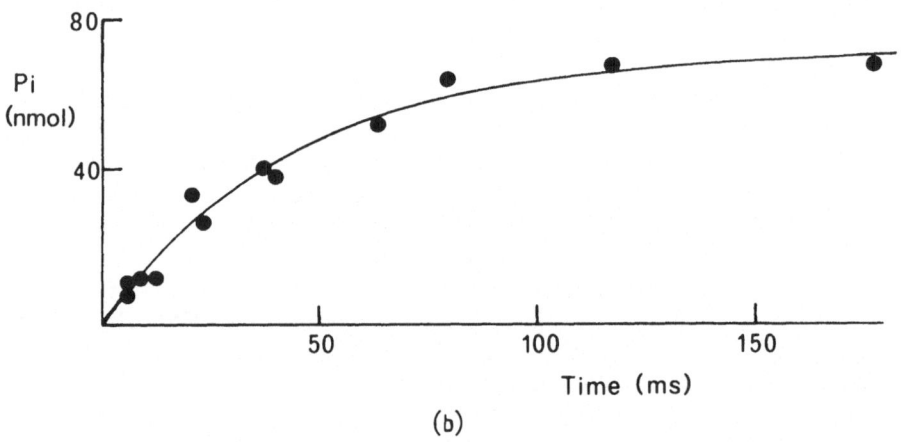

(b)

Figure 6.8   Coupling of electron transfer from Kp2 to Kp1 protein with MgATP hydrolysis.
(a) Stopped-flow oscillograph of the absorbance change at 420 nm associated with the
MgATP (5 mM)-induced oxidation of Kp2 (50 μM) by Kp1 (10 μM) at pH 7.4, 10°C in the
presence of $Na_2S_2O_4$ (10 mM). τ = 42 ± 3 ms. (b) Time course for appearance of P_i under
identical conditions obtained using the rapid quench technique with trichloroacetic acid
(30% w/v) as quenching agent followed by colorimetric analysis for P_i. τ = 44 ± 4 ms. At 23°
the same reaction occurs with τ = 7 ms

genase action is the dissociation of oxidised Kp1 from reduced Kp1 (k_3 = 6.4 ± 0.8 s^{-1} at 23°). Stopped-flow spectrophotometry was used to measure the rate of reduction of $Kp2_{ox}(MgATP)_2$ by SO_2·$^-$ as a function of Kp1 concentration. At a high concentration of Kp1 the rate of reduction is limited by the dissociation reaction k_{-3}. At low Kp1 concentrations when the two proteins are largely dissociated, the rate is that observed for $Kp2_{ox}(mgADP)_2$ in the absence of Kp1, i.e. k_4.

The last reaction in the Fe protein cycle is the reduction of $Kp2_{ox}(MgADP)_2$ by SO_2·$^-$ which occurs with k_4 = 3 × 10^6 M^{-1} s^{-1} at 23°. The detailed mechanism of this reaction is not known. Studies with Ac nitrogenase showed that in the absence of MgADP, $Ac2_{ox}$ is reduced by SO_2·$^-$ with $k > 10^8$ M^{-1} s^{-1} (Thorneley *et al.*, 1976). Therefore MgADP must either induce a conformation change in the protein or inhibit the access of SO_2·$^-$ to the site of reduction (Yates *et al.*, 1975).

Scheme 1 can be used to show how the rate limiting step of nitrogenase changes with protein concentration. When [Kp1] ≃ [Kp2] ≪ 5 μM, the association of reduced $Kp2(MgATP)_2$ with Kp1 (k_1) is rate limiting. This explains the non-linear dependence of acetylene reduction activity on component protein concentration ('dilution effect') (see Thorneley *et al.*, 1975). When the component proteins are in the concentration range 1 μM < [Kp1] ≃ [Kp2] < 10 μM, the dissociation of $Kp2_{ox}(MgADP)_2$ from Kp1(k_{-3}) becomes rate limiting. At concentrations of [Kp1] ≃ [Kp2] > 10 μM, k_{-3} is still rate-limiting, but the activity is inhibited due to the contribution of the back reaction (k_{+3}) (figure 6.9). Thus at the concentrations of Kp1 and Kp2 present *in vivo* (*c.* 100 μM) (Eady *et al.*, 1978b; Roberts *et al.*, 1978; Hennecke and Shanmugam, 1979) substrate reduction is inhibited. The mechanism discussed below suggests that this inhibition is unavoidable if nitrogenase is to use the available reducing power and MgATP to reduce N_2 to NH_4^+ rather than $2H^+$ to H_2.

Steady-state analysis assays with low concentrations of nitrogenase proteins have shown that MgADP inhibits the reduction of substrates (Moustafa and Mortenson, 1967; Davis and Orme-Johnson, 1976; Orme-Johnson and Davis, 1976, Thorneley and Cornish-Bowden, 1977; Hageman *et al.*, 1980). This inhibition is not fully described by scheme 1 but presumably is implicit in k_4.

A PROPOSED MECHANISM FOR NITROGENASE

The principal observations which any mechanism for nitrogenase must explain are listed below:

(1) The H_2 :NH_3 product ratio (with saturating N_2) increases from a limiting value of 1:2 as the electron flux is decreased. This has been achieved by using (a) Fe:MoFe protein ratios < 1:1 (b) limiting MgATP or (c) MgADP inhibition (Silverstein and Bulen, 1970; Hageman and Burris, 1980; Wherland *et al.*, 1981).

Table 6.5 Rate constants used in the simulation of the kinetics of Kp nitrogenase at 23°C, pH 7.4

Rate constant	Value	Comment	Reference
k_1	$5 \times 10^7 M^{-1} s^{-1}$*	Responsible for dilution effect	Lowe and Thorneley (1984)
k_{-1}	$15\ s^{-1}$*		
k_2	$200\ s^{-1}$	Electron transfer coupled to MgATP hydrolysis	Thorneley (1975)
k_3	$4.4 \times 10^6 M^{-1} s^{-1}$	Responsible for inhibition at high protein concentrations	
k_{-3}	$6.4\ s^{-1}$	Rate limiting when Kp2 and substrates are saturating	
k_4	$3.0 \times 10^6 M^{-1} s^{-1}$	Rate of reduction of $Kp2_{ox}(MgADP)_2$ by SO_2^{-}	Thorneley and Lowe (1983)
k_5	$4.4 \times 10^6 M^{-1} s^{-1}$*	Responsible for inhibition of H_2 evolution	
k_{-5}	$6.4\ s^{-1}$*	when MgATP but not reductants is limiting	
k_6	$1.2 \times 10^9 M^{-1} s^{-1}$	$S_2O_4^{2-} \overset{k_{-6}}{\underset{k_6}{\rightleftharpoons}} 2SO_2^{\cdot -}$	
k_{-6}	$1.7\ s^{-1}$		
k_7	$250\ s^{-1}$†	Responsible for enhanced H_2 evolution at low e^- flux	
k_8	$8.0\ s^{-1}$†	Slow in order to maximise N_2 binding to E_3	
k_9	$400\ s^{-1}$†‡	Rapid H_2 evolution from the most reduced hydridic species	
k_{10}	$5 \times 10^5 M^{-1} s^{-1}$	Determined from $K_m^{H_2}$ at low e^- flux	Lowe and Thorneley (1984)
k_{-10}	$2.0 \times 10^5 M^{-1} s^{-1}$	Determined from $K_i^{H_2}$ at low e^- flux	
k_{11}	$2.2 \times 10^6 M^{-1} s^{-1}$‡	Determined from $K_m^{N_2}$ at high e^- flux	
k_{-11}	$4.0 \times 10^5 M^{-1} s^{-1}$	Determined from $K_i^{H_2}$ at high e^- flux	

* Kp1–Kp2 association–dissociation rates assumed to be independent of Kp1 oxidation level.
† H_2 evolution rates. These depend on small differences between large numbers and are subject to errors of factors of about two.
‡ Since these rate constants determine K_ms and K_is, only their ratios are absolute values.

N.B. The differences between the values of some of the rate constants shown and those reported previously (Lowe *et al.*, 1984) are due to the more accurate determination of k_1.

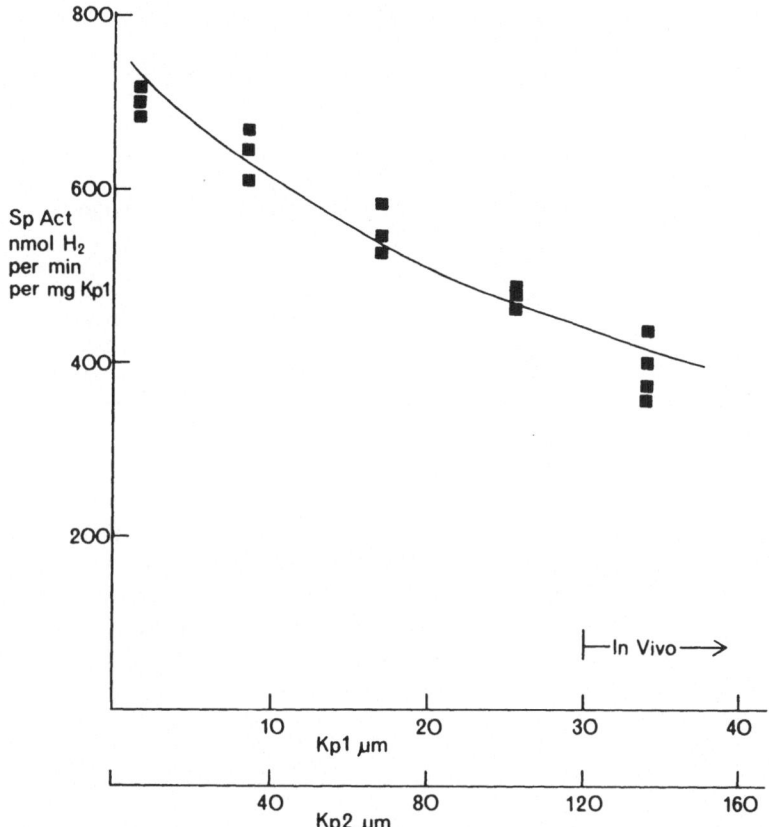

Figure 6.9 Inhibition of nitrogenase activity at high protein concentrations, 23°, pH 7.4, The inhibition is due to the equilibrium k_3/k_{-3} (Scheme 2) being displaced in favour of the complex $Kp2_{ox}(MgADP)_2$ Kp1 as the protein concentration increases. A potential advantage of this inhibition is that the efficiency of N_2 reduction relative to H_2 evolution may be improved (see text)

(2) The non-linear dependence of substrate reduction activity on protein concentration (see above).

(3) The catalysis by nitrogenase of HD formation from D_2, a reaction that is enhanced by N_2 (Bulen, 1976; Burgess *et al.*, 1981).

(4) The pre-steady state kinetics of H_2, N_2H_4 (from an enzyme-bound intermediate thought to be $=N—NH_2$) and NH_3 formation (table 6.6). The time course for N_2H_4 shows an overshoot and oscillation as the steady state is approached (figure 6.7) (Thorneley and Lowe, 1981, 1982).

The rate constants for scheme 1 must be consistent with the steady state rates of substrate reduction over a wide range of protein concentrations (see 2 above).

A mechanism which satisfies all these criteria is shown in scheme 2 which is essentially that of Lowe and Thorneley (1984). The species E_n represents free

Table 6.6　Pre-steady state kinetics of intermediate and product formation by nitrogenase

Substrate	Intermediate or product	Lag (ms)[a]	Burst (ms)[b]
H^+ (under Ar)	H_2	100	linear
H^+ (under N_2)	H_2	100	1000
N_2	N_2H_4	250	1500 (25% overshoot)
N_2	NH_3	400	1500
N_3^-	N_2H_4	180	linear
N_3^-	NH_3	100	800
C_2H_2	C_2H_4	150	linear

[a]linear extrapolation to time axis
[b]time when steady state obtained.

Kp1 protein that has gone round the electron transfer cycle (scheme 1) 'n' times. The three reactions represented by the arrows linking E_n with E_{n+1} are reactions 1, 2 and 3 of scheme 1 respectively. Hence the subscript 'n' indicates the number of rate limiting protein dissociation steps (k_{-3}) that Kp1 has undergone as well as the number of electron equivalents by which Kp1 has been reduced. The initial state, E_0, is that of Kp1 as isolated in the presence of

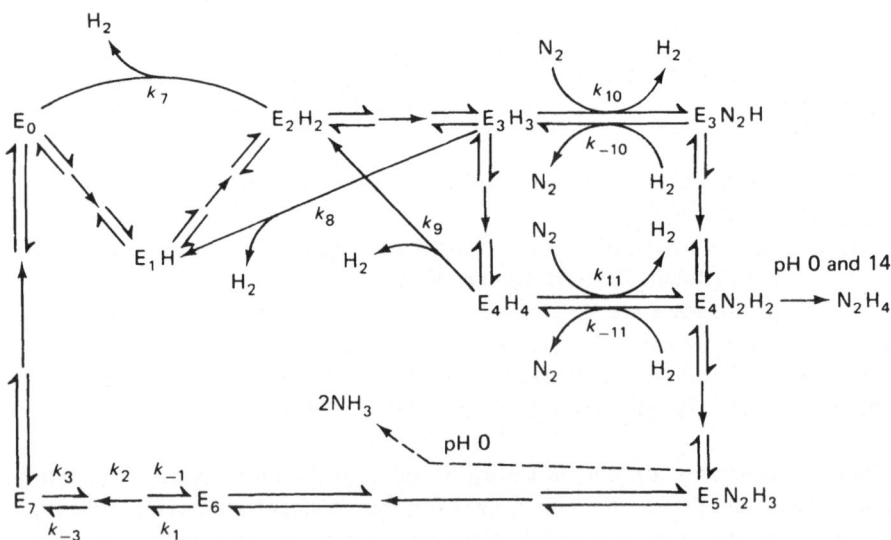

Scheme 2　MoFe protein cycle for nitrgenase

dithionite. The kinetics of the electron transfer cycle (scheme 1) are assumed to be independent of the level of reduction of Kp1.

The positions in scheme 2 at which H_2, N_2H_4 and NH_3 appear are those which give the best fits to the pre-steady state kinetic data using the independently measured rate constants from table 6.5.

Hydrogen evolution

Hydrogen is evolved from species E_2 and is the first reduction product released by nitrogenase. Since H_2 evolution is a waste of reducing equivalents and ATP, the fate of E_2 is crucial for the efficiency of nitrogenase. It is the minimisation of H_2 evolution that Thorneley and Lowe (1982) and Lowe and Thorneley (1983) consider to be the reason for nitrogenase being such a slow enzyme. The rate of wasteful H_2 evolution is proportional to the concentration of E_2. In order to decrease the concentration of E_2, its rate of production is minimised by the slow dissociation of oxidised Fe protein ($k_{-3} = 6.4 \pm 0.8$ s^{-1}). In addition the rate of conversion to E_3, the N_2 binding species, is maximised by reduced Fe protein reacting at close to the diffusion controlled rate ($k_1 = 5 \times 10^7$ M^{-1} s^{-1}). If nitrogenase were a faster enzyme (with $k_{-3} \gg 6.4$ s^{-1}) it would be much less efficient at N_2 reduction and would have increased hydrogenase activity. This mechanism predicts that high protein concentrations favour N_2 reduction over H_2 evolution and could therefore explain the very high *in vivo* concentrations of nitrogenase. The efficiency of nitrogen fixation *in vivo* is also increased by the presence of an 'uptake' hydrogenase which recycles reducing equivalents derived from H_2 (Yates and Walker, 1980).

At low electron flux, the rate of conversion of E_2 to E_3 is decreased and the evolution of H_2, by conversion of E_2 to E_0, is favoured. Thus criterion 1 above is satisfied.

The 'dilution effect' (criterion 2 above) is explained by postulating a change in rate limiting step when protein concentrations decrease below 5 μM. Under these conditions the second order association of reduced Fe protein with MoFe protein (k_1) becomes slower than the first order dissociation of oxidised Fe protein from reduced MoFe protein (k_{-3}).

Nitrogen binding and HD formation

Chatt (1980) has suggested that N_2 could bind to nitrogenase by displacement of H_2 from a metal hydride. A precedent for this type of chemistry is shown in equation (6.3) (Frigo *et al.*, 1971).

$$[MoH_4(dppe)_2] + 2N_2 \rightleftharpoons trans\text{-}[Mo(N_2)_2(dppe)_2] + 2H_2 \qquad (6.3)$$

$$(dppe = Ph_2PCH_2CH_2PPh_2)$$

Lowe and Thorneley (1984) propose that this occurs at the three and four electron-reduced forms of the MoFe protein in scheme 2. It must be stressed that there is no direct evidence, at present, for molybdenum-hydride intermediates being formed at the active-site of nitrogenase. However the hypothesis that E_3 is a hydridic species does provide an elegant explanation and quantitative description of the HD formation reaction, H_2 inhibition of N_2 reduction and the limiting stoicheiometry of one H_2 evolved for each N_2 reduced (criteria 1 and 3 above).

Scheme 3 shows a mechanism for N_2-catalysed HD formation from D_2 that gives no D_2 when nitrogenase functions under N_2 + HD and no D^+ in solution. E_3H_3 is regarded as a metal dihydride with a proton bound to an adjacent group. This proton can react with the dihydride to yield H_2 but not with the metal centre to yield a trihydride. An analogous scheme based on the other N_2 binding species E_4 can be constructed using the dihydride intermediate, $E_4{}^{2-}H_2(2H^+)$.

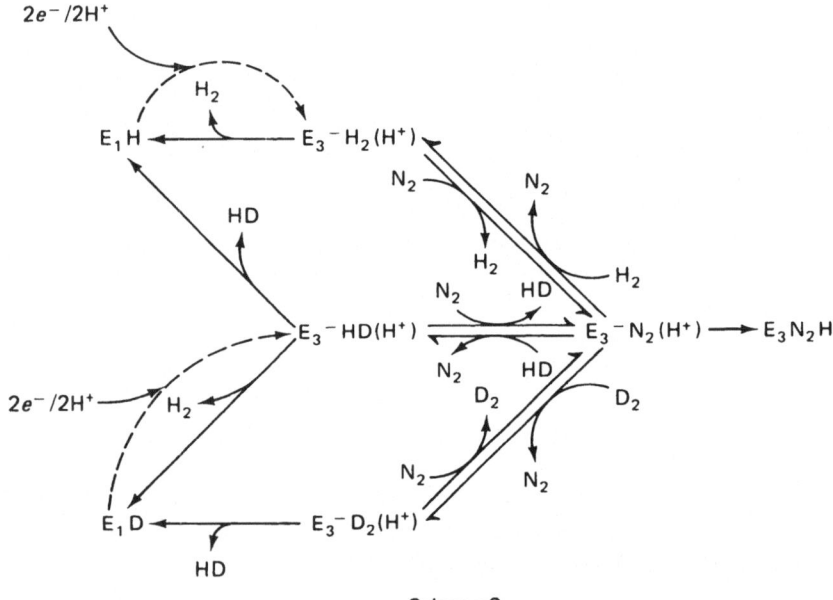

Scheme 3

Equation (6.4), which has been omitted from scheme 3 in the interest of clarity, can explain N_2-independent HD formation.

$$E_3{}^-H_2(H^+) + D_2 \qquad E_3{}^-D_2(H^+) + H_2 \tag{6.4}$$

The hydrazido (2-) intermediate ($=N-NH_2$)

Hydrazine is obtained from nitrogenase which is reducing N_2 when the enzyme is quenched with acid or alkali (Thorneley *et al.*, 1978). It does not accumulate in solution during N_2 reduction under physiological conditions. Hydrazine arises from the reaction of protons or solvent, in an acid/base catalysed reaction, with a dinitrogen hydride, which is an intermediate in N_2 reduction. Scheme 4 shows how this occurs. p-dimethylaminobenzaldehyde was used to detect spectrophotometrically the very low concentrations of N_2H_4 produced.

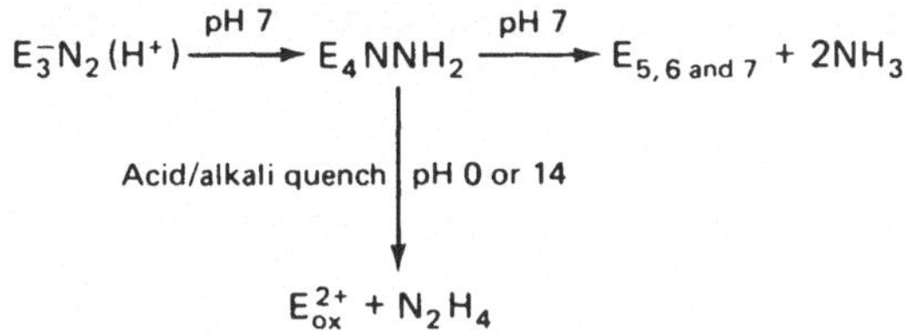

$$E_3^-N_2(H^+) \xrightarrow{\text{pH 7}} E_4NNH_2 \xrightarrow{\text{pH 7}} E_{5,6\text{ and }7} + 2NH_3$$

Acid/alkali quench | pH 0 or 14

$$E_{ox}^{2+} + N_2H_4$$

Scheme 4

The time course for N_2H_4 formation is shown in figure 6.7. The dinitrogen hydride intermediate appears in scheme 2 after four slow steps at E_4. This is consistent with the suggestion of Thorneley *et al.* (1978) that it is a hydrazido(2-) species which on quenching yields hydrazine. In scheme 2, two electrons are used in the evolution of H_2 when N_2 is bound and two more electrons reduce the bound N_2 to the hydrazido(2-) level. The two additional electrons required for conversion to hydrazine, a four electron reduction product of N_2, are obtained by oxidation of the metal centres in the protein on quenching with acid or alkali. The chemistry of dinitrogen and dinitrogen hydrides coordinated to molybdenum and tungsten (Chatt *et al.*, 1978; Richards, 1980) has shown that the hydrazido(2-) species [M=N—NH_2, M = Mo or W] have relatively high stability and do yield hydrazine when treated with acid or alkali.

Ammonia formation

Two moles of NH_3 are released on quenching with acid after only five or six slow steps in scheme 2 at species E_5 or E_6. Since two electrons have been used to produce H_2 on N_2 binding, this is equivalent to a three or four electron reduction of Kp1. Thorneley and Lowe (1982) suggested that these electrons are used to cleave the nitrogen–nitrogen triple bond to yield NH_3 and an enzyme bound nitrido ($\equiv$N) or imido (=NH) intermediate. This intermediate would rapidly hydrolyse, under the acid conditions employed in the rapid quench experiment, to yield the second NH_3 and an oxidised protein. Under physiological conditions the nitride or imido intermediate may have to be reduced further, before the second NH_3 is released at E_6 or E_7. An important conclusion is that at least one NH_3 is formed after only three or at most four electron reduction of the bound dinitrogen. This strongly supports the view of Chatt (1980) that N_2 is reduced by progressive weakening of the nitrogen–nitrogen triple bond consequent on an increase in multiple bonding character between the metal ion and the β-nitrogen atom. Protonation of the β-nitrogen and subsequent cleavage yield NH_3 and a nitrido ($\equiv N$) or imido (=NH) group bound to the metal. Nitride has also been proposed as an enzyme bound intermediate in the nitrogenase catalysed reduction of azide to hydrazine and ammonia (Dilworth and Thorneley, 1981).

MECHANISM OF THE REDUCTION OF SOME OTHER SUBSTRATES

Azide

Azide is reduced by nitrogenase according to equations (6.7-6.9),

$$N_3{}^- + 3H^+ \xrightarrow{\ 2e^-\ } N_2 + NH_3 \tag{6.5}$$

$$N_3{}^- + 7H^+ \xrightarrow{\ 6e^-\ } N_2H_4 + NH_3 \tag{6.6}$$

$$N_3{}^- + 9H^+ \xrightarrow{\ 8e^-\ } 3NH_3 \tag{6.7}$$

Reactions (6.5) and (6.6) were studied by Hardy and Knight (1967), Schöllhorn and Burris (1967b), Parejko and Wilson (1971) and Hwang *et al.* (1973). Reaction (6.7) was more recently discovered by Dilworth and Thorneley (1981). A comparison of the time courses for hydrazine formation from an intermediate in N_2 reduction (figure 6.7) and as a product of azide reduction (figure 6.10) clearly shows how in the former case the hydrazine concentration remains constant once the steady state has been obtained whereas in the latter case hydrazine accumulates in solution. The lag of 180 ms (three slow steps, k_{-3}, in scheme 2) for hydrazine production from azide (figure 6.10) suggests that the first three electrons are used to produce N_2H_4. This implies that if the α-nitrogen atom of azide is coordinated to a metal centre, the β and γ nitrogen atoms are reduced to give hydrazine. This results in a nitrido ($\equiv N$) or imido (=NH) group remaining coordinated to the metal centre and subsequently being reduced to ammonia. Whether the oxidation level of the metal centre binding this nitrido or imido intermediate in azide reduction is the same as that involved in N_2 reduction is not known. However N_2 is a competitive inhibitor of azide reduction to hydrazine; this is consistent with N_2 and azide binding at the same metal centre although the oxidation level may differ. Azide can also be reduced without concomitant H_2 evolution. Thus azide, unlike N_2, can bind without having to displace H_2 from a metal site.

Hydrogen cyanide

Cyanide reduction by nitrogenase was first demonstrated by Hardy and Knight (1967) with methane, ammonia and methylamine as the products. Li *et al.* (1982) in a detailed steady state kinetic study with Av nitrogenase confirmed these products and showed that HCN was the substrate (K_m = 4.5 mM). The reduction of HCN is represented by equations (6.8-6.10).

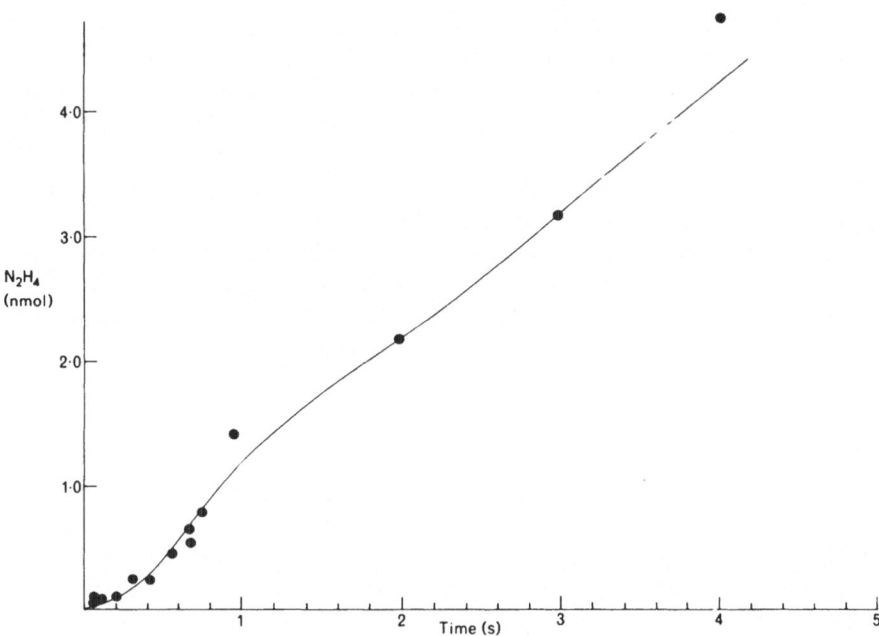

Figure 6.1 Pre-steady state kinetics of azide reduction to give hydrazine at 23° pH 7.5. The data were obtained by using the rapid quench technique. One syringe contained 31 μM-Kp1, 155 μM-Kp2, 10 mM-MgCl$_2$, 1 mM-Na$_2$S$_2$O$_4$ and 25 mM-Hepes pH 7.5. The other contained 18 mM-ATP, 10 mM MgCl$_2$, 20 mM-Na$_2$S$_2$O$_4$, 25 mM-Hepes, pH 7.5 and 40 mM-NaN$_3$. Samples were shot into 1.6 ml of p-dimethyl-aminobenzaldehyde reagent, after centrifugation the absorbance at 458 nm in a 4 cm light path cuvette was determined. The curve is that computed for three slow steps prior to hydrazine formation

$$ HCN + 2H^+ \xrightarrow{\text{2e}^-/\text{H}_2\text{O}} NH_3 + (HCHO \text{ not identified}) \qquad (6.8) $$

$$ HCN + 4H^+ \xrightarrow{\text{4e}^-} CH_3NH_2 \qquad (6.9) $$

$$ HCN + 6H^+ \xrightarrow{\text{6e}^-} NH_3 + CH_4 \qquad (6.10) $$

Cyanide ion inhibits HCN reduction and concomitant H$_2$ evolution. This inhibition is reversed by CO and partially by azide, implying a common binding site. Since HCN, like azide can completely suppress H$_2$ evolution, it probably binds to nitrogenase at the relatively high oxidation levels of systems E$_0$, E$_1$ or E$_2$ in scheme 2.

Acetylene

The two electron reduction of acetylene to ethylene is a convenient and extremely sensitive assay (c. 0.01 nmol C$_2$H$_2$ per ml of sample by gas chromatography) for nitrogenase activity (Dilworth, 1966; Schöllhorn and Burris, 1967b; Eady,

1980a,b; Turner and Gibson, 1980). The reduction is stereospecific since *cis*-1-2-dideuteroethylene is formed when the reaction occurs in D_2O (Dilworth, 1966).

Acetylene is a non-competitive inhibitor of N_2 reduction whereas N_2 is competitive with acetylene reduction (Rivera-Ortiz and Burris, 1975). Since ethylene is produced after a lag of 150 ms (table 6.4), a cycle involving only E_0, E_1, E_2, and E_3 can reduce acetylene. This is consistent with the complete suppression of H_2 evolution by acetylene (Hageman and Burris, 1980).

Although N_2 fixation rates have often been estimated from acetylene reduction data by using a ratio of *c*. 4:1 for C_2H_2:N_2 reduced, extreme caution is advised, since under certain conditions this relationship is not correct (Thorneley and Eady, 1977; Smith *et al.*, 1976b; Turner and Gibson, 1980).

Cyclopropene

Ring strain in cyclopropene causes it to have chemical reactivity intermediate between that of the nitrogenase substrate acetylene and the non-substrate ethylene. The V_{max} of cyclopropene is about half that of acetylene with a ratio of 2:1 for the products propene and cyclopropane, equation (6.11) (McKenna *et al.*, 1980).

$$\begin{array}{c} CH_2 \\ / \quad \backslash \\ HC = CH \end{array} + 2H^+ \xrightarrow{2e^-} \frac{1}{3} \begin{array}{c} CH_2 \\ / \quad \backslash \\ CH_2 - CH_2 \end{array} + \frac{2}{3} CH_3 - CH = CH_2 \qquad (6.11)$$

The K_m with respect to either product is *c*. 0.01 atm (McKenna and Huang, 1979). Reduction of cyclopropene by nitrogenase in D_2O produces ten times more *cis*-d_2 than *trans*-d_2 isomer of cyclopropane (McKenna *et al.*, 1979). The characteristic ratio of the products, cyclopropane and propene, may be useful for the evaluation of chemical systems intended to model nitrogenase catalysis.

FUTURE PROSPECTS

The last decade has seen considerable advances in our understanding of nitrogenase. We anticipate significant developments in the near future. X-ray diffraction studies on the MoFe protein should provide a much better understanding of the structures of metal clusters. These in turn should provide insight into the way in which N_2 is activated. Further mechanistic information can be expected from the more sophisticated application of spectroscopic and kinetic techniques to the problem of energy transduction involving ATP hydrolysis, the substrate reduction pathway and protein–protein interactions. The use of mutant and genetically engineered bacterial strains is beginning to contribute to the solution of these problems and to our knowledge of the functions of the other *nif* gene products. This basic research is a prerequisite for the successful transfer and maintenance of the ability to fix nitrogen in other organisms of potential agronomic importance.

REFERENCES

Antonio, M. R., Teo, B-K., Orme-Johnson, W. H., Nelson, M. J., Groh, S. E., Lindahl, P. A., Kauzlarich, S. M. and Averill, B. A. (1982) *J. Am. chem. Soc.*, **104**, 4703

Ausubel, F. M. and Cannon, F. C. (1981) *Cold Spring Harbor Symposia on Quantitative Biology*, **XLV**, 487

Averill, B. A., Bale, J. R. and Orme-Johnson, W. H. (1978) *J. Am. chem. Soc.*, **100**, 3034

Bishop, P. E., Jarlenski, D. M. L. and Hetherington, D. R. (1980) *Proc. nat. Acad. Sci. U.S.A.*, **77**, 7342

Bishop, P. E., Jarlenski, D. M. L. and Hetherington, D. R. (1982) *J. Bacteriol.*, **150**, 1244

Bogusz, D., Houmard, J. and Aubert, J-P. (1981) *Eur. J. Biochem.*, **120**, 421

Braaksma, A., Haaker, H., Grande, H. J. and Veeger, C. (1982) *Eur. J. Biochem.*, **121**, 483

Buchanan-Wollaston, V., Cannon, M. C., Beynon, J. L. and Cannon, F. C. (1981a) *Nature, Lond.*, **294**, 776

Buchanan-Wollaston, V., Cannon, M. C. and Cannon, F. C. (1981b) *Mol. Gen. Genet.*, **184**, 102

Bulen, W. A. (1976) In: *Proceedings of First International Symposium on Nitrogen Fixation* (ed. Newton, W. E. and Nyman, C. J.), Washington University Press, Pullman, U.S.A., p. 177

Burgess, B. K., Jacobs, D. B. and Stiefel, E. I. (1980a) *Biochim. Biophys. Acta*, **614**, 196

Burgess, B. K., Stiefel, E. I. and Newton, W. E. (1980b) *J. biol. Chem.*, **255**, 353

Burgess, B. K., Wherland, S., Stiefel, E. I. and Newton, W. E. (1981) *Biochemistry*, **20**, 5141

Burris, R. H. (1980) In: *Nitrogen Fixation, Vol. 1*, (ed. Newton, W. E. and Orme-Johnson, W. H.) University Park Press, Baltimore, U.S.A., p. 7

Cannon, F. C., Reidel, G. E. and Ausubel, F. M. (1979) *Molec. Gen. Genet.*, **174**, 59

Chatt, J., Dilworth, J. R. and Richards, R. L. (1978) *Chem. Rev.*, **78**, 589

Chatt, J. (1980) In: *Nitrogen Fixation* (ed. Stewart, W. D. P. and Gallon, J. R.), Academic Press, London, p. 1

Coughlan, M. (1980) *Molybdenum and Molybdenum Containing Enzymes*, Pergamon Press, Oxford

Cramer, S. P., Gillum, W. O., Hodgson, K. O., Mortenson, L. E., Steifel, E. I., Chisnell, J. R., Brill, W. J. and Shah, V. K. (1978) *J. Am. chem. Soc.*, **100**, 3814

Davis, L. C. and Orme-Johnson, W. H. (1976) *Biochim. Biophys. Acta*, **452**, 42

Davis, L. C., Henzl, M. T., Burris, R. H. and Orme-Johnson, W. H. (1979) *Biochemistry*, **18**, 4860

Dilworth, M. J. (1966) *Biochim. Biophys. Acta*, **127**, 285

Dilworth, M. J. and Thorneley, R. N. F. (1981) *Biochem. J.*, **193**, 971

Dixon, R. A., Kennedy, C., Kondorosi, A., Krishnapillai, V. and Merrick, M. (1977) *Molec. Gen. Genet.*, **157**, 189

Dixon, R. A., Eady, R. R., Espin, G., Hill, S., Iaccarino, M., Kahn, D. and Merrick, M. (1980) *Nature, Lond.*, **286**, 128

Eady, R. R., Lowe, D. J. and Thorneley, R. N. F. (1978a) *FEBS Lett.*, **95**, 2111

Eady, R. R., Issack, R., Kennedy, C., Postgate, J. R. and Ratcliffe, H. D. (1978b) *J. gen. Microbiol.*, **104**, 277

Eady, R. R. and Smith, B. E. (1979) In: *A Treatise on Dinitrogen Fixation* (ed. Hardy, R. F., Bottomely, F. and Burns, R. C.), Sections I and II, Inorganic and Physical Chemistry and Biochemistry, Wiley, New York, p. 394

Eady, R. R. (1980a) In: *Methods in Enzymology, Vol. 69, part C* (ed. A. San Pietro), Academic Press, London, New York, p. 751

Eady, R. R. (1980b) In: *Methods for Evaluating Biological Nitrogen Fixation* (ed. F. J. Bergersen), Wiley, New York, p. 213

Eady, R. R. (1984) In: *Molecular Biology of Nitrogen Fixation* (ed. Broughton, W. J. and Pühler, A.), Oxford University Press, Oxford, in press

Elmerich, C., Houmard, J., Sibold, L., Manheimer, I. and Charpin, N. (1978) *Mol. Gen. Genet.*, **165**, 181

Frigo, A., Puosi, G. and Turco, A. (1971) *Gazz. Chim. Ital.*, **101**, 637

Gibson, A. H. and Newton, W. E. (1981) *Current Perspectives in Nitrogen Fixation*, Australian Academy of Sciences, Canberra

Gutteridge, S., Tanner, S. J. and Bray, R. C. (1978) *Biochem. J.*, **175**, 869

Haaker, H., and Veeger, C. (1977) *Eur. J. Biochem.*, **77**, 1

Hageman, R. V. and Burris, R. H. (1980) *Biochim. Biophys. Acta*, **591**, 63

Hageman, R. V., Orme-Johnson, W. H. and Burris, R. H. (1980) *Biochemistry*, **19**, 2333

Hardy, R. W. F. and Knight, E. (1967) *Biochim. Biophys. Acta*, **139**, 69

Harker, A. R. and Wullstein, L. H. (1981) *J. biol. Chem.*, **256**, 11983

Hase, T., Nakano, T., Mutzubara, H. and Zumft, W. G. (1981) *J. Biochem.*, **90**, 295

Hawkes, T. R. (1981) D.Phil. thesis, University of Sussex

Hawkes, T. R. and Smith, B. E. (1983) *Biochem. J.*, **209**, 43

Hawkes, T. R., Lowe, D. J. and Smith, B. E. (1983) *Biochem. J.*, **211**, 495

Hawkes, T. R., McLean, P. A. and Smith, B. E. (1984) *Biochem. J.*, **217**, 317

Hennecke, H. and Shanmugam, K. T. (1979) *Arch. Microbiol.*, **123**, 259

Hill, S. and Kavanagh, E. P. (1980) *J. Bacteriol.*, **141**, 470

Hill, S., Kennedy, C., Kavanagh, E. P., Goldberg, R. and Hanau, R. (1981) *Nature, Lond.*, **290**, 424

Hoffman, B. M., Venters, R. A., Roberts, J. E., Nelson, M. and Orme-Johnson, W. H. (1982a) *J. Am. chem. Soc.*, **104**, 4711

Hoffman, B. M., Roberts, J. E. and Orme-Johnson, W. H. (1982b) *J. Am. chem. Soc.*, **104**, 860

Holm, R. (1977) *Acc. Chem. Res.*, **10**, 427

Holm, R. (1981) *Chem. Soc. Rev.*, **10**, 455

Houmard, J., Bogusz, D., Bigault, R. and Elmerich, C. (1980) *Biochemie*, **62**, 267

Huynh, B. H., Münck, E. and Orme-Johnson, W. H. (1979) *Biochim. Biophys. Acta*, **527**, 192

Huynh, B. H., Henzl, M. T., Christner, J. A., Zimmermann, R., Orme-Johnson, W. H. and Münck, E. (1980) *Biochim. Biophys. Acta*, **623**, 124

Hwang, J. C. and Burris, R. H. (1972) *Biochim. Biophys. Acta*, **283**, 339

Hwang, J. C., Chen, C. H. and Burris, R. H. (1973) *Biochim. Biophys. Acta*, **292**, 256

Imam, S. and Eady, R. R. (1980) *FEBS. Lett.*, **110**, 35

Johnson, M. K., Thomson, A. J., Robinson, A. E. and Smith, B. E. (1981) *Biochim. Biophys. Acta*, **671**, 61

Kennedy, C., Eady, R. R., Kondorosi, E. and ReKosh, D. K. (1976) *Biochem. J.*, **155**, 383

Kennedy, C. (1977) *Molec. Gen. Genet.*, **157**, 199

Kennedy, C., Cannon, F. C., Cannon, M., Dixon, R. A., Hill, S., Jensen, J., Kumar, S., McLean, P. A., Merrick, M., Robson, R. and Postgate, J. R. (1981) In: *Current Perspectives in Nitrogen Fixation* (ed. Gibson, A. H. and Newton, W. E.), Elsevier/North Holland, Australian Academy of Sciences, Canberra, p. 146

Kimber, S. J., Bishop, E. O. and Smith, B. E. (1982) *Biochim. Biophys. Acta*, **705**, 385

Kurtz, D. M., McMillan, R. S., Burgess, B. K., Mortenson, L. E. and Holm, R. H. (1979) *Proc. nat. Acad. Sci. U.S.A.*, **76**, 4986

Levchenko, L. A., Roschupkina, O. S., Sadkov, A. P., Marakushev, S. A., Mikhailov, G. M. and Borod'ko, Y. G. (1980) *Biochem. Biophys. Res. Commun.*, **96**, 1384

Li, J-G., Burgess, B. K. and Corbin, J. L. (1982) *Biochemistry*, **21**, 4393

Ljones, T. and Burris, R. H. (1978a) *Biochemistry*, **17**, 1866

Ljones, T. and Burris, R. H. (1978b) *Biochem. Biophys. Res. Commun.*, **80**, 22

Ljones, T. (1979) *FEBS Lett.*, **98**, 1

Lowe, D. J., Eady, R. R. and Thorneley, R. N. F. (1978) *Biochem. J.*, **173**, 277

Lowe, D. J. (1978) *Biochem. J.*, **175**, 955

Lowe, D. J., Smith, B. E. and Eady, R. R. (1980) In: *Recent Advances in Biological Nitrogen Fixation* (ed. Subba-Rao, N. S.), Arnold, London, p. 34
Lowe, D. J. and Thorneley, R. N. F. (1984) *Biochem. J.* (in press)
Lowe, D. J., Thorneley, R. N. F. and Postgate, J. R. (1984) In: *Advances in Nitrogen Fixation Research* (ed. Veeger, C. and Newton, W. E.), Nijhoff-Junk, The Hague, p. 133
Ludden, P. W. and Burris, R. H. (1979) *Proc. nat. Acad. Sci. U.S.A.*, 76, 6201
Ludden, P. W., Preston, G. G. and Dowling, T. E. (1982) *Biochem. J.*, 203, 663
Lundell, D. J. and Howard, J. B. (1978) *J. biol. Chem.*, 253, 3422
Lundell, D. J. and Howard, J. B. (1981) *J. biol. Chem.*, 256, 6385
MacNeil, T., MacNeil, D., Roberts, G. P., Supiano, M. A. and Brill, W. J. (1978) *J. Bacteriol.*, 136, 253
McKenna, C. E. and Huang, C. W. (1979) *Nature, Lond.*, 280, 609
McKenna, C. E., Jones, J. B., Haratyan, E. and Huang, C. W. (1979) *Nature, Lond.*, 280, 611
McKenna, C. E., McKenna, M-C. and Huang, C. W. (1979) *Proc. nat. Acad. Sci. U.S.A.*, 76, 4773
McKenna, C. E., Huang, C. W., Jones, J. B., McKenna, M.-C., Nakajima, T. and Nguyen, H. T. (1980) In: *Nitrogen Fixation, Vol. 1* (ed. Newton, W. E. and Orme-Johnson, W. H.), University Park Press, Baltimore, p. 223
McLean, P. A. and Dixon, R. A. (1981) *Nature, Lond.*, 292, 655
McLean, P. A., Smith, B. E. and Dixon, R. A. (1983) *Biochem. J.*, 211, 589
Mazur, B. J. and Chui, C-F. (1982) *Proc. nat. Acad. Sci. U.S.A.*, 79, 6782
Merrick, M., Filser, M., Kennedy, C. and Dixon, R. A. (1978) *Molec. Gen. Genet.*, 165, 103
Merrick, M., Filser, M., Dixon, R., Elmerich, C., Sibold, J. and Houmard, J. (1980) *J. Gen. Microbiol.*, 117, 509
Merrick, M., Hill, S., Hennecke, J., Hahn, M., Dixon, R. A. and Kennedy, C. (1982) *Molec. Gen. Genet.*, 185, 75
Mevarech, M., Rice, D. and Haselkorn, R. (1980) *Proc. nat. Acad. Sci. U.S.A.*, 77, 6476
Miller, R. W., Robson, R. L., Yates, M. G. and Eady, R. R. (1980) *Can. J. Biochem.*, 58, 542
Mortenson, L. E. and Thorneley, R. N. F. (1979) *Ann. Rev. Biochem.*, 48, 387
Moustafa, E. and Mortenson, L. E. (1967) *Nature, Lond.*, 216, 1241
Münck, E., Rhodes, H., Orme-Johnson, W. H., Davis, L. C., Brill, W. J. and Shah, V. K. (1975) *Biochim. Biophys. Acta*, 400, 32
Nason, A., Lee, K. Y., Pan, S. S., Ketchum, P. A., Lamberti, A. and De Vries, J. (1971) *Proc. nat. Acad. Sci.*, 68, 3242
Nelson, M. J., Lundahl, P. A. and Orme-Johnson, W. H. (1982) *Adv. inorg. Biochem.*, 4, 1
Nelson, M. J., Levy, M. A. and Orme-Johnson, W. H. (1983) *Proc. nat. Acad. Sci. U.S.A.*, 80, 147
Newton, W. E., Burgess, B. K. and Stiefel, E. I. (1980) In: *Molybdenum Chemistry of Biological Significance* (ed. Newton, W. E. and Otsuka, S.), New York and London, Plenum Press, p. 1
Newton, W. E., McDonald, J. W., Friesen, G. D., Burgess, B. K., Conradson, S. D. and Hodgson, K. O. (1981) In: *Current Perspectives in Nitrogen Fixation* (ed. Gibson, A. H. and Newton, W. E.), Australian Academy of Sciences, Canberra, p. 30
Newton, W. E. and Orme-Johnson, W. H. (1980) *Nitrogen Fixation*, University Park Press, Baltimore
Nieva-Gomez, D., Roberts, G. P., Klevickis, S. and Brill, W. J. (1980) *Proc. nat. Acad. Sci. U.S.A.*, 77, 2555
O'Donnell, M. J. and Smith, B. E. (1978) *Biochem. J.*, 173, 831
Orme-Johnson, W. H. and Davis, L. C. (1976) In: *Iron–Sulfur Proteins*, Vol. III (ed. Lovenberg, W.), Academic Press, London, p. 15
Orme-Johnson, W. H., Hamilton, W. D., Ljones, T., Tso, M.Y-W., Burris, R. H., Shah, V. K. and Brill, W. J. (1972) *Proc. nat. Acad. Sci. U.S.A.*, 69, 3142
Orme-Johnson, W. H., Davis, L. C., Henzl, M. T., Averill, B. A., Orme-Johnson, N. R., Münck, E. and Zimmermann, R. (1977) In: *Recent Developments in Nitrogen Fixation* (ed. W. E. Newton, J. R. Postgate, and C. Rodriquez-Barrueco), Academic Press, London, p. 131
Orme-Johnson, W. H., Lindahl, P., Meade, J., Warren, W., Nelson, M., Groh, S., Orme-Johnson, N. R., Münck, E., Huynh, B. H., Emptage, M., Rawlings, J., Smith, J., Roberts,

J., Hoffmann, B. and Mims, W. B. (1981) In: *Current Perspectives in Nitrogen Fixation* (ed. Gibson, A. H. and Newton, W. E.), Australian Academy of Sciences, Canberra, p. 79

Palmer, G., Multani, J-S., Cretney, W. C., Zumft, W. G. and Mortenson, L. E. (1972) *Arch. Biochem. Biophys.*, 153, 325

Parejko, R. A. and Wilson, P. W. (1971) *Proc. nat. Acad. Sci. U.S.A.*, 68, 2016

Peisach, J., Orme-Johnson, N. R., Mims, W. R. and Broquist, H. P. (1977) *J. biol. Chem.*, 252, 5643

Pickett, C. J. (1983) In: *Specialist Periodical Report*, Vol. 8, *Electrochemistry* (ed. Pletcher, D.), Royal Society of Chemistry, London, p. 81

Pienkos, P. T., Shah, V. K. and Brill, W. J. (1977) *Proc. nat. Acad. Sci. U.S.A.*, 74, 5468

Postgate, J. R. (1981) In: *Current Perspectives in Nitrogen Fixation* (ed. Gibson, A. H. and Newton, W. E.), Australian Academy of Sciences, Canberra, p. 217

Postgate, J. R. and Cannon, F. C. (1981) *Phil. Trans. Roy. Soc. Lond. B*, 292, 589

Preston, G. G. and Ludden, P. W. (1982) *Biochem. J.*, 205, 489

Pühler, A. and Klipp, W. (1981) In: *Biology of Inorganic Nitrogen and Sulphur* (ed. Boethe, H. and Trebst, A.), Springer-Verlag, Berlin, Heidelberg, New York, p. 276

Rawlings, J., Shah, V. K., Chisnell, J. R., Brill, W. J., Zimmermann, R., Münck, E. and Orme-Johnson, W. H. (1978) *J. biol. Chem.*, 253, 1001

Richards, R. L. (1980) In: *Trends in the Chemistry of Nitrogen Fixation* (ed. Chatt, J., da Câmara Pina, L. M. and Richards, R. L.), Academic Press, London, p. 199

Rivera-Ortiz, J. M. and Burris, R. H. (1975) *J. Bacteriol.*, 123, 537

Roberts, G. P., MacNeil, P. T., MacNeil, P. D. and Brill, W. J. (1978) *J. Bacteriol.*, 136, 267

Roberts, G. P. and Brill, W. J. (1980) *J. Bacteriol.*, 144, 210

Roberts, G. P. and Brill, W. J. (1981) *Ann. Rev. Microbiol.*, 35, 207

Robson, R. L. (1979) *Biochem. J.*, 181, 569

Robson, R. L. and Postgate, J. R. (1980) *Ann. Rev. Microbiol.*, 34, 183

Robson, R., Kennedy, C. and Postgate, J. R. (1983) *Can. J. Microbiol.*, 29, 954

Ruvkun, G. B. and Ausubel, F. M. (1980) *Proc. nat. Acad. Sci. U.S.A.*, 77, 191

Schöllhorn, R. and Burris, R. H. (1967a) *Proc. nat. Acad. Sci. U.S.A.*, 57, 1317

Schöllhorn, R. and Burris, R. H. (1967b) *Proc. nat. Acad. Sci. U.S.A.*, 58, 213

Scott, D. B., Hennecke, H. and Lim, S. T. (1979) *Biochim. Biophys. Acta*, 565, 365

Scott, K. F., Rolfe, B. G. and Shine, J. (1981) *J. mol. appl. Genet.*, 1, 71

Shah, V. K. and Brill, W. J. (1977) *Proc. nat. Acad. Sci. U.S.A.*, 74, 3249

Shah, V. K., Chisnell, J. R. and Brill, W. J. (1978) *Biochem. Biophys. Res. Commun.*, 81, 232

Shah, V. K. and Brill, W. J. (1981) *Proc. nat. Acad. Sci. U.S.A.*, 78, 3438

Sibold, L., Ouviger, B., Charpin, A. and Elmerich, C. (1983) *Biochimie*, 65, 53

Silverstein, R. and Bulen, W. A. (1970) *Biochemistry*, 9, 3809

Smith, B. E., Lowe, D. J. and Bray, R. C. (1972) *Biochem. J.*, 130, 641

Smith, B. E., Lowe, D. J. and Bray, R. C. (1973) *Biochem. J.*, 135, 331

Smith, B. E. and Lang, G. (1974) *Biochem. J.*, 137, 169

Smith, B. E., Thorneley, R. N. F., Yates, M. G., Eady, R. R., and Postgate, J. R. (1976a) *Proc. 1st Internat. Symp. Nitrogen Fixation*, Vol. 1 (ed. Newton, W. E. and Nyman, C. J.), Washington University Press, p. 150

Smith, B. E., Thorneley, R. N. F., Eady, R. R. and Mortenson, L. E. (1976b) *Biochem. J.*, 157, 439

Smith, B. E. (1977) *J. Less Common Metals*, 54, 465

Smith, B. E. (1980) In: *Molybdenum Chemistry of Biological Significance* (ed. Newton, W. E. and Otsuka, S.), Plenum Press, New York and London, p. 179

Smith, B. E., O'Donnell, M. J., Lang, G. and Spartalian, K. (1980) *Biochem. J.*, 191, 449

Smith, B. E. (1983) In: *Nitrogen Fixation: The Chemical Biochemical and Genetical Interfaces* (ed. Müller, A. and Newton, W. E.), Plenum Press, New York, p. 23

Smith, B. E., Lowe, D. J., Chen, G-X., O'Donnell, M. J. and Hawkes, T. R. (1983) *Biochem. J.*, 209, 207

Smith, J., Emptage, M. H. and Orme-Johnson, W. H. (1982) *J. biol. Chem.*, 257, 2310

Starr, M. P., Stolp, H., Trüper, H. G., Balows, A. and Schlegel, H. G. (1981) *The Prokaryotes, Vol. I*, Springer-Verlag, Berlin, Heidelberg, New York

Stephens, P. J., McKenna, C. E., Smith, B. E., Nguyen, H. T., McKenna, M-C., Thomas, A. J., Devlin, F. and Jones, J. B. (1979) *Proc. nat. Acad. Sci. U.S.A.*, 76, 2585

Stephens, P. J., McKenna, C. E., McKenna, M-C., Nguyen, H. T., Morgan, T. V. and Devlin, F. (1981) In: *Current Perspectives in Nitrogen Fixation* (ed. Gibson, A. H. and Newton, W. E.), Australian Academy of Sciences, Canberra, p. 357

Stephens, P. J., McKenna, C. E., McKenna, M-C., Nguyen, H. T. and Lowe, D. J. (1983) In: *Interaction between Iron and Proteins in Oxygen and Electron Transport* (ed. Ho, C.), Proceedings of the Airlie House Symposium, Elsevier/North-Holland, Amsterdam, p.405

Stewart, W. D. P., Rowell, P. and Rai, A. N. (1980) In: *Nitrogen Fixation* (ed. Stewart, W. D. P. and Gallon, J. R.), Academic Press, London, p. 239

St. John, R. T., Johnston, H. M., Seidman, C., Garfinkel, D., Gordon, J. K., Shah, V. K. and Brill, W. J. (1975) *J. Bacteriol.*, **121**, 759

Sundaresan, V. and Ausubel, F. M. (1981) *J. biol. Chem.*, **256**, 2808

Thorneley, R. N. F. and Eady, R. R. (1973) *Biochem. J.*, **133**, 405

Thorneley, R. N. F. and Willison, K. R. (1974) *Biochem. J.*, **139**, 211

Thorneley, R. N. F. (1975) *Biochem. J.*, **145**, 391

Thorneley, R. N. F., Eady, R. R. and Yates, M. G. (1975) *Biochim. Biophys. Acta*, **403**, 269

Thorneley, R. N. F., Yates, M. G. and Lowe, D. J. (1976) *Biochem. J.*, **155**, 137

Thorneley, R. N. F. and Eady, R. R. (1977) *Biochem. J.*, **167**, 457

Thorneley, R. N. F. and Yates, M. G. (1977) *Biochem. J.*, **167**, 457

Thorneley, R. N. F. and Cornish-Bowden, A. (1977) *Biochem. J.*, **165**, 255

Thorneley, R. N. F., Eady, R. R. and Lowe, D. J. (1978) *Nature, Lond.*, **272**, 557

Thorneley, R. N. F. and Lowe, D. J. (1981) In: *Current Perspectives in Nitrogen Fixation* (ed. Gibson, A. H. and Newton, W. E.), Australian Academy of Sciences, Canberra, p. 360

Thorneley, R. N. F. and Lowe, D. J. (1982) *Israel J. Botany*, **32**, 61

Thorneley, R. N. F. and Lowe, D. J. (1983) *Biochem. J.*, **215**, 393

Török, I. and Kondorosi, A. (1981) *Nucleic Acid. Res.*, **9**, 5711

Turner, G. L. and Gibson, A. H. (1980) In: *Methods for Evaluating Biological Nitrogen Fixation* (ed. Bergersen, F. J.), Wiley, New York, p. 111

Veeger, C., Laane, C., Scherings, G., Matz, L., Haaker, H. and Van Zeeland-Wolbers, L. (1980) In: *Nitrogen Fixation, Vol. I* (ed. Newton, W. E. and Orme-Johnson, W. H.), University Park Press, Baltimore, p. 111

Walker, G. A. and Mortenson, L. E. (1973) *Biochem. Biophys. Res. Commun.*, **53**, 904

Watt, G. D., Burns, A., Lough, S. and Tennent, D. L. (1980) *Biochemistry*, **19**, 4926

Watt, G. D., Burns, A. and Tennent, D. L. (1981) *Biochemistry*, **20**, 7272

Weininger, M. S. and Mortenson, L. E. (1982) *Proc. nat. Acad. Sci. U.S.A.*, **79**, 378

Wherland, S., Burgess, B. K., Stiefel, E. I. and Newton, W. E. (1981) *Biochemistry*, **20**, 5132

Wong, G. B., Kurtz, D. M., Holm, R. H., Mortenson, L. E. and Upchurch, R. G. (1979) *J. Am. chem. Soc.*, **101**, 3078

Yamane, T., Weininger, M. S., Mortenson, L. E. and Rossman, M. G. (1982) *J. biol. Chem.*, **257**, 1221

Yang, S-S., Pan, W-H., Frisen, G. D., Burgess, B. K., Corbin, J. L., Stiefel, E. I. and Newton, W. E. (1982) *J. biol. Chem.*, **257**, 8042

Yates, M. G. and Jones, C. W. (1974) *Adv. Microb. Physiol.*, **11**, 97

Yates, M. G., Thorneley, R. N. F. and Lowe, D. J. (1975) *FEBS Lett.*, **60**, 89

Yates, M. G. and Walker, C. (1980) In: *Nitrogen Fixation, Vol. 1*, (ed. Newton, W. E. and Orme-Johnson, W. H.), University Park Press, Baltimore, p. 95

Yates, M. G. (1980) In: *The Biochemistry of Plants, Vol. 5* (ed. Miflin, B.), Academic Press, New York, p. 1

Zimmermann, R., Münck, E., Brill, W. J., Shah, V. K., Henzl, M. T., Rawlings, J. and Orme-Johnson, W. H. (1978) *Biochim. Biophys. Acta*, **537**, 185

Zumft, W. G., Cretney, W. C., Huang, T. C., Mortenson, L. E. and Palmer, G. (1972) *Biochem. Biophys. Res. Commun.*, **48**, 1525

Zumft, W. G., Palmer, G. and Mortenson, L. E. (1973) *Biochim. Biophys. Acta*, **292**, 413

Zumft, W. G., Mortenson, L. E. and Palmer, G. (1974) *Eur. J. Biochem.*, **46**, 525

Index

Acetylene, reduction by nitrogenase, 243-244
Achromobacter cyclolastes, bacterial cupredoxins, 231
Achromobacter fischeri, nitrite reductase, 75
Aconitase, 100, 106, 108-110
  EPR spectroscopy, 102, 109-110
  extended X-ray absorption fine structure spectroscopy, 102
  as iron-sulphur protein, 108-110
  Mossbauer spectroscopy, 109-110
Adrenodoxin, iron-sulphur cluster of, 82
Alcaligines, sp. NC1B11015, 10
Alcaligines denitrificans, 2, 4, 17
  azurin, 2, 4, 9, 12, 18
Alcaligines faecalis, 2, 4, 15, 17, 31, 33
  azurin, 34
Amicyanin, 4
Amino acid composition
  of bacterial cupredoxins, 31-34
  of nitrogenase Fe protein, 212-214
  of plant cupredoxins, 32
Ammonia, formation by nitrogenase, 241-242
Anabaena 7120, nitrogenase, 207
Anabaena cylindrica, nitrogenase, 207
Antiferromagnetic coupling, in iron-sulphur protein, 89
Ascorbate oxidase, as blue copper protein, 1
Azide, reduction by nitrogenase, 242
Azotobacter vinelandii
  ferrodoxin, 97, 99, 103, 106, 108, 110, 224
  nitrogenase, 207, 212-214, 219
Azospirillum brasilense, nitrogenase, 207
Azurin, 1-2, 4-19, 22
  of *Alcaligines denitrificans*, 4, 9, 12, 18
  of *Alcaligines faecalis*, 34
  beta sheet, 7
  chromium complexes, 15, 26
  common structural features of, 6
  cytochrome *c* oxidase interaction, 14, 18, 55-58

cytochrome c_{551} interactions with plastocyanin, 7, 16, 18
  EPR spectroscopy, 11
  folding topology, 6
  histidine 35, 11-13, 17
  hydrophobic surfaces, 9
  nickel complexes of, 12-13
  nitrite reductase interactions, 70
  NMR spectroscopy of, 11-12, 14
  optical absorption spectroscopy of, 10
  plastocyanin structural comparison, 6-7, 21-23
  of *Pseudomonas aeruginosa*, 4-5, 9, 12, 16, 31, 33
  redox reactions of, 13-19, 54
  resonance Raman spectroscopy, 10-11
  sequence alignment with plastocyanin, 10-11
  sources, 4
  stellacyanin common features, 35
  structure, 4-10

Beta sheets
  of azurin, 7, 20-22
  of plastocyanin, 7, 20-22
Blue copper centres
  in ascorbate oxidase, 1
  in azurin, 10
  in caeruloplasmin, 1
  in laccase, 1
  in plastocyanin, 25
  in stellacyanin, 35
Blue copper proteins, 1-42
  bacterial, 31-34
  in plants, 34-36
  properties of, 1-4

251